Human Agro-Energy Optimization for Business and Industry

Pandian Vasant
MERLIN, Faculty of Electrical and Electronics Engineering, Ton Duc Thang University, Vietnam

Roman Rodríguez-Aguilar
Panamerican University, Mexico

Igor Litvinchev
Universidad Autónoma de Nuevo León, Mexico

Jose Antonio Marmolejo-Saucedo
Panamerican University, Mexico

A volume in the Advances in Environmental Engineering and Green Technologies (AEEGT) Book Series

IGI Global
PUBLISHER of TIMELY KNOWLEDGE

Published in the United States of America by
IGI Global
Engineering Science Reference (an imprint of IGI Global)
701 E. Chocolate Avenue
Hershey PA, USA 17033
Tel: 717-533-8845
Fax: 717-533-8661
E-mail: cust@igi-global.com
Web site: http://www.igi-global.com

Library of Congress Cataloging-in-Publication Data

Names: Vasant, Pandian, editor. | Rodriguez-Aguilar, Roman, editor. | Litvinchev, Igor, editor. | Marmolejo-Saucedo, Jose Antonio editor.
Title: Human agro-energy optimization for business and industry / edited by Pandian Vasant, Roman Rodriguez-Aguilar, Igor Litvinchev, Jose Antonio Marmolejo-Saucedo.
Description: Hershey, PA : Engineering Science Reference, [2023] | Includes bibliographical references and index. | Summary: "The edited book on humanized optimization approaches for Smart Energy and Agro-Business Industry is a critical scholarly resource that examines efficient use of modern smart farming and renewable energy sources which have a positive impact on sustainable development. Featuring coverage on cogeneration thermal modules, photovoltaic (pv) solar, and renewable energy systems (RES) application practices, this publication is geared towards academics, advocates, government officials, policy makers, humanized managers, practitioners, professionals, and graduates interested in the latest research on smart human energy and smart optimization for sustainable rural development"-- Provided by publisher.
Identifiers: LCCN 2022043711 (print) | LCCN 2022043712 (ebook) | ISBN 9781668441183 (hardcover) | ISBN 9781668441190 (paperback) | ISBN 9781668441206 (ebook)
Subjects: LCSH: Agriculture and energy. | Energy development. | Power resources.
Classification: LCC S494.5.E5 H86 2023 (print) | LCC S494.5.E5 (ebook) | DDC 338.1/6--dc23/eng/20220928
LC record available at https://lccn.loc.gov/2022043711
LC ebook record available at https://lccn.loc.gov/2022043712

This book is published in the IGI Global book series Advances in Environmental Engineering and Green Technologies (AEEGT) (ISSN: 2326-9162; eISSN: 2326-9170)

British Cataloguing in Publication Data
A Cataloguing in Publication record for this book is available from the British Library.

Advances in Environmental Engineering and Green Technologies (AEEGT) Book Series

ISSN:2326-9162
EISSN:2326-9170

MISSION

Growing awareness and an increased focus on environmental issues such as climate change, energy use, and loss of non-renewable resources have brought about a greater need for research that provides potential solutions to these problems. Research in environmental science and engineering continues to play a vital role in uncovering new opportunities for a "green" future.

The **Advances in Environmental Engineering and Green Technologies (AEEGT)** book series is a mouthpiece for research in all aspects of environmental science, earth science, and green initiatives. This series supports the ongoing research in this field through publishing books that discuss topics within environmental engineering or that deal with the interdisciplinary field of green technologies.

COVERAGE

- Pollution Management
- Contaminated Site Remediation
- Alternative Power Sources
- Biofilters and Biofiltration
- Water Supply and Treatment
- Cleantech
- Radioactive Waste Treatment
- Renewable Energy
- Industrial Waste Management and Minimization
- Green Technology

IGI Global is currently accepting manuscripts for publication within this series. To submit a proposal for a volume in this series, please contact our Acquisition Editors at Acquisitions@igi-global.com or visit: http://www.igi-global.com/publish/.

The Advances in Environmental Engineering and Green Technologies (AEEGT) Book Series (ISSN 2326-9162) is published by IGI Global, 701 E. Chocolate Avenue, Hershey, PA 17033-1240, USA, www.igi-global.com. This series is composed of titles available for purchase individually; each title is edited to be contextually exclusive from any other title within the series. For pricing and ordering information please visit http://www.igi-global.com/book-series/advances-environmental-engineering-green-technologies/73679. Postmaster: Send all address changes to above address.

Titles in this Series

For a list of additional titles in this series, please visit:
www.igi-global.com/book-series/advances-environmental-engineering-green-technologies/73679

Applying Drone Technologies and Robotics for Agricultural Sustainability
Pethuru Raj (Reliance Jio Platforms Ltd., Bangalore, India) Kavita Saini (Galgotias University, India) and Vinicius Pacheco (Office of the Prosecutor General, Brazil)
Engineering Science Reference • © 2023 • 318pp • H/C (ISBN: 9781668464137) • US $250.00

Intelligent Solutions for Optimizing Agriculture and Tackling Climate Change Current and Future Dimensions
Ahmed Karmaoui (Moulay Ismail University, Meknès, Morocco & Moroccan Center for Culture and Science, Morocco)
Engineering Science Reference • © 2023 • 300pp • H/C (ISBN: 9781668446492) • US $215.00

Recent Advances in Global Meliponiculture
Shamsul Bahri Abd Razak (Universiti Malaysia Terengganu, Malaysia) Tuan Zainazor Tuan Chilek (Universiti Malaysia Terengganu, Malaysia) and Jumadil Saputra (Universiti Malaysia Terengganu, Malaysia)
Engineering Science Reference • © 2023 • 250pp • H/C (ISBN: 9781668462652) • US $215.00

Biomass and Bioenergy Solutions for Climate Change Mitigation and Sustainability
Ashok Kumar Rathoure (M/s Akone Services, India) and Shankar Mukundrao Khade (Ajeenkya D.Y. Patil University, India)
Engineering Science Reference • © 2023 • 412pp • H/C (ISBN: 9781668452691) • US $250.00

Handbook of Research on Building Greener Economics and Adopting Digital Tools in the Era of Climate Change
Patricia Ordóñez de Pablos (The University of Oviedo, Spain)
Engineering Science Reference • © 2022 • 391pp • H/C (ISBN: 9781668446102) • US $350.00

IGI Global
PUBLISHER of TIMELY KNOWLEDGE

701 East Chocolate Avenue, Hershey, PA 17033, USA
Tel: 717-533-8845 x100 • Fax: 717-533-8661
E-Mail: cust@igi-global.com • www.igi-global.com

Editorial Advisory Board

Table of Contents

Detailed Table of Contents

There is an increasing trend across the globe in establishing solar power plants in water ways and dams. This chapter presents, for the first time, the design and analysis of a typical floating solar power plant on the water surface of the Goreagab dam located in Namibia. Engineering design of the components is carried out in Google-SketchUp, and the standard System Advisory Model software is used for performance analysis. Three different case scenarios have been considered to study the possible extents of plant sizes and capacities. Design, simulations, and the analysis strongly favor the possibilities of establishing a floating solar power plant in the Goreagab dam. Additional benefits can be realized if an appropriate agrivoltaic ecosystem is established.

Hydroponics is becoming increasingly popular because it can be used in unconventional locations such as inside warehouses, marginal lands, and water-scarce regions. Hydrogen (pH), nutrient level, water level, temperature, and other parameters must be monitored on a daily basis in a typical hydroponic design. A hydroponic setup that is not dependent on the environment outside has been developed. The technology consumes less water and fertiliser as compared to a soil-based system. This system uses the idea of a relay and alarm connected to a microcontroller to make sure that nutrient solution is always available. The creation of an internet of things (IoT) network enabled remote parameter monitoring and control, which increased system stability. In addition to measuring and controlling the water level, pH, water stream rate, and EC at this time, the numerous sensors attached to the NFT framework support gathering fundamental system information.

Due to increasing power demand and the utilisation of natural resources in growing countries, the use of biomass and biofuels is emerging. The role of thermophilic ligninolytic bacterial enzymes in the biomass-to-biofuel conversion process is discussed in this chapter. Various elements of biomass feedstocks, compositions, and viabilities of lignocellulosic biomass are illustrated. The lignocellulosic biomass pre-treatment methods, inhibitors during the pre-treatment process, hydrolysis methods, and bacteria production processes have been explained. The processes for isolation, screening, and maintenance of cellulolytic bacteria are exemplified through suitable schematics. The utilization of agro-waste in the ethanal and biofuel production processes is also exemplified.

In this chapter, the various terminologies of soilless Agro-cultivation systems have been explained. Various factors have been used to explain the classification of soilless cultivation systems. Different fabrication elements of soilless, aquaponic

segmentation is suggested. The various parameters of the aquaponics system and the sensors used in the IoT and digital twin framework system are demonstrated. The controlling strategies, various smart technologies, and implementation procedures are also illustrated.

Energy is one of the most basic parts of the framework, and very essential for the national growth and economic development, of any country. The sustainable new sources of energy reproduction from traditional mediums like coal, lignite, petroleum, hydro, and nuclear capacity to non-traditional medium like wind, solar, agriculture, and home waste. Energy interest in the nation was prolonged very fast and is relied upon to rise further in the years to come. To fulfill the expanding need for energy, solar- or sun-based energy will play a real job in helping to this interest. The current scenario indicates that 60 GW of conventional power/energy capacity has been add over the last three years, which is believed to be the highest addition so far. As of April 2021, India has a thermal installed ability of 234 GW, and 53% of the part is covered by coal, diesel, and natural gas. The private sector produces 47.4% of thermal energy, although states and centres generate 27.1% and 25.5% individually. As of now, solar energy in India is exceptionally good.

In this chapter, the various sustainable development activities of hybrid floating solar photovoltaic (FSPV) technology with various emerging technologies are discussed. The basic principles, benefits, and important components of emerging FSPV techniques are illustrated. The developments of various environments, recent developments, classifications, and performance improvements by integrating various existing renewable energy resources have been elaborated. FSPV hybridization with hydro-power plants, hydro-storage, offshore, wave, and air energy storage are also

Ethanol can be made from lignocellulosic biomass, a promising replacement for fossil fuels. The bioethanol production process is divided into three basic unit operations. Conventional and well-established corn-to-ethanol technology can be cost competitive with the cellulosic ethanol process. For the pre-treatment of bagasse, steam explosion without a catalyst and a mixture of sulfuric and oxalic acids were used. The slab was submitted to enzymatic hydrolysis in the first scheme without any additional treatments like washing or solid-liquid separation. According to a study, post-treating bagasse with a pressure filtering process improved the amount of sugar that might be produced for ethanol. The total cellulosic ethanol process can be made successful. The effect of varying concentrations of sodium sulphite and its effect on both process configurations were also studied. The study provided an 8-10% increase in overall enzymatic efficiency as compared to the control substrate.

In this chapter, the general procedures for the biomass and bio-ethanal manufacturing processes are explained. The various physical and chemical pre-treating methods for biomass are elaborated. The various acid hydrolysis methods, enzyme immobilisation methods, fermentation methods, and optimising the parameters and factors for hydrolysis have been exemplified. The stage-by-stage bioethanol production processes from Sunn hemp seed has been illustrated. Biomass preparation, pre-treatment processes, hydrolysis, synthesis of ferrites, and study of bioethanol characterizations

Foreword

In our time in which ecological issues, the necessity to cover the energy needed in industries and households, and the new managerial concepts to be implemented remain at the forefront of Operational Research (OR), *Energy* and especially *Renewable Energy* has become an all-important field of study today. This is all the more true with *Agriculture*, which produces and consumes energy in so multiple ways, and more and more fosters our natural environments. Furthermore, while *Smart Technologies* continue to emerge and at the same time becoming refined, their applications widen and grow in their possible or promised impacts of triggering regime switching or paradigm shifting, even revolutionize sights and whole projects on energy production, energy consumption and sustainability. This potential and promise can only be realized fully with a thorough understanding of newest breakthroughs in the areas of *Energy Optimization for Business and Industry* in and of themselves as well as in their rapidly growing methodologies from mathematics and statistics, analytics and machine learning, probability theory and stochastics, OR and AI (Artificial Intelligence). With the present book, the Editors and Chapter Authors even went one step further than Energy Optimization for Business and Industry, namely by *Human Agro-Energy Optimization.*

The objective of this special Book is to explore latest developments of mathematical ideas and techniques in modeling and simulation, related with *(i)* Modeling Agro-Energy and its links with Business and Industry, including their foundations in physics, mathematics, statistics, and related elements of optimization, optimal control and game theory, *(ii)* Simulation of the models from agenda or program step (i), including simulation foundations in probability theory and stochastics, stochastic calculus and stochastic dynamics, numerical, computational and discrete mathematics and algorithms, graph and network theory, and *(iii)* Optimization of the models and simulation results received in the agenda or program steps (i) and (ii). They all can be applied in economics, finance and biology, as it becomes illustrated in this book which focuses its attention on modern and future *Agro-energy* and its utilization in emerging *Businesses* and *Industries*. Chapters on recently evolving topics of agriculture, business and industry have been welcomed especially. The editor of the book invited experts and researchers, practitioners and educators from all around the globe to prepare and submit high-quality innovative research chapters and critical review papers on a variety of subjects.

In recent decades, the analytical-algebraical and statistical-stochastic toolbox, the methods, instruments and programs of computer science and informatics, of agriculture and landscape architecture, of applied mathematics and statistics, of rising OR and MS (management science), of analytics and AI have attracted the attention of many scientists and engineers, users and consumers worldwide, causing a strong impact on and influx in the information sciences, natural sciences, high-tech and development, economics and finance and, in recent times, the health and security sectors, e.g., in storage, handling and transfer of data and energy, services and of goods, such as from agriculture, in traffic and production and transport, logistics and supply chains, inventory and waste management, etc. Key technologies based on Big Data, on Health- and Safety-awareness are known more and more, and used in study fields of science and engineering, design and technology, of ergonomics and medicine, marketing and trade. Modern trends in applied research challenges of those fields are also extended into the realms of economy and finance and not least into food industry and lifestyle sectors, hospitality and tourism, travel as well as urban and rural management, into healthcare and hygiene, education and continued education, worldwide.

This edited compendious is established and focuses on attractive, interesting and carefully selected chapters in a broad diversity of research topics. Well-known and emerging methodologies of modeling and simulation, optimization and optimal control are used to approach and solve some of the very complex, difficult and striking problems connected with the considered scope of investigations.

The authors of accepted contributed or invited chapters of this edited compendium are experienced and qualified scientists from universities or research institutes and experts from the applications, coming from all over the world. They designed, refined and used the deep model-based methods of mathematics and the less model-based, also so-called intelligent or smart algorithms and codes, and combined them with their handy, practical counterparts from in the engineering, computer science and informatics. The latter ones are often referred to as heuristics or model-free. In fact, they are regarded as less rigorous, freed from rigor of formal mathematics, to benefit from nature- and specifically bio-inspired approaches, to overcome and approximately solve hard problems in efficient ways. The rise of those algorithms from AI happened simultaneously to the powerful progress in model-based mathematics. In the last decades, some of the methodologies were called *Statistical*, *Machine* or *Deep Learning*, *Theory of Inverse Problems*, *Meta-* and *Matheuristic*s, *Analytics*, *AI*, and *OR*. All these and further model-free or model-based streams, traditions and philosophies are brought together to exchange and collaborate in all over the world, to leave behind misunderstandings and misperceptions between these two academic streamlines, to benefit from their joined forces and synergy effects, to jointly prepare scientific progress and give a common contribution to the solution of urgent real-life problems. These big problems occur in any field of the modern world and academics, not least in business and industry, in the environment and the energy sector, in agriculture and landscape architecture, in life- or human sciences, in neuroscience and medicine, and for the overcoming of conflicts and antagonisms, in developing and improving of living conditions, in ensuring fairness and providing of chances for the following generations. Here the fields of human-aided intelligent computing, human resource management, aggregate production planning and logistics, inventory, supply chain and waste management, pension fund systems, actuarial sciences and disaster relief deserve an extra mentioning.

In our modern world, many problems lead to mathematical descriptions and solutions, and even complex problems are approximated by mathematical and heuristic formulations. Modern optimization, optimal control and game theory, be they deterministic or stochastic, continuous, discrete or hybrid, constitute a mathematical framework to address issues with conflicting or cooperating parties, with trade-offs and negotiations, herewith having an important part in various decision-oriented real-life challenges. Some of the purposes of this remarkable book have been to explore and represent latest developments of model-based and model-free ansatzes, concepts and techniques in modeling, simulation, and optimization. This all requires novel and intelligent ideas which provide fresh insights into emerging challenges in our world, and strategies for solving as many as possible of them.

This compendium, *Human Agro-Energy Optimization for Business and Industry*, is an international collection and selection of innovative research that explores the recent steps forward on the ways of smart applications in sustainable modern and future businesses and industries, economies and societies. While it covers a wide scope of subjects, including energy assessment and control, uncertainty modeling and fuzzy logic or other approaches, biogeography and ecology, this work is designed and well-suited for academicians and investigators, developer and students, ecomomists and businessmen, industryleaders and policy-makers, engineers and educators, implementers and multipliers, multipliers and artists related with the subject fields. This compendium provides a *"golden opportunity"* to the worldwide community to familiarize and deepen, interact and discuss, distribute and share their newest and best theoretical findings, methodological developments, and applied creations and inventions, discoveries and results with their colleagues and students, friends and the public.

In all of these points, modern *Optimization* and *Optimal Control*, *Statistics* and *Data Science, Statistical* and *Machine Learning, Probability Theory* and *Stochastics*, *AI*, and *OR* play together as *key technologies* of modeling, model regularization and selection, of pre- and post-processing, of simulation, validation and testing, of guiding, supplying and refreshing, of continuous interest, respecting and encouraging.

Having been kindly invited by the Editor Prof. Dr. Pandian Vasant (MERLIN Research Centre, Ton Duc Thang University, Vietnam), we foreword authors express our gratitude to him and our hope that the selected topics and chapters of this book will also in the readers' eyes represent a core choice of today's international scientific research which copes with the emerging and complex problems of *Human Agro-Energy Optimization for Business and Industry*. We are very thankful to the house of IGI Global publishing house and its decision makers for hosting this precious compendium as a pioneering scientific agenda. We extend particular thanks to the book's editor for his dedication, interest, confidence, and support which he spent on this work from the beginning for anyone related, and to the valuable and kind leadership, management, and staff of IGI Global publishing house. We thank all the chapter authors for their diligent, intelligent, and tasteful work and their agreement to share their state-of-the-art works with our whole community, and the dedicated anonymous referees. Eventually, we wish that the authors' research will inspire and initiate collaboration and joined forces at highest stages for the sake of humanity and entire creation.

Gerhard-Wilhelm Weber
FEM, Poznan University of Technology, Poland & IAM, Middle East Technical University, Turkey

Selma Gütmen
FEM, Poznan University of Technology, Poland

Marta Kanczurzewska
Institute of Mathematics, Poznan University of Technology, Poland

NOTE

This foreword and related study was supported by the Grant by the Rector of Poznan University of Technology for its Institute of Mathematics.

Preface

This compendium addresses original, innovative, creative as well as recreative scientific and practical results from smart optimization, classical and novel computing, artificial intelligence and operational research, renewable energy and sustainability, economics and business, city or rural planning, meta-heuristics and big-data analytics, life quality and health, modern industries and emerging technologies. Indeed, it consists of carefully selected and very valuable chapters which have been reviewed and chosen by the experienced editor Prof. Dr. Pandian Vasant in order to become published under the highly innovative and rich name *Human Agro-Energy Optimization for Business and Industry* in the prestigious publishing house of IGI Global.

In the current era, there are many environmental and energy challenges facing the agricultural sector, which negatively impact the climate, business, industry, and society. Through modern technologies and engineering. However, these challenges are now able to be met with solutions. There is a high need for research in this area so that the industry can continue to be sustainable.

Human Agro-Energy Optimization for Business and Industry presents research on humanized optimization approaches for smart energy and the agro-business industry. It is a critical scholarly resource that examines the efficient use of modern smart farming and renewable energy sources, which have a positive impact on sustainable development. Covering topics such as biomass characterization, energy efficiency, and sustainable development, this premier reference source is an essential resource for agricultural scientists, engineers, government officials, software developers, managers, business leaders, executive officers, students and educators of higher education, librarians, researchers, and academicians.

COVERAGE

The many academic areas covered in this publication include, but are not limited to:

Aeroponics,
Aquaponics,
Biomass Characterization,
Energy Efficiency,
Floating Solar Power Plant,
Hydroponics,
IoT-Based Smart Farming,
Renewable Energy Systems,
Sustainable Development,
Weather Characterization.

This original compendium on *Human Agro-Energy Optimization for Business and Industry* reviews newest technologies in intelligent computing, operational research (OR) and artificial intelligence (AI) which are adjusting and, if needed, reducing the dimensions of data as well as variables related. This handbook discusses codes of modeling and simulation, intelligent optimization and optimal control which could be applied in numerous branches of energy engineering whenever uncertainty and stochasticity is a major issue. From an energy perspective, special attention is paid to the role of Agriculture, Farming and corresponding modern fields of landscape architecture and design as well as rural and city planning. Including emerging methods and techniques of OR and AI, and applying them together with advanced numerical codes and evolutionary algorithms to real-life problems for strategical and daily applications, this compendium works out distributed energy and hybrid renewable energy systems which employ OR and AI methodologies, also in view of new chances of blockchain technology for smart energy. By the rich variety of this work's contributions, the state-of-the-art developments and trends in ever changing key technologies are smartly covered. In these ways, this new reference work should become very helpful for students and researchers, developers and scientists, decision makers and users of science and technology, business and industry, management and economy, social sciences and the arts, for humanitarian helpers and preparers of the world of tomorrow.

CHALLENGES

Studies on agriculture and energy, farming and agro-energy, energy production and energy consumption, replacement and substitution effects, environmental protection and coverage of energy needs, science and technology, justice and safety, peace and freedom have been on the rise in recent decades. Any attempt to handle the many issues on their interfaces demands a deep understanding of the challenges which we face in our millennium. In view of the subjects of this compendium, such challenges may be classified into four classes:

The challenge of understanding the needs of our nations' citizens, individually and the whole population, of their economies and industries, and of their natural environments, including the antagonisms and the chances between those needs.
The challenge of understanding the modern energy sectors, of energy production and energy consumption, with a special focus on the role of the agriculture with its changing functions and roles in modern societies.
The challenge of finding pathways of Human Agro-Energy Optimization for Business and Industry into the future, especially in the light of the antagonisms and chances, functions and roles which we mentioned in items 1-2.
The challenge of advancing our scientific and technological, neuro-cognitive and bio-medical instruments. Methodologies and tools in order to solve the challenges stated in items 1-3.

This novel compendium looked for creative techniques in OR and AI, including fields of data science (DC) and deep learning (DL), inverse problems (IP) and machine learning (ML), neuroscience (NS) and statistical learning (SL), to intelligently condense, arrange and reduce the dimensions both of input variables and of datasets. It includes OR and AI, simulation tools and heuristics, optimization and optimal control, probabilistic and stochastic methodologies, e.g., by enhanced hybrid swarm-based optimization algorithms. The new book offers examples and tables, figures and charts for illustrating, better understanding, code-writing and implementing newest results and methodologies of modern Mathematics for the study subjects. This state-of-the-art work includes newest advancements in theory, methods and efficient implementations of green energy resources, renewable resources, biomass-, biofuel, geothermal or hydrogen, solar, tidal-, wave- or wind-power, for energy production and utilization, electrical instruments, devices and vehicles, industries and households, in OR and AI, DC and DL, IP and ML, NS and SL, in intelligent and soft computing, optimization and optimal control, mathematics and their modern applications.

The 13 chapters of this innovative book on *Human Agro-Energy Optimization for Business and Industry* are as follows:

Chapter 1, "Design and Simulation of a Floating Solar Power Plant for Goreagab Dam, Namibia" by Matheus Nadhipite Matheus and K.S. Sastry Musti, presents a design and analysis of a typical floating solar power plant on the water surface of the Goreagab Dam in Namibia. Three different case scenarios are considered to study any extents of plant sizes and capacities. Design, simulations, and the analysis strongly favor the possibilities of establishing a floating solar power plant in the Goreagab Dam.

Chapter 2, "Experimental Study on Reliable Smart Hydroponics System," by Sampath B., states that hydroponics is becoming increasingly popular because it can be used in unconventional locations. It develops a hydroponic setup which is not dependent on the environment outside, consuming less water and fertilizer compared to a soil-based system. IoT network enabled remote parameter monitoring and control increased system stability. Numerous sensors attached to the NFT framework support gathering fundamental system information.

Chapter 3, "A Study on Conversion of Ligninolytic Biomass to Biofuels," by Lakshmana Kishore T. and Haribalaji V., discusses thermophilic ligninolytic bacterial enzymes in biomass-to-biofuel conversion and illustrates elements of biomass feedstocks, compositions, and viabilities of lignocellulosic biomass. The latter one's pre-treatment, inhibitors during the pre-treatment process, hydrolysis methods and processes of bacteria production and for isolation, screening, and maintenance of cellulolytic bacteria and utilization of agro-waste in ethanal and biofuel production are exemplified.

Chapter 4, "Sustainable Developments of Modern Soil-Less Agro-Cultivation Systems," by Sampath Boopathi, Praveen Kumar Siva Kumar, Radhey Shyam Meena, Samson Isaac J., Suresh Kumar P., and M. Sudhakar, explains various terminologies of soilless agro-cultivation systems, a classification of soilless cultivation systems, and illustrates various fabrication elements of soil-less aquaponic systems with bio-fish and plant waste components. Implementation of the IoT in the aquaponic system is illustrated by microcontroller devices.

Chapter 5, "IoT-Based Smart Farming," by Pawan Kumar Singh, informs that agriculture has not been untouched by IoT enabled solutions. Along with more data, farmers predict more accurately climate changes and take better decisions in real time to save losses and optimize yield. IoT-based solutions in agriculture enable scientist and farmers to understand and innovate the farming techniques for a growing food demand of an increasing human population.

Chapter 6, "Sustainable Development in Modern Aquaponics Cultivation Systems Using IoT Technologies," by Ravi Samikannu, Ashok Kumar Koshariya, E. Poornima, S. Ramesh, Ashok Kumar, and S. Boopathi, helps in the development of a workable commercial aquaponics system, and looks at the IoT and smart systems. A suggested tool models the relationship between crop size and fresh weight and estimates crop size as it grows using image segmentation. Various parameters of the aquaponics system and the sensors used in the IoT and digital twin framework system are demonstrated.

Chapter 7, "Energy Efficiency in the Indian Scenario," by Vartika Singh, explains that solar or sun based energy will play a real role in helping to fulfill the expanding need for energy. As of April 2021, India has a thermal installed ability of 234 GW, and 53% of the part is covered by coal, diesel, and natural gas. The private sector produces 47.4% of thermal energy although states and centers individually generate 27.1% and 25.5%. Today, solar energy in India is exceptionally good.

Chapter 8, "Sustainable Developments of Hybrid Floating Solar Power Plants," by Vinay Kumar Domakonda, Shaik Farooq, Sekhar Chinthamreddy, R. Puviarasi, M. Sudhakar, and S. Boopathi, discusses various sustainable development activities of hybrid floating solar photovoltaic (FSPV) technology with various emerging technologies. Integrating various existing renewable energy resources, basic principles, benefits, and components of emerging FSPV are illustrated and developments of environments, recent developments, classifications, and performance improvements are elaborated.

Chapter 9, "Comprehensive Characterisation of Biomass and Weather for Deployment of the Renewable Energy Systems," by Yusto Mugisha Yustas and Valerian Cosmas Silayo, informs that inefficient use of traditional biomass energy sources for cooking and scarce electricity are challenges in least developed countries such that renewable energy (RE) related resources and sources are considered. A new approach's effectiveness is demonstrated involving sampling and laboratory analysis of cow dung, data collection and analysis of solar irradiance, wind speed, and ambient temperature. The results show that development of RE systems is feasible in the areas.

Chapter 10, "Artificial Intelligence Techniques in Water Purification and Utilization," by Suthanthira Vanitha, Radhika K., and Sampath Boopathi, explains that use of AI and cutting-edge technologies such as data analytics, regression models and algorithms, facilitating water management. AI-based systems will be sustainable, cost-effective, and capable of predicting potential damage to optimize water management with minimal waste. Machine Learning supports water management and improves accuracy of sensor-based systems in agriculture.

Chapter 11, "New Strategies in Treatment and Enzymatic Processes," by Sampath B., states that ethanol can be made from lignocellulosic biomass, a promising replacement for fossil fuels. Conventional and well-established corn-to-ethanol technology can be cost competitive with the cellulosic ethanol process. For the pre-treatment of bagasse, steam explosion without a catalyst and a mixture of sulfuric and oxalic acids were used. The total cellulosic ethanol process can be made successful. The study provided an 8-10% increase in overall enzymatic efficiency when compared to the control substrate.

Chapter 12, "Bioethanol Production From Biomass Extracted From Sunn Hemp Seed," by Ashok Kumar Koshariya, J. Madhusudhanan, M. Jinnah Sheik Mohamed, S. S. Kerur, Srinivas Viswanth Valeti, and V. Haribalaji, explains the general procedures for biomass and bio-ethanal manufacturing processes. The microstructure of untreated and treated hemp biomass are displayed. Incorporation of optimization parameters for producing quality bioethanol is also interpreted.

Chapter 13, "A Glance at Trends Manufacturing Systems Modeling and Optimization," by Gilberto Pérez-Lechuga, José Francisco Martínez-Sánchez, and Tomás Gómez-Rodríguez, analyzes main optimization methods used in flexible manufacturing systems (FMS). The results are obtained from a random sample of more than one hundred documents published from 1986 to 2020 dedicated to the optimization of FMS. The results are presented, and the virtual future of the FMS projected, as well as the new challenges and opportunities to venture into this important field of human activity. These are highly representative.

We are very grateful to the editor of this book, Professor Pandian Vasant, for his dedicated hard-work. Moreover, we express our sincere thanks to all the authors of the chapters and the many anonymous referees. Without them this book project would not have been realized. Eventually, we would like express our heartfelt thanks to the editors, officers, assistants, and staff members at all levels of IGI Global for their collaboration in publishing this remarkable compendium.

Gerhard-Wilhelm Weber
Poznan University of Technology, Poland & IAM, Middle East Technical University, Turkey

Selma Gütmen
FEM, Poznan University of Technology, Poland

Marta Kanczurzewska
Institute of Mathematics, Poznan University of Technology, Poland

NOTE

This preface and related study was supported by the Grant by the Rector of Poznan University of Technology for its Institute of Mathematics.

Acknowledgment

The editors would like to sincerely acknowledge the help and support of all the people involved in this project and, more specifically, to the authors and reviewers that took part in the review process. Without their effort and hard work, this book would not have become a reality.

First, the editors would like to thank each one of the authors for their excellent contributions. Our sincere gratitude goes to the chapter's authors who contributed their time and expertise to this book.

Second, the editors wish to acknowledge the valuable contributions of the reviewers regarding the improvement of quality, coherence, clarity, and content presentation of chapters. Most of the authors also served as referees; we highly appreciate their double task.

Pandian Vasant
MERLIN, Faculty of Electrical and Electronics Engineering, Ton Duc Thang University, Vietnam

Chapter 1
Design and Simulation of a Floating Solar Power Plant for Goreagab Dam, Namibia

Matheus Nadhipite Matheus
Namibia University of Science and Technology, Namibia

K. S. Sastry Musti
https://orcid.org/0000-0003-4384-7933
Namibia University of Science and Technology, Namibia

ABSTRACT

There is an increasing trend across the globe in establishing solar power plants in water ways and dams. This chapter presents, for the first time, the design and analysis of a typical floating solar power plant on the water surface of the Goreagab dam located in Namibia. Engineering design of the components is carried out in Google-SketchUp, and the standard System Advisory Model software is used for performance analysis. Three different case scenarios have been considered to study the possible extents of plant sizes and capacities. Design, simulations, and the analysis strongly favor the possibilities of establishing a floating solar power plant in the Goreagab dam. Additional benefits can be realized if an appropriate agrivoltaic ecosystem is established.

INTRODUCTION

It is a well-known fact that energy production requires land, water, fuel, and heavy financial investments. Thus, there is a nexus between several aspects such as land,

DOI: 10.4018/978-1-6684-4118-3.ch001

fuel, water and even food (Musti, 2021; Al-Widyan et al, 2021). Almost all nations in the world are opting for solar energy due to financial savings, however availability of land is a major issue. Though, solar energy is inexhaustible, economical, renewable and natural; it requires a lot of land when compared to fossil fuel based energy generation (Malu et al, 2017). Sub-Saharan countries are facing shortages of energy and food and a greater degree of resource nexus can be seen. Thus, establishing economical energy production systems with reduced conflict between resources is imperative, especially for countries like Namibia (Musti, 2021).

Of late, several nations have started setting up solar based energy plants in waterbodies like rivers, canals and even in the ocean (Tarigan, 2021; Radhiansyah et al., 2021; PRNewswire, 2021; Laila, 2021; AgriSolar, 2022). Though there are a good number of solar plants that were set up on waterbodies, two major projects one from Netherland (EnergyLive, 2019) and the other one from Singapore (PRNewswire, 2021) are purpose built with distinct advantages (Ranjbaran et al., 2019). These are typically known as Floating Solar Photo Voltaic Systems (FSPV Systems). The novelty of these FSPV systems brings in several advantages including effective usage water resources, reduction of water evaporation losses, off-shore energy production that can save use of regular on-shore land etc. However, it should be noted that capacity addition requires in-depth investigations such as techno-economic studies, environmental impact studies, resource conflicts and overall benefits to the society in terms of providing employment etc. (Azran & Majid, 2014). Different advances and disadvantages are associated with different energy production with different fuels. In case of solar energy, the major challenge is the large volumes of land requirement, which may result in a conflict with agricultural/ food production (Pearce and Dinesh, 2016). Several other parameters such current grid operating conditions, supply-demand-gap, demand forecast, ongoing Demand Side Management (DSM) policies, intended location of new plant etc., also play their own role in feasibility studies (Hauwanga & Musti, 2022). Dominance of solar energy in the overall energy pool can lead to duck curve phenomena (Pitra & Musti, 2021). Liberal DSM incentives to reduced energy use can lower the overall demand as well (Musti et al, 2020). On the other hand, several developed nations encouraging conserving the energy through circular economy principles. Smart cities based on circular economies may have localized energy generation and may be less dependent on the regular power grids (Musti, 2020).

From the above, it can be seen that several options exist for energy production, however there are several constraints as well. FSPV plants do offer distinct advantages if designed carefully. Since this is relatively new and niche area, it is important that proper design and techno-economic studies need to carried out in the first place (Solomin, 2021). This chapter presents an engineering design framework for such a plant in Namibia then undertakes the economical evaluation for different options. The content in this chapter is divided into several sections. Firstly, this a comprehensive literature review is presented. This review considers different aspects such as – social, contemporary, environmental and technical developments as related to FSPV plants. Then design of a typical FSPV plant based on Google-SketchUp platform is presented, followed by simulation studies using SAM software to determine the energy production levels. The next section takes up three different case scenarios, and financial studies for all the cases have been presented. Discussion and scope for further work are provided in the following section. Then the concluding remarks are provided.

LITERATURE REVIEW

In the recent past, several authors have presented various aspects are related to FSPV plants including – narration of various such plants around the world; advantages and disadvantages; environmental impact studies, design and technical studies etc. This section looks at a few important aspects that are necessary for designing a new plant for a specific location (Hauwanga &. Musti, 2022), dedicated loads through single buyer mechanisms with known demand response characteristics (Musti, 2022; Musti, 2020). A few authors have presented various aspects related to the direct nexus between energy production with the use of water resources specifically in Africa (Gulati et al, 2013; Sanchez et al, 2020; Musti 2021). This naturally warrants a thorough understanding of water evaporation with and without FSPV plants.

Basics of Fspv Systems

Floating Solar Photo Voltaic Systems (FSPV Systems) contains solar PV panels that are designed to stay on top of waterbodies such as reservoirs, hydroelectric dams, industrial ponds, water treatment ponds, mining ponds, lakes, and lagoons (Pearce et al, 2016). These are also known as floatovoltaics (Solomin, 2021). Different floating type structures are used to keep these panels floating. Usually, floating structure is anchored and moored.

Terminology

A good number of new words or phrases that are used in this area. Some of them are – Agrivoltaics, Agri-PV systems, Agri-solar systems, floatovoltaics, Floating Solar PV systems (FSPV systems) etc. The simultaneous use of sections of land for both solar photovoltaic power generation and agriculture is known as agrivoltaics (Pearce et. al, 2016). The term agrivoltaics in general, refers to the integration of solar

PV projects within an agricultural activity. It includes a variety of business models, co-designed between the solar and the agricultural sectors such as Agrivoltaic (or Agri-PV) systems, the deployment of PV on the roof of agricultural buildings or, powering agricultural machinery with solar power.

Agri-PV systems are Agri-solar solutions where a PV installation and a sustainable agricultural activity are co-located, and light management is performed. In other words, Agri-PV projects optimize the quantity and quality of light that reaches crops and the PV cells. Significantly, in certain instances agricultural productivity increases as a result of the solar panels' shadow reducing some of the stress that high temperatures and UV ray damage have on plants. In terms of efficiency, agricultural land is best suited for solar farms (Pearce and Dinesh, 2016). Solar industry can make more money/power by replacing farming land with fields of solar panels, as opposed to using bare lands. This is mainly because farms have typically been established in damp places, and solar systems generally lose efficiency at higher temperatures. Vapor pressure's cooling effects are a key role in improving panel efficiency. Several articles indicated that establishing agrivoltaics with FSPV systems is indeed a win-win situation (EUScienceHub, 2021; Almeida et. al, 2022),

ADVANTAGES OF FSPV SYSTEMS

Agri-solar systems do offer several advantages. According to Pearce et al. (2016) and Gozálvez et al. (2012), agrisolar systems enable a more efficient use of land, through the co-location of agriculture and PV. Agrisolar systems enable a more responsible use of water in agriculture. If Agrisolar was deployed on only 1% of Europe's arable land, its technical capacity would amount to more than 900 GW, more than 6 times the current installed PV capacity (EUScienceHub, 2021). Agrisolar offers a turnkey solution to reduce greenhouse gas emissions from the energy sector, deploy additional solar capacity, promote more sustainable agricultural practices, reduce the environmental impacts of agriculture, and drive rural development. Considering the drawbacks Agrivoltaics will result in a decrease in productivity because some land will be sacrificed for mounting structures and system equipment, which is one of the drawbacks of permitting some agriculture to take place on the solar power plant (Pearce and Dinesh, 2016). However, because photovoltaic systems are technologically sophisticated, farmers won't be able to repair anything that might malfunction or be damaged, necessitating a large enough workforce of experts. Additionally, a substantial investment is needed.

Besides the inherent benefits of installing PV on a water surface, such as avoided land-energy conflicts, reduced shading on modules from surrounding objects, reduced module operating temperature resulting in a better performance of the PV plant, or the potential improvement on water quality through decreased algae growth etc (Akole and Jadhav, 2018; Al-Widyan et. al, 2021). Thus, there are some specific potential advantages that can be realized from hybridizing FPV with an established hydro power facility (Laila et. al, 2021; Akole et. al(2018). In recent years, many theoretical and experimental studies have been carried out to design cooling systems for PV panels (Solomin, 2021; Hermanu et al, 2019). FSPV systems are also regarded as one of the solutions for this problem. When the solar power plant is built on waterbody, the air on the surface is cooler and overcome the problem of overheating solar panel surface (Gozálvez et al, 2012). Tarigan (2021) points to the fact that FSPV plants can be approximately 11% more efficient than conventional solar plants based on the research conducted at Marine Pertamina in Palembang, South Sumatra, Indonesia.

PV modules in deep and shallow water are proposed as a solution for cleaning panels, reducing reflection losses and efficiency improvement by elimination of thermal drift (Junianto, 2020). This system can be used to power underwater devices, as well as swimming pools and decorative pools and fountains. On the other hand, hybrid energy harvesting system employing thermoelectric generators integrated with solar panel are designed to harness both light as well as thermal energy. The thermoelectric generators are employed at the rear side of the solar panel so as to protect the solar panel from getting heated up and at the same time harvesting the wasted thermal energy (Singh et. al., 2016). Floating pontoon or floating support structure needs to be carefully designed. A pontoon is a floating device made up of polymer and has enough buoyancy to float on water with heavy load placed on it. The platform is design to hold suitable number of modules in series parallel combination according to the requirement and space availability (Giri et al., 2019). Multiple plastic hollow floats with effective buoyancy to self-weight ratio are connected in series such as to form a huge pontoon. They are basically made of HDPE (high density poly-ethylene), known for its tensile strength, maintenance free, UV and non-corrosive. Glass fiber reinforced plastic (GRP) can also be used for construction of floating platform (Solomin, 2021; Hermanu et al, 2019). Other aspects of FSPV to enhance the overall efficiency and energy production have also been taken up. Tilt angle of the PV modules is crucial in determining how much wind drag is created on the mounting framework (Giri et al., 2019). This is especially crucial in the case of FSPV plants since an erroneous tilt angle can increase drag forces and seriously harm the overall plant structure. The work of Singh et. al.(2016), examines the relationship between the tilt angle of a PV module and wind speed, finding that a high wind speed necessitates a smaller tilt angle. This work also looks into the plot size how it affects the drag force and possible solutions to keep a single plot size small and symmetrical. In addition to the conventional monofacial, Bifacial Floating Solar PV (BFSPV) is a cutting-edge technology that collects solar energy from both sides of the solar panel (Widayat et al., 2020). When used in FSPV, more energy can be produced, as BFSPVs have two receiving sides, which allows them to receive sunlight reflection from the water surface on the lower side of the PV, increasing the amount of electrical energy produced. Widayat et al. (2020) presented the comparison of bifacial floating Solar PV and monofacial floating Solar PV.

FSPV Plants around the World

Floating solar parks provide a wide range of options for using economically exploited bodies of water such as reservoirs, fish farming waters or lakes on former open-cast lignite mines – in two different ways. Technical studies have identified a potential for Europe of 20 GW if only 1% of the surface of man-made freshwater reservoirs will be used (AgriSolar, 2022; EUScienceHub, 2021). These studies also indicate that floating solar energy is technically manageable and comes with only slightly higher cost than ground-mounted systems. In order to optimize the Levelized Cost of Electricity (LCOE), the stability and long-life durability of the systems are absolutely key. Interestingly, the developers of Sekdoom project clearly indicated that the basis for these floating installations is the nature of the mounting system, specifically in terms of design and quality (EnergyLive, 2019).

During 2019, The Netherlands has completed a floating solar array facility, located near the city of Zwolle (EnergyLive, 2019). This is popularly known as the Sekdoom Project. With its total capacity of 14.5MW, this project can supply electricity to approximately 4,000 households (EnergyLive, 2019). Construction of the Sekdoom project took only a few months' time and it is one of the biggest floating solar farms in Europe. Construction of this plant has used typical installations that are extension to ground-mounted facilities and a smart contribution to improve the so called 'double function' applications for solar, such as AgriSolar, Carports, building integrated PV and rooftop PV. The Sekdoom project developers indicated that more than 100MW of floating PV capacity will be added in immediate future in Germany, France, Italy, and Spain. According to the developers, the potential for floating solar farms in Europe is indeed significant (EnergyLive, 2019).

Singapore also adopted offshore FSPV plants as a part of it journey towards carbon neutrality (PRNewswire, 2021). The market for solar energy is heating up worldwide, with more and more countries joining the Race to Zero, United Nations states that "a global campaign to rally leadership and support from businesses, cities, regions, and investors for a healthy, resilient, zero carbon recovery". Odetoye et. al (2022) presented a case study for a large scale FSPV This is a concerted, global effort to fight against climate change. Singapore, a small and resource-scarce city-state, is no exception, and the island is now home to a number of offshore floating solar farms deployed in the Straits of Johor.

Table 1. FSPV plant around the world

Name of Location, Country	FPV size (sq km), capacity (MW)	Country's Total MWh Electricity Production
Huainan city, China	8 ,40MW	5,882,943,540 MWh
Traits of Johor, Singapore	0.05 ,5MW	7,400,000 MWh
Walton-on-Thames, United Kingdom	0.057 ,6.3 MW	1,651,000,000 MWh
Zwolle, Netherland	0.18 ,14.5MW	119,000,000 MWh
Ichihara, Chiba Prefecture, Japan	0.18 , 13.7MW	989,335,520 MWh
Khandwa, India	100 , 600 MW	1,386,436,420 MWh
Huainan City, china	0.45, 150 MW	5,882,943,540 MWh
California, USA	0.0607 , 4.8 MW	3,902,306,027 MWh

Restricted by its small size, Singapore lacks access to natural and renewable energy sources such as wind and hydroelectric power. Thus, Singapore has considered offshore floating solar energy option as a viable alternative. In doing so, it has achieved both solar energy production and also has constructed an undersea cable system connecting the floating platform to the mainland, creating a 22 KV electrical distribution network. Singapore's Green Plan 2030 has set ambitious targets of achieving 1.5 GW-peak by 2025, with further plans to reach 2 GW-peak by 2030. Table1 summarizes various projects all around the world.

Though the recent literature and several projects point to the successful experiences of FSPV systems, addition of any structure in large waterbodies requires the active involvement of the state. Establishing a floating solar PV plant is part of energy capacity addition, it is typically done at the national level in any country. For this comprehensive engineering and cost benefit studies have to be undertaken. There are different factors that need to be carefully assumed as part of engineering analysis. For this reliable information on existing power plants is required in some instances exercise solar generation can lead to duck curve phenomenon (Pitra & Musti, 2021). Location of any privately owned power plants are subjected to risks and location of point of coupling is one of the major contributing factors (Hauwanga & Musti, 2022). If floating solar plant is of small-scale capacity than network reticulation is desirable to avoid protection issues. Thus, radial type networks are desirable with concentrated loads and this in turn reduces the complexity of the network topology (Mks, 2005). In the design process of localized, islanded grids demand response need to be considered in sizing the capacity of the plant.

From the above, several aspects are clear. There is significant interest in FSPV systems. However, there are several challenges that come with Agrivoltaics, specifically in setting up the FSPV systems. While design of solar panels itself is very well matured, the engineering design of floats, support structures and overall system orientation etc. are not very well discussed. Waterbodies do exist in different sizes and water buoyancy and turbulence can vary from place to place, just as the solar radiation. Further, just as commissioning of any other power plant, FSPV systems also need careful engineering evaluation in terms of overall costs, greenhouse gas emissions, social and environmental impact. Such evaluation should be undertaken for a specific location, as parameters such as solar radiation, social conditions and costs vary from place to place. Then how the energy produced can be utilized, how the plant can be connected to the main grid, if at all it is; how the wheeling charges are impacting the plant operational economics etc. have to be also studied, just as for any other plant.

Since, FSPV systems need to be designed, analyzed carefully, this chapter undertakes a systematic engineering approach for design and analysis for a typical FSPV system. Though the approach suggested in this work is general in nature, Goreagab Dam, Namibia is considered for undertaking a case study. The objectives of this chapter will be:

- Design and simulate a floating solar PV system for Goreagab Dam, Namibia using Google Sketchup for different support structures
- Undertake comprehensive techno-economic analysis with different case studies with SAM (System Advisor Model)
- Estimation of energy output of FSPV systems when compared to land-based systems
- Illustrate how FSPV systems contribute to demand side management, circular economy of the local energy markets.

Some of the computations that this chapter considers include – efficiency of the energy conversion, number of panels and modules; size of the overall system that contribute to the technical specification of the plant and guidance on reticulating the network to avoid known technical and operational challenges; and water saving potential.

FLOATING SOLAR PV PLANT DESIGN

The design process of any FSPV needs to start with available water body space or area of water surface, as this is the major factor in limiting the overall size and capacity of the plant. Then typical components of the FSPV that need to be designed are:

Figure 1. Water surface area of Goreangab Dam on Google Map

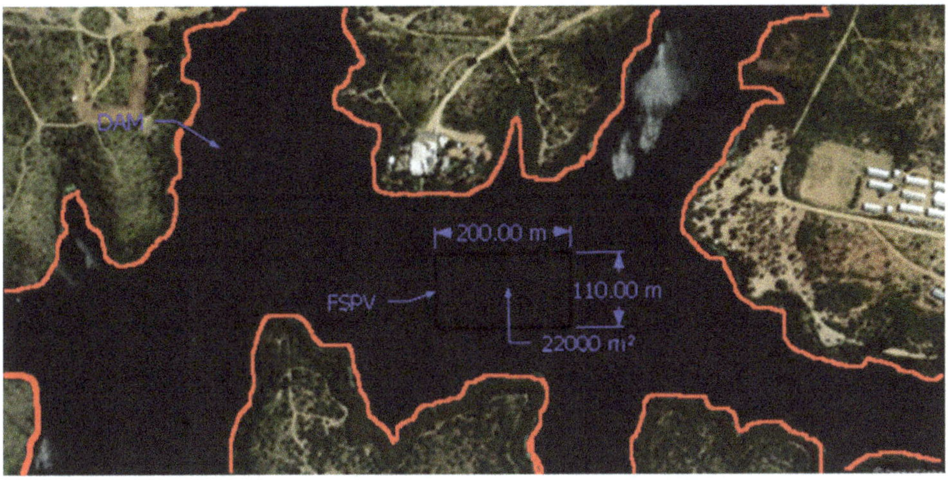

- Anchoring and Mooring System
- Floating Platform
- PV Modules
- Inverters

Since this chapter aims to design the FSPV plant for Goreagab Dam, Namibia, specific parameters related that facility need to be considered. The water surface area of the dam is estimated to be around 1,100 000 . Significantly, for this simulation the total capacity of the FSPV is 5.01 which approximately corresponds to a dam area of 22,000 which is 2% of the total water surface area of the dam. The dimensions of the FSPV are shown in Figure 1.

The design of the Floating Solar PV is identical to that of a traditional solar PV system, with the exception that it requires a unique configuration to float on the water surface. A standard floating framework supports the PV system. Arrays, inverters, combiner boxes, lighting arresters, and other components will be on a single board, the bed will be made of fiber-reinforced plastic and floats (FRP) or metal structures, or high-density polyethylene (HDPE).

Anchoring and Mooring

The floating platform's anchoring and mooring mechanism keeps it in place and provides the mechanical stability it needs for the duration of its operation. The system's most important components are anchors and mooring lines. FPV platforms can be fastened to the bottom of a body of water or to the shore's bank. Figure 2

Figure 2. Concrete block – Google Sketchup design

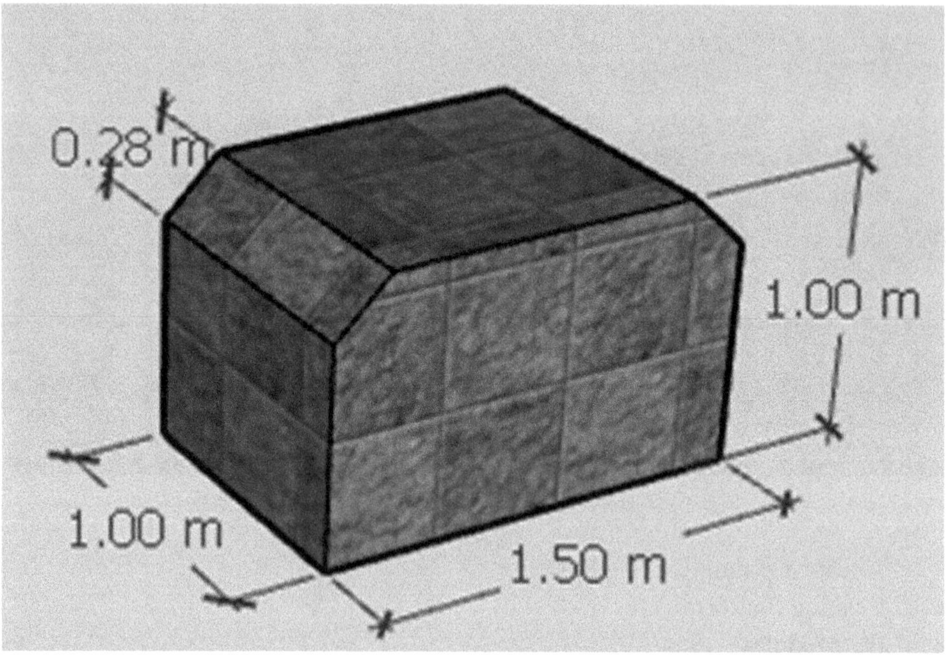

displays the size and Google Sketchup design of the concrete block. The concrete block is intended to be 1.5 in area, and 1.50 m, 1 m, and 1 m in length, width, and height, respectively. As per the current design, the floating platform is fastened to the water's bottom as seen in figure 4. Dead weights like concrete blocks and helical anchors that are fastened into piles on the ground are two common types of anchors used in bottom anchoring.

Floating Platform

A floating platform made out of several high-grade plastic fiber floats tied together, which is also known as pontoon is generally used to hold the solar panels. This float platform is very sturdy structure, but is allowed to move freely as with water buoyancy, while holding the solar panels on top of it. For this design, the pontoon is intended to have an area of 0.23 , with the dimensions given in figure 3 being 0.50 m in length, 0.50 m in breadth, and 0.20 m in height.

Table 2 presents the number of pontoons, and the associated number of PV modules. 183,141 pontoons are used to support the 10773 solar PV modules that make up this design's whole PV module count.

Figure 3. A section of floating Pontoon - Google SketchUp design

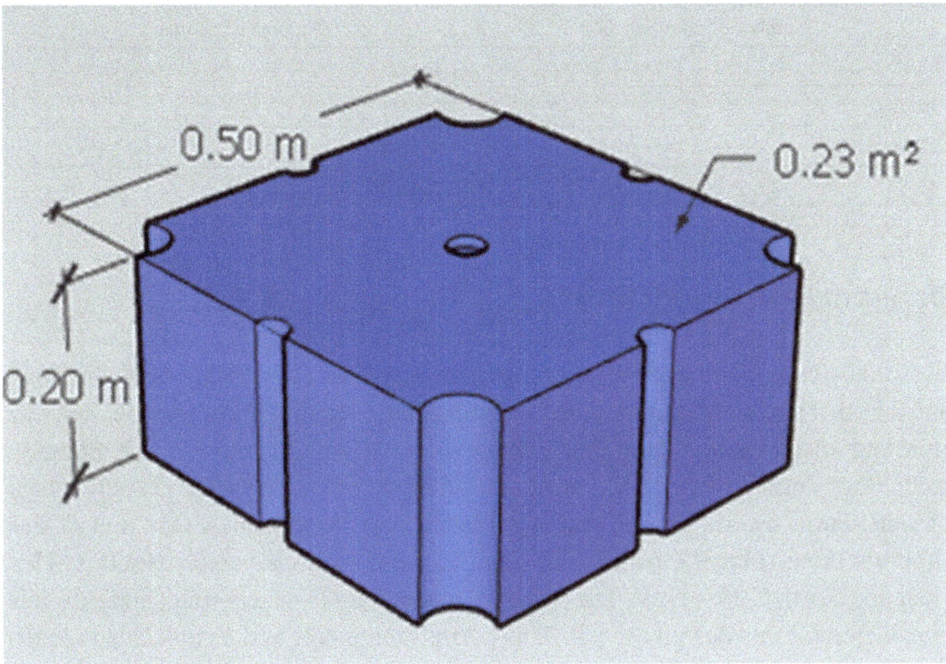

Figure 4. Standard onshore mooring with floating solar panels - Google Skechup design

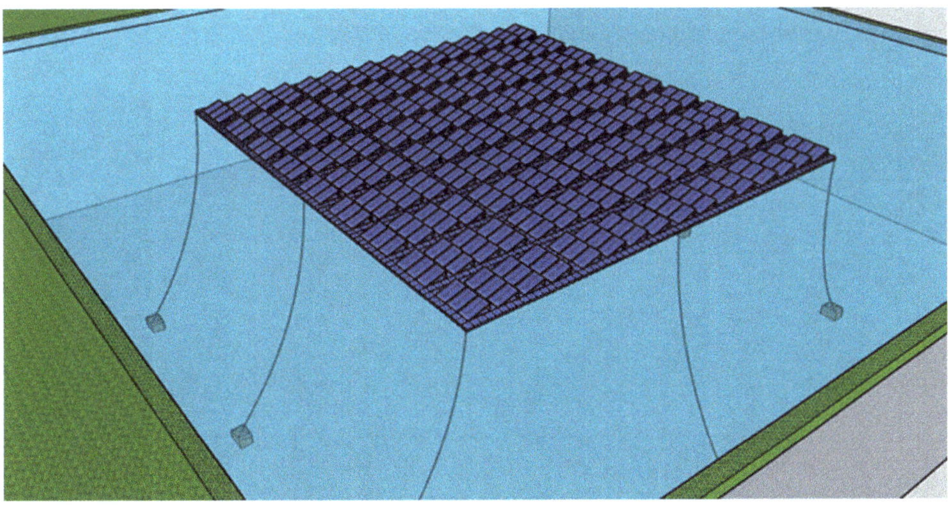

Table 2. Number of Pontoons and solar panels

Number of solar panels	Number of pontoons
8	136
10773	183141
22596	384132
44682	759594

Different Floating Platforms

Essentially two design alternatives can be considered one with Seahorse technology and the other one with Stingray technology. Seahorse technology offers one walkway float and solar PV panel floats that hold the solar PV panels in a single unit. Stingray technology option tries to reduce evaporation losses by covering a larger percentage of water surfaces. Stingray is best suitable for large waterbodies. One unit of this supports three solar PV panels and has two float sections namely South C-Hex float and North C-Hex float. The seahorse technology is widely used, and thus it is considered for this design as well. This permits engineers and technicians to reach the PV panels for easy system maintenance because it includes a walkway. Figure 5 shows the side and top view of the design in Google Sketchup.

The solar panels are slanted at a 20-degree angle when viewed through the beam of the floating platform, which supports them while they are on the floats. The

Figure 5. Seahorse technology - Google SketchUp design

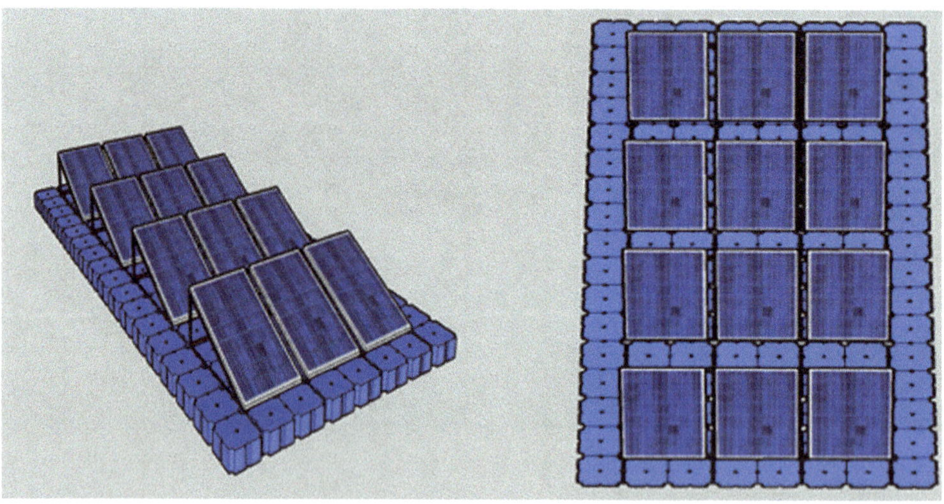

Figure 6. beam - Google SketchUp design

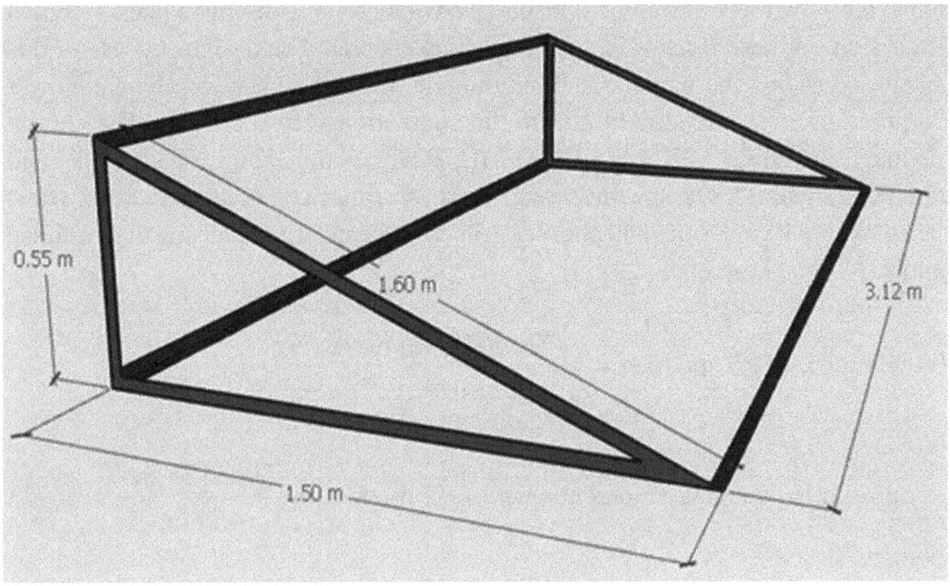

beam has the proportions depicted in figure 5. The beam can support a maximum of three solar PV modules.

SOLAR PV PANELS

The specific energy output of a PV system which is defined as the amount of energy output of a PV system in comparison with the total solar irradiation received by the solar panels under operating conditions. It is calculated by comparing PV system energy output E_{out}, in kWh to maximum power capacity, P_{max} in kW_p under STC, specified by the standard formula.

$$SpecificEnergyOutput = \frac{E_{out}}{P_{max}}$$

Where E_{out} is the energy output for the actual condition and P_{max} is the power capacity under standard test conditions. However, it should be noted that the efficiency of the PV module depends on the amount of solar radiation that can be converted into electrical energy. Besides the temperature, type of material and type of solar panels also influence specific energy output. The most common types polycrystalline and monocrystalline type panels have been considered in general for most installations, due to their own sets of technical features, advantages and disadvantages. Detailed discussion on these aspects is avoided in this chapter. The design process in this chapter considers the monocrystalline module, from the manufacturer SunPower Corporation as it is readily built into the simulation software SAM. The chosen module is SunPower SPR-A425-COM-MLSD with nominal power of 464.6 W, and an efficiency of 22.89% for the overall total DC capacity of 5005135.8Wp. Now, the number of PV modules and parallel strings can be determined using the formulae below (Radhiansyah et al., 2021)

$$\text{Total number of PV modules} = \frac{TotalPVplantcapacity}{RatedpowerofaPVmodule}$$

$$\text{Substituting the values, total number of PV modules} = \frac{5005135.8\wp}{464.6\wp} = 10773$$

modules

$$\text{Then, total Number of strings in parallel} = \frac{TotalnumberofPVmodules}{NumberofPVmodules \in astringmodule}$$

$$\text{Substituting the values, total Number of strings in parallel} = \frac{10773}{21} = 513$$

Inverter sizing and DC to AC Ratio

A typical commercial inverter in solar energy industry is a power converting and conditioning device that is made up of IGBT switches with reverse current blocking diodes and it converts PV array output DC voltage into AC voltage. As a result, the inverter rating represents its ability to process the power of the PV array. SAM has made available several standard inverter models. The chosen module is DYNAPOWER COMPANY LLC: CPS-1500 Gen4, with a maximum AC power of 1498660 W and a CEC weighted efficiency of 97.167%. Significantly, the DC/AC ratio is the relationship between the amount of DC power of the modules linked to the AC power of the inverters. To get the best PV system performance, the inverter rated capacity and the rated PV array capacity must be matched (Faranda et al, 2015; Meneghel et al, 2018). An appropriate DC/AC ratio design must take into account the specific project requirements. Irradiation, system loss, inverter lifetime, efficiency, operational voltage range, module mounting angle, etc. are among the primary determinant (Giri et al., 2019).

Figure 7. Overall schematic for the FSPV

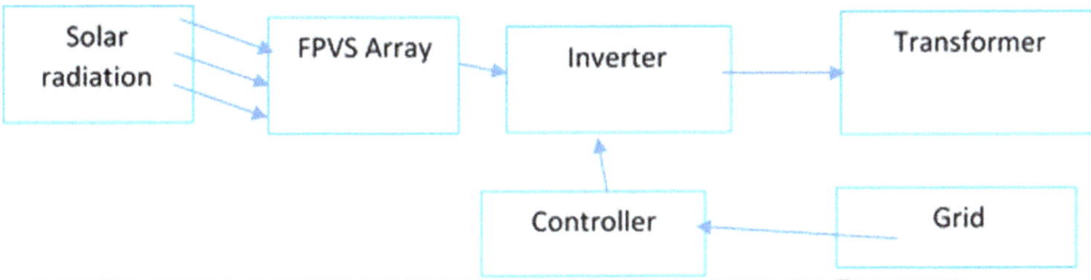

The DC output of solar arrays is typically specified higher than the AC output of inverters because solar arrays rarely produce their rated DC output. First off, Standard Test Conditions (STC), which are strictly controlled laboratory environments used to determine a solar panel's nameplate wattage, are considerably dissimilar from the actual working circumstances of commercial solar installations. A matched AC capacity inverter will end up being large for solar systems since they operate below their rated DC output. This can be expensive as the nameplate capacity of inverters heavily influences their pricing. The use of an inverter with a rated AC output lower than the nameplate capacity of the DC array is a fairly trivial solution. The DC/AC ratio or inverter load ratio is calculated by dividing the array capacity (kW DC) over the inverter capacity in kW AC (Giri et al., 2019). With three RES-BESS60DP-480 inverters, the overall inverter capacity for this design is 200.707 (3 inverters with s 59.789 kW each), resulting in a DC/AC ratio of 1.12.

$$\text{DC/AC ratio} = \frac{arraycapacity\left(kWDC\right)}{invertercapacity\left(kWAC\right)} = \frac{5005.136kW}{1498.660 \times 3kW} = 1.12$$

A high DC/AC ratio has the benefit of allowing you to increase energy output without raising the expense of the solar installation's AC component. We can only use this ratio to a certain degree before reaching a limit. Inverter clipping is the result of a high DC to AC ratio. During periods of peak output, when the solar panels generate more energy than the inverter's maximum AC power, this occurs. The inverter "clips" the excess solar energy during these times (Giri et al., 2019). However, in general, 1.12 is considered the standard industry AC/DC ratio in majority field installations.

Table 3. Input Parameters for Simulation

Parameter	Input
Site name	Goreangab, Namibia
Latitude	-22.59
Longitude	17.06
Tilt(deg)	20
Azimuth (deg)	180
Installed power (kW)	5,000
Ground coverage ratio	0.3
Mounting system	fixed mounting, free standing
Number of modules	10773
Number of inverters	3
DC/AC ratio	1.12

On-grid FSPV System

In a typical on-grid scheme, the FSPV system the output from the panels is connected to the state-owned transmission network without using energy storage components as governed by REFIT program policies.

A properly sized solar charge controller is necessary for the solar power system to regulate the voltage and current. A maximum power point tracking (MPPT) charge controller with an efficiency between 94% and 98% can be used. Two different inverter layout configurations - Central-inverter and string-inverter configurations are considered in the study, however detailed discussion on this is avoided to save space. Central-inverter configuration will have the inverters sit on floating platforms at the center of the plant to minimize cable runs and electrical losses. Whereas the string-inverter configuration will have the inverters usually placed alongside each platform on secondary or multifunction floats, which are designed to hold electrical equipment or serve as maintenance walkways. The central-inverter variants for the inverter architecture is found to be more suitable for this design since they decrease cable runs and electrical losses, apart from other benefits.

Energy Output Simulations

SAM (System Advisor Model) software is utilized for the simulations to the overall costing and effectiveness of renewable energy systems. SAM provides a set of input variables for the system under simulation based on the location and a few initial

Table 4. Simulation results and metrics

Metric	Value
Annual energy (year 1)	7,251,508 kWh
DC capacity factor (year 1)	16.50%
Energy yield (year 1)	1,449 kWh/ kW
Performance ratio (year 1)	0.79
Electricity bill without system (year 1)	$104,614.00
Electricity bill wit system (year 1)	$47,268.00
Net saving with system	$57,347.00
Net capital cost	$8,605,086.00
Debt	$8,605,086.00

specifications. For this simulation, Windhoek, Namibia, weather data is collected from the National Solar Radiation Data Base (NSRDB) for simulation purposes. The results of the simulation are extrapolated based on prior works by Tarigan (2021) and Ranjbaran et al. (2019) to produce the results if the system is installed on floats to float on water surface. The simulation is performed using an open ground area as the location of solar panels. Table 3 presents the input parameter for the simulation.

According to the data provided by NSRDB, Widhoek has a very high potential for solar energy in terms of global irradiation. Direct, diffuse, and reflected radiation are all components of the global radiation. The range of the daily global irradiance is 20 to 400 kW/m2, with a daily average of roughly 210 kW/m2. From October to March, the global irradiance predominates. For a grounded-based PV system with an energy yield of 1,449 kWh/kW year, the annual energy in Windhoek will be 7,251,508 kWh with the input parameters listed in table 2. The corresponding results and metrics for this set of input values are shown in table 4.

A majority of the designed system's power is generated between the first three months and the last three months, according to an analysis of how much power the system generates. This is mostly because Namibia has three distinct climate types that have an impact on the landscape. With 300 days of sunshine on average each year, it is undoubtedly a sunny nation. The summer season lasts from October to April, and highs of 40 °C (104 °F) are not uncommon in Namibia.

Based on the simulations, it can be seen that the monthly net electricity output ranges from 325,000 to 912,500kWh, with January having the highest and June having the lowest amounts. Column 2 in table 5 displays the annual monthly net electricity to grid. The PV system mounting structures come in a variety of types, and the software can mimic them all. The goal of choosing a mounting system type is to maximize the amount of solar radiation that hits the panel surface, which translates to maximize the PV system's energy output. A PV system's mounting method has a big impact on how much power it produces specifically, but it also determines how much it costs to install and maintain. Though the required simulations are carried out, details are omitted in this discussion. SAM simulations indicated that the annual specific energy is approximately 1450 kWh/kWp and corresponding annual energy of 7.2 million units; for installations with mounting standoff higher than 3.5 inches.

Estimation of Energy Production from Fspv Plant

Since there is no specific software that can model FSPV plant, SAM cannot be used directly to determine the energy output. However, several studies have widely acknowledged that FSPV plants yield higher energy than their land-based PV counterparts. The studies carried out by Tarigan (2021), Ranjbaran et al. (2019) and Azran and Majid (2014) clearly indicated that an FSPV system produces more energy than a ground-based PV system, by roughly 10% - 15%. Based on this, it is possible to estimate the AC energy and the specific energy produced by the FSPV system that is designed in this study. A lower side, conservative projection of 10% difference in energy gain is applied to determine the output. Table 5 shows the monthly energy output of land-based and FSPV energy plants under simulated conditions. As can be observed, the Windhoek FPV system's Monthly Net Electricity to Grid ranges from 357,500 kWh to 1,003,750 kWh over the course of a year with an annual energy production at 7,976,658.8 kWh. The annual specific energy production is 1593.9 kWh/kWp.

Financial Analysis

SAM simulations provide financial estimates based on a comprehensive analysis. For the 5MW FSPV, the overall installation cost will be roughly $8.6 million, with an installed cost per capacity of $1.72/Wdc. The sum of the indirect, sales tax, and direct costs is the total installation cost.

Case Scenarios with Different Capacities

Since this chapter proposes a technical proposal for a new FSPV plant in the dam, it is important to carryout investigations for different capacities. However, capacity of FSPV itself is limited by the available water surface area, which is specific to the dam under question. Also, in most cases, the small to medium size solar power plants may have limits imposed by the regulatory bodies on the capacity. Setting such limit is very important to ensure fair market conditions and also to provide level-playing conditions in the market for several, different stakeholders, particularly

Table 5. Monthly net electricity to grid in year 1

Month	AC energy (kWh)	
	Land based	FSPV
January	912500	1003750
February	637876	701663.6
March	562878	619165.8
April	537875	591662.5
May	387879	426666.9
June	325000	357500
July	362500	398750
August	512500	563750
September	662500	728750
October	775000	852500
November	825000	907500
December	750000	825000
Yearly	7251508	7976658.8

the owners of independent power producers (IPPs). In case of Namibia, 20MW is the capacity limit for a solar power plant if any IPP wants to invest in. Thus, two additional case scenarios for capacities 10MW and 20MW are also considered.

Table 6. Summary of the three FSPV case scenarios

FSPV capacity	5 MW	10 MW	20 MW
Annual Energy (kWh)	7976658.8	26167455.6	33400571.6
Annual Performance Ratio	0.79	0.79	0.79
Annual Energy Yield (kWh/ kW)	1593.9	2492.6	1609.3
Annual DC Capacity Factor (%)	16.5	25.9	16.7
Area (m^2)	22000	45900	90720
Number of Modules	10773	22596	44682
Module power (W)	464.6	464.6	464.6
Number of Inverters	3	3	4
Inverter power (kW)	1498.66	2844.7	4654.42
DC/AC Ratio	1.11	1.23	1.12
Installation Cost ($)	8605085	18048873	35690377

The same framework as presented in the earlier sections is applied and the results are obtained. Table 6 summarizes the overall results for the three case scenarios.

Water Loss Due to Evaporation

Evaporation is one of the most significant environmental processes that can lower the quality and quantity of water available for industrial, agricultural, and domestic uses; as a result, evaporative losses become a significant obstacle to the effectiveness of water supply and storage in many regions of the world. Numerous techniques have been tried and developed to prevent water from evaporating, including physical, chemical, and biological means of reducing evaporation from water surfaces. Physical solutions for decreasing evaporation include floating covers, suspended covers and injection of air bubbles into water (bubble plume). Those that employ coverings that are suspended or floating can save a lot of water (between 70 and 95 percent). This puts FSPV plants into a win-win situation since a greater deal of shade is naturally provided due to the presence of floating platform and the solar panels. Some of the known technologies to contain the evaporation losses are presented here under.

Stingray technology - One South C-Hex float and one North C-Hex float make up the stingray technology design. The solar panels are encased between the two floats as they float on the water's surface. With this design, a greater portion of the water's surface is covered, attempting to minimize evaporation losses. This technology is best suitable for large waterbodies (Sanchez et. al, 2021). Ultra-thin monolayer technology is a potentially cost-effective method of reducing evaporation. Current commercial monolayer materials, however, perform poorly, are easily disturbed by wind, and require repeated application. Significantly, to lessen the effects of wind on monolayer films and lower water storage evaporation losses, monolayer solutions can be employed in combination with physical barriers. According to estimates, between 20 and 40 percent of the water kept on cotton fields is lost due to evaporation losses from agricultural water storages. Despite the fact that there are currently available solutions to decrease evaporation, combination of monolayer technology with inexpensive physical barriers result in a cheaper capital costs (Malu et. al, 2017; Sanchez et. al, 2020). This technology helps farmers to buy less water and keeps more water available for irrigation. The economic advantages are highest in years with low water allocations (EnergyLive, 2019).

Physical Methods of evaporation reduction that use floating covers and suspended covers also have been tried. Two types of floating covers exist - continuous and modular covers. To lessen evaporation, floating continuous coverings often create an impermeable barrier that floats on the water's surface. While composite materials such as polystyrene, wax, and foam have all been studied as potential materials for floating continuous covers. Polyethylene plastic has proven to be the most popular and long-lasting option. On the other hand, modular covers are separate, free-floating pieces that do not completely enclose the water's surface. As a result, they improve the flow of dissolved oxygen through the spaces that are left between them. Examples of commonly used floating modular covers include aqua-caps and shade balls (Sanchez et al, 2021).

While suspended covers are suspended structures that are attached to water surfaces using steel cables and supporting poles. This cover lessens wind flow and blocks sun radiation, which lowers the rate of evaporation from water surfaces. With this model, the solar photovoltaics PV cover will be examined. FSPV plants, by design should significantly reduce evaporation. On the other hand, using solar panels above water features has the added benefit of cooling effect, which improves the overall efficiency as atmospheric thermal stress is less. A study examined potential of FSPV plants as an alternate option for land-based installations that would be put in place above irrigation canals and on the downstream face of dams in the Mediterranean islands. Typically, each MWp produced is predicted to save 3000 cubic meters of water (Sanchez et al, 2021). Based on this proposition, the 5 MW Floating FSPV plant is estimated to save around 15,000 cubic meters of water. Overall, through this literature review the development of local energy development is quite possible because the resources are available in Namibia therefore, existing solar power generation in Namibia can be improved which can Decrease the cost of electrical energy in Namibia.

DISCUSSION AND SCOPE FOR FURTHER WORK

From the above sections, it can be seen that FSPV systems are gaining popularity around the world and also in Africa. It is clear that the nexus between resources in Africa is well identified and environmental aspects are given high consideration as can be seen from literature as well from the rising numbers of renewable energy capacity in the recent times. FSPV systems provide a good opportunity to resolve the nexus between land, energy and water. However, with the addition of large-scale symbiotic systems close to the FSPV system is likely to result in more benefits as this contributes to food production.

However, it is important to consider several aspects before coming to any sort of conclusion. Some of those include –approval processes for FSPV plants; possible ecological and environmental impact; location of the plant itself; energy evacuation processes and market design and general public opinion. An attempt will be made to discuss these issues briefly. In general, it is known that hydropower resources and water dams are generally located very far from urban areas. They are not only Cost-effective, but also environmentally friendly. Thus, the state and regulator need to identify a special purpose vehicle to launch FSPV power plants through appropriate changes in policy and/or in approval methodologies.

Any dam however small or big; can be a permanent host to wide variety of birds, fisheries and water flows. Any disruption can impact the ecosystem of the dam adversely. Thus, in-depth studies have to be undertaken to determine the possible impacts and then remedial measures need to be identified. Then, electrical power plants always are purpose-built based on the forecasted demand. Utilization of the energy depends upon demand response. System configuration and the power grid topology will influence energy evacuation process. Location of the plant type of the fuel nature of ownership all will play their own rules in getting environmental clearances and necessary approvals. If the location is far from the power grid, then a good amount of energy will be lost in the form of transmission losses. Hence a lot of feasibility studies will be carried out before setting up any power plant.

In general, transmission losses can be up to 10%; this may not favor FSPV plants that are located very far unless the scale of energy production is very high. Another major challenge is the duck curve phenomena in which peak time solar energy cannot be used. This happens when the power grid is connected to too many existing solar power plants that might contribute a significant portion of energy. This results in a condition where in any further capacity addition in solar segment cannot be favored.

Thus, there are a good number of factors that may adversely affect FSPV plants as the scale of energy can be low when compared to large-scale hydropower plants. To understand the overall cost benefits and issues related the nexus between the resources, exhaustive studies need to be carried out considering the overall system with different parameters and operating conditions.

Interestingly symbiotic systems (combining aquatic and plant based agricultural systems) can function well in terms of taking the peak energy. The loads in symbiotic systems are predominantly water pumping/ circulation, lighting etc.; and these loads can be supplied satisfactorily by small scale FSPV plants. On the other hand, agricultural farming loads and self-contained symbiotic systems can be located near to FSPV plants with appropriate policy interventions. In other words, both water and energy sources will be close to the loads; thus saving precious resources, which is critical for countries such as Namibia. Colocation reduces both water and energy transportation losses. This means, proposals for FSPV plants need to include a dedicated set of loads that can be co-located nearby. Such dedicated loads, especially through single buyer model can offer several advantages. Firstly, they are by and large commercial or industrial loads, which means higher tariffs can be set on equitable basis. Secondly, such consumers are aware of safety and regulation aspects and thus resulting in better customer relations and free flow of information on demand and supply schedules. This is very important, particularly when energy sources are intermittent in nature. FSPV plants by nature are intermittent for obvious reasons.

However, studies are required to determine the benefits of co-locating or adding symbiotic facilities to the FSPV. In other words, engineering, techno-economic studies of agrivoltaic systems need to be undertaken.

CONCLUSION

Design and simulation studies for a new, FPVS power plant for Goreagab Dam, Namibia, have been carried out. The total area of the water surface of the dam is about 1 100 000 m^2. By considering 2% of the total area would be used for the FPV system, about is 5.01 MW_p solar panels can be installed. The annual specific energy production is 1593.9 kWh/kWp. Based on the design and simulation studies, this 5MW FSP can deliver around 8million units of electrical energy annually. Further, a significant saving potential can be achieved in terms of reduced evaporation losses. The two more case scenarios for 10 and 20 MW based on the surface area, are also considered and found to be feasible. On the hindsight, it is possible to consider setting up a FSPV plant that can deliver multiple benefits, subjected to other in-depth investigations related ecology and environmental studies.

REFERENCES

AgriSolar. (2022). *Agrisolar SolarPower Europe*. Agrisolar. https://www.solarpowereurope.org/interests/ agrisolar

Akole, H. U., & Jadhav, P. S. (2018). Floating Solar Plant. *3rd Int. Conf. Converg. Technol. I2CT* (pp. 7–10). IEEE. 10.1109/I2CT.2018.8529479

Al-Widyan, M., Khasawneh, M., & Abu-Dalo, M. (2021). Potential of floating photovoltaic technology and their effects on energy output, water quality and supply in Jordan. *Energies*, *14*(24), 1–13. doi:10.3390/en14248417

Almeida, R., Schmitt, R., Grodsky, S., Flecker, A., Gomes, C., Zhao, L., Liu, H., Barros, N., Kelman, R., & McIntyre, P. (2022). Floating solar power could help fight climate change— Let's get it right. *Nature*, *606*(7913), 246–249. doi:10.1038/d41586-022-01525-1 PMID:35672509

Azran, Z., & Majid, A. (2014). Study on Performance of 80 Watt Floating Photovoltaic Panel. *Journal of Mechanical Engineering Science*, *7*, 1150–1156. doi:10.15282/jmes.7.2014.14.0112

Chico Hermanu, B. A., & Santoso, B., & Wicaksono, F. X. R. (2019). Design of 1 MWp floating solar photovoltaic (FSPV) power plant in Indonesia. *AIP Conference Proceedings*. doi:10.1063/1.5098188

EnergyLive (2019). Netherlands' largest floating solar array switched on. *Energy Live News.*

EUScienceHub. (2021). Floating solar panels in African hydropower reservoirs: a win-win for electricity generation and water savings. EU Science Hub. https://joint-research-centre.ec.europa.eu/jrc-news/floating-solar-panels-african-hydropower-reservoirs-2021-02-26_en,

Faranda, R. S., Hafezi, H., Leva, S., Mussetta, M., & Ogliari, E. (2015). The optimum PV plant for a given solar DC/AC converter. *Energies*, *8*(6), 4853–4870. doi:10.3390/en8064853

Giri, S., Ahmad, A., & Katiyar, V. (2019). Avoiding Power Clipping Losses by Inverter having High DC-to-AC Loading Ratio in Grid Connected Solar PV Plant Using Battery Energy Storage System, 213-219. PEEIC. doi:. doi:10.1109/PEEIC47157.2019.8976581

Gozálvez, J. J. F. (2012). Covering Reservoirs With a System of Floating Solar Panels: Technical and Financial Analysis. *Sel. Proc. from 16th Int. Congr. Proj. Eng.*, (pp. 177–187). IEEE.

Gulati. Jacobs, M., Joostem, A., Naidoob, D., & Fakir, S. (2013). The Water–energy–food Security Nexus: Challenges and Opportunities for Food Security in South Africa. *Aquatic Procedia*, *1*, 150–164. doi:10.1016/j.aqpro.2013.07.013

Hauwanga, A. S., & Musti, K. S. S. (2022). Evaluation of Energy Losses and Costs Based on the Location of PCC in Smart Grids. IEEE 11th International Conference on Communication Systems and Network Technologies (CSNT), (pp. 76-82). IEEE. 10.1109/CSNT54456.2022.9787642

Junianto, B., Dewi, T., & Sitompul, C. R. (2020). Development and Feasibility Analysis of Floating Solar Panel Application in Palembang, South Sumatra. *Journal of Physics: Conference Series*, *1500*(1), 012016. doi:10.1088/1742-6596/1500/1/012016

Malu, P. R., Sharma, U. S., & Pearce, J. M. (2017). Agrivoltaic potential on grape farms in India. *Sustainable Energy Technologies and Assessments*, *23*, 104–110. doi:10.1016/j.seta.2017.08.004

Meneghel, D., Bortoni, E., & Karimi, A. (2018). *Boosting DC/AC Ratio of PV Plant for BESS Integration on DC side,* 1-4. Semantic Scholar. doi:. doi:10.1109/SusTech.2018.8671354

Mks, S. (2005). Simplified algorithm to determine break point realys & relay coordination based on network topology [for realys read relays. 2005 IEEE International Symposium on Circuits and Systems (ISCAS), (*vol. 1,* pp. 772-775). IEEE. doi:10.1109/ISCAS.2005.1464702

Musti, K. S. (2020). Circular economy in energizing smart cities. In Handbook of Research on Entrepreneurship Development and Opportunities in Circular Economy (pp. 251-269). IGI Global. doi:10.4018/978-1-7998-5116-5.ch013

Musti, K. S. (2021). *Industry 4.0-Based Large-Scale Symbiotic Systems for Sustainable Food Security in Namibia.: Impacts of Climate Change on Agriculture and Aquaculture.* IGI Global. doi:10.4018/978-1-7998-3343-7.ch009

Musti, K. S., Iileka, H., & Shidhika, F. (2020). Industry 4.0-based enterprise information system for demand-side management and energy efficiency. In Novel Approaches to Information Systems Design (pp. 137-163). IGI Global. doi:10.4018/978-1-7998-2975-1.ch007

Musti, K. S. S. (2020). Quantification of Demand Response in Smart Grids. *IEEE Int. Conf. (INDISCON),* (pp. 278-282). IEEE. 10.1109/INDISCON50162.2020.00063

Odetoye, O. A., Ibikunle, F. A., Olulope, P. K., Onyemenam, J. O., & Okeke, U. N. (2022). Large-Scale Solar Power in Nigeria: The Case for Floating Photovoltaics. *IEEE Int. Conf. on Disruptive Technologies for Sustainable Development (NIGERCON),* (pp. 1-5). IEEE. 10.1109/NIGERCON54645.2022.9803156

Pearce, J. M., Harshavardhan Dinesh. (2016). The potential of agrivoltaic systems. *Renewable & Sustainable Energy Reviews*, *308*, 299–308. doi:10.1016/j.rser.2015.10.024

Pitra, G. M., & Musti, K. S. S. (2021). Duck Curve with Renewable Energies and Storage Technologies. 13th International Conference on Computational Intelligence and Communication Networks (CICN), (pp. 66-71). IEEE. 10.1109/CICN51697.2021.9574671

PRNewswire. (2021) Where the Sun Meets the Sea: Offshore Floating-PV Powers Singapore's Journey Toward Carbon Neutrality. PR Newswire.

Radhiansyah, T. D., & Hamdani, D. (2021). Performance Analysis of Offshore Floating PV Systems in Isolated Area. *2021 3rd Int. Conf. High Volt. Eng. Power System,* (pp. 651–655). IEEE. doi:10.1109/ICHVEPS53178.2021.9600926

Ranjbaran, P., Yousefi, H., Gharehpetian, B., & Astaraei, F. R. (2019). A review on floating photovoltaic (FPV) power generation units. *Renewable & Sustainable Energy Reviews, 110*, 332–347. doi:10.1016/j.rser.2019.05.015

Sanchez, R. G., Kougias, L., Fahl, F., & Jäger-Waldau, A. (2021). Assessment of floating solar photovoltaics potential in existing hydropower reservoirs in Africa. *Renewable Energy, 169*, 687–699. doi:10.1016/j.renene.2021.01.041

Sanchez, R.G., Seliger, R, Fahl, F, De Felice, L., & Taha, B.M.J., Fabio. (2020). Freshwater use of the energy sector in Africa. *Applied Energy, 270*. doi:10.1016/j.apenergy.2020.115171

Singh, J., Kuchroo, P., Bhatia, H., & Sidhu, E. (2016). Floating TEG based solar energy harvesting system. *Doi, 763-766*, 763–766. doi:10.1109/ICACDOT.2016.7877689

Solomin, E., Sirotkin, E., Cuce, E., Selvanathan, S. P., & Kumarasamy, S. (2021). Hybrid Floating Solar Plant Designs: A Review. *Energies, 14*(10), 2751. doi:10.3390/en14102751

Tarigan, E. (2021). Simulation of the Floating PV System to Supply Electricity Demand for the City of Surabaya, Indonesia. *Int. conf. on on Renewable Energy: Generation and Applications (ICREGA),* (pp. 48–53). IEEE. doi: . doi:10.1109/ICREGA50506.2021.9388219

Zareef, L., Jalali, M., & Sadat, S. M. (2021). Analysis of the potential for use of floating solar panels on Naghlo hydropower dam. *Int. Jnl of Innovative Research and Scientific Studies, 4*(2), 126–132. doi:10.53894/ijirss.v4i2.66

Chapter 2
Experimental Study on a Reliable Smart Hydroponics System

Varuna Kumara
Moodlakatte Institute of Technology, India

T. A. Mohanaprakash
Panimalar Engineering College, India

Shaik Fairooz
Malla Reddy Engineering College, India

K. Jamal
GRIET, India

T. Babu
St. Joseph's College of Engineering, India

Sampath B.
Mythayammal Engineering College, India

ABSTRACT

Hydroponics is becoming increasingly popular because it can be used in unconventional locations such as inside warehouses, marginal lands, and water-scarce regions. Hydrogen (pH), nutrient level, water level, temperature, and other parameters must be monitored on a daily basis in a typical hydroponic design. A hydroponic setup that is not dependent on the environment outside has been developed. The technology consumes less water and fertiliser as compared to a soil-based system. This system uses the idea of a relay and alarm connected to a microcontroller to make sure that nutrient solution is always available. The creation of an internet of things (IoT) network enabled remote parameter monitoring and control, which increased system stability. In addition to measuring and controlling the water level, pH, water stream rate, and EC at this time, the numerous sensors attached to the NFT framework support gathering fundamental system information.

DOI: 10.4018/978-1-6684-4118-3.ch002

INTRODUCTION

Hydroponics is a form of soil-free plant growth that uses mineral nutrient solutions in water. By automatically supplying nutritional solutions to the plant roots, the system will produce superior results. This research was carried out to develop an effective hydroponic tomato crop production method. In hydroponic methods, nutrients are fed directly to the roots of the plants while they are submerged in a growing medium. Plants can develop their roots in inert media or just nutrient solution. Plants can be grown more effectively and quickly using the hydroponic method than in soil. Without the use of soil or chemical fertilisers, hydroponic farming allows plants to grow without the risk of pest infestation. Compared to traditional soil gardening, hydroponic farming enables farmers to grow more food in a smaller area. It is possible to cultivate flowers, plants, and vegetables on roofs of houses. Multiple crops can be grown at once using a hydroponic system(Kumar et al., 2023; Sampath et al., 2023). Depending on the needs of the plant, nutrients are provided in the form of a nutrient solution that contains a precise combination of the necessary ingredients for plant growth. This stops the growth of algae in the fertiliser solution, which the crops are susceptible to. A method for overcoming these issues has been chosen by the developed system, which groups resources logically and prioritises their needs. The goal is to develop a hydroponic system based on NFT that anyone can use to grow their own plants. The user won't have to deal with tedious setup or ongoing maintenance thanks to this method. Through a mobile interface connected to the microcontroller powering the device, the user will be able to specify the plants they want to cultivate. This system will carry out all daily testing required, modify system levels, such as the nutrient solution and water level, as appropriate, alert the user of a problem that needs to be addressed, and log all testing data for analysis. This chapter covered the development of a mobile application for a smart phone that will collect data from numerous sensors and controllers. The Arduino Uno, sensors, and Internet of Things (IoT) technology will all be used to complete this assignment. The majority of hydroponic systems on the market today are either highly expensive or don't have complete control over all the factors required for healthy plant growth(Ezzahoui et al., 2022; Sathyanarayana et al., 2022). This newly created system offers total automation with monitoring and management of all plant growth-related variables, including soil moisture, pH, and temperature. Two Arduino boards are used by the system to analyse and regulate the data that is received. It also makes use of a mobile app to let users of Android-compatible phones and other smart devices monitor and operate the system. Various expansion boards (shields) and other circuits can be interfaced to an Arduino Uno microcontroller board's sets of digital and analogue input/output (I/O) pins. In this project, a Raspberry Pi and monitoring sensors are also utilised. The Raspberry Pi is a little single-board computer that has an inbuilt Internet of Things (IoT) architecture that enables it to communicate with sensors and actuators over the internet and exchange data with them. The hydroponic Nutrient Film Technique system that was discussed in the preceding chapter is automated with the aid of an embedded controller. With transparent plastic coverings on either side, the table has been erected inside of a tent. The integration of NFT systems with IoT technology, which configures the readings from numerous sensors to the Arduino microcontroller, is the main topic of this chapter. The NFT system was covered in the chapter before; in this one, we'll look at how these sensors communicate with one another online. These activities involve characterising the equipment, demonstrating programming, integrating the equipment and programming, conducting investigations, and accumulating data and results. A machine-to-machine (M2M) connection method uses an Internet of Things (IoT) network to link various devices, such as sensors, actuators, and other heterogeneous items/gadgets. The client's message or data is transported from one end to the other via the internet, for example, from a system to an online server where anybody can access the information. End users can view and even control the IoT framework in the cloud, or they can imagine the cloud acting as the IoT's central control hub.

HYDROPONIC GROWING MEDIUMS

Instead of using soil to support a plant and absorb nutritional solutions, hydroponics uses a growing medium. The plant lives in the soil and looks for food and water. Because the nutrient solution easily satisfies their requirements, hydroponic plants do not have to work as hard as those that grow in soil. The following are some growing mediums used in hydroponic farming: Hydro-rockwool corn, coco fiber, or coco chips Oasis Cube, Perlite, and Vermiculite; Floral Foam; Poly Foam Insulation; and Rice Hulls In addition, the growing medium must be such that it can be cleaned easily, have the ability to drain easily, and be chemically inert (Calone & Orsini, 2022).

Crop Selection: Premium quality crops are grown in hydroponic greenhouses, where they fetch a premium price even for everyday veggies. To ascertain the types of crops that are in demand, it is crucial to conduct market research. Common veggies including lettuce, tomatoes, and cucumbers are grown hydroponically. It is recommended to start with one crop then add the others one at a time if you wish to cultivate a range of crops. Every crop has unique dietary, environmental, and cultivation requirements. One should wait until a crop is fully matured before beginning another to avoid confusion or incorrect crop maintenance.

Tomatoes: Many novice gardeners pick tomatoes as their crop of choice for hydroponic production. The producer of indeterminate tomato plants must use several culturing procedures to each plant on a weekly basis. These include harvesting, leaning, lowering, suckering, leaf and cluster pruning, cutting, and suckering. During their fruiting stages, tomatoes require a minimum of eight hours of light each day. A farmer can either bring in a specific bee hive or perform this daily by hand. The bees pollinate more effectively and eliminate the need for manual labour. The majority of hydroponic tomato growers propagate their plants during the winter. It is impossible to produce during this time without artificial lighting in the majority of India's climes. This is due to the fact that purchasing and using artificial lights for a commercial application would be too expensive and time-consuming. As long as they are not refrigerated, vine-ripened tomatoes have a far longer shelf life than green-picked tomatoes. The optimal temperature range for tomatoes is between 20 and 30 °C. Although fruiting cannot be accomplished at the decreased light levels, early development can.

Fancy Lettuce: Fresh lettuce and lettuce blends are in high demand right now. A lettuce farm should be taken into consideration for a region where it is difficult to find fresh, high-quality leaf crops. In order to keep the lettuce fresh, hydroponic lettuce farmers typically sell a fresh bagged mix or a whole leaf harvest. The possibility of plant disease, fungi, or pests is significantly decreased since fancy lettuce grows from seedling to harvest in 4 to 12 weeks. The planting of sequenced cluster tomatoes allows for continuous harvesting of leaf crops. The majority of leaf crop growers employ hybrid types created for tasty and rapid output.

Cucumbers: Cucumbers cultivated in hydroponic systems have incredibly soft, sweet skin and great flavour. The majority of hydroponically grown cucumbers are European seedless types developed for greenhouse farming. Since the skin is so delicate, the majority of farmers individually wrap the cucumbers to guard against dents and scratches. A cucumber plant is cultured similarly to a tomato plant. The majority of hydroponic cucumber farmers cultivate two crops each year. When compared to tomatoes, it takes less time—around six weeks—from seed to first harvest. Tomatoes and cucumbers may both be produced in the same greenhouse under identical conditions.

Pepper: Although there is a demand for pepper farmers, there aren't many of them and there isn't a lot of information on the subject.

Carrots: Vermiculite and perlite are examples of materials that can be used to cultivate hydroponic plants underground. For the plants' roots to penetrate the water deeply, a DWC-based system is required. Water should be able to be sieved through holes drilled into the bottom. For the purpose of retaining water, the container must be sat in first. In six to ten days, the carrots will begin to sprout. In the world of subterranean hydroponics, carrots are grown.

Figure 1. Various nutrients for hydroponic cultivation systems

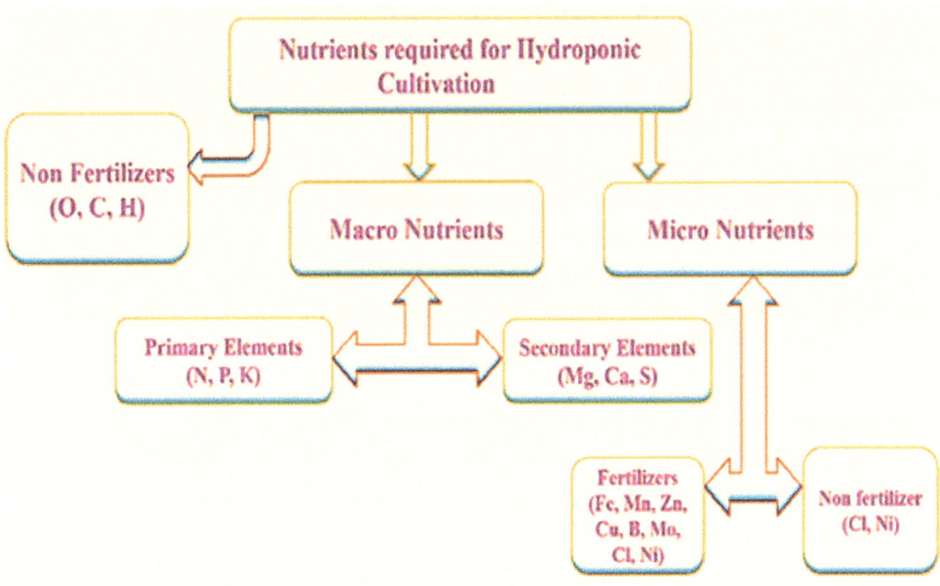

NUTRIENT SOLUTIONS

Various nutrients for hydroponic cultivation systems are illustrated in Figure 1. A farming method known as hydroponics involves growing plants in highly oxygenated, nutrient-rich water without the use of soil. The key to a good hydroponic system is fertiliser solution control. The hydroponic fertiliser solution is a mixture of water, oxygen, and soluble forms of the necessary mineral components. Organic matter in the soil is broken down into basic nutrients via biological decomposition, providing food for plants. These nutrients are subsequently dissolved in water and absorbed by the roots. About seventeen different components are necessary for plants to grow properly. There are nine of them: Calcium (Ca), Magnesium (Mg), Potassium (K), Nitrogen (N), Oxygen (O_2), Sulfur (S), Phosphorus (P), and Hydrogen (H). The creation of the remaining eight elements, known as micronutrients or trace elements, necessitates very small concentrations of these substances(Taha et al., 2022).

PHYSICAL COMPONENTS OF HYDROPONIC SYSTEM

The hydroponic system arrangements are illustrated in the Figure 2. The various physical components of hydroponic system is described below (Das & Sahu, 2020).

- **Grow Tray**: Growing medium and plants are housed in the hydroponic system's grow tray. In the tray, the plant roots develop.
- **Reservoir**: The component of the system that stores the nutrient solution is called the reservoir. The nutrient solution is made up of plant nutrients mixed with water that has the following dimensions: 61" wide, 41" long, and 30" high.
- **Submersible Pump**: By immediately pumping the solution from the reservoir to the tray or plants, submersible pumps distribute nutrient solutions. To prevent unwelcome bacteria from growing in the solution in a hydroponic system, it is crucial that the solution be cycled. Pumps can be used in a hydroponic system in a variety of ways.
- **Delivery Tube**: The delivery tube is used to deliver the nutrient / water solution from the reservoir to the grow chamber tray.
- **Automated Timer**: An automated timer is simple timer that allows hydroponic unit to be activated on a predetermined schedule.
- **Oxygen Pump**: oxygen pump is used to oxygenate the nutrient solution; it needs to make sure the plants are getting proper oxygen level to thrive. A simple way to ensure for enough oxygen is in the nutrient solutions is by using an air pump connected to an air stone in the hydroponic solution reservoir.

IOT COMPONENTS OF NFT SYSTEM

An Arduino Mega, a 5-volt relay, a 12-volt DC water pump, a humidity and temperature sensor (a DHT22), a PH sensor (a SENO161), a 12-volt battery, a 6-litre plastic container, and plastic pipe are among the nine distinct types of sensors and hardware that were used(Espinal & Matulić, 2019).

Arduino: A microcontroller board based on the ATmega2560 is called the Arduino Mega 2560. It contains a 16 MHz crystal oscillator, 4 UARTs (hardware serial ports), 16 analogue inputs, and a total of 54 digital input/output pins, 14 of which can be utilised as PWM outputs. The ATmega2560 serves as the foundation for the microcontroller board known as the Arduino Mega 2560. It has 54 digital input/output pins, 14 of which can be used as PWM outputs, 16 analogue inputs, 4 UARTs (hardware serial ports), a 16 MHz crystal oscillator, and 16 analogue outputs.

Water Pump (12V DC): It can be used for a variety of domestic chores, including cooking, cleaning, bathing, heating up a room, and watering flowers, among others. The motor and pump controller for the smart hydroponic system are depicted in Figure 3.

Figure 2. Physical layout of NFT system

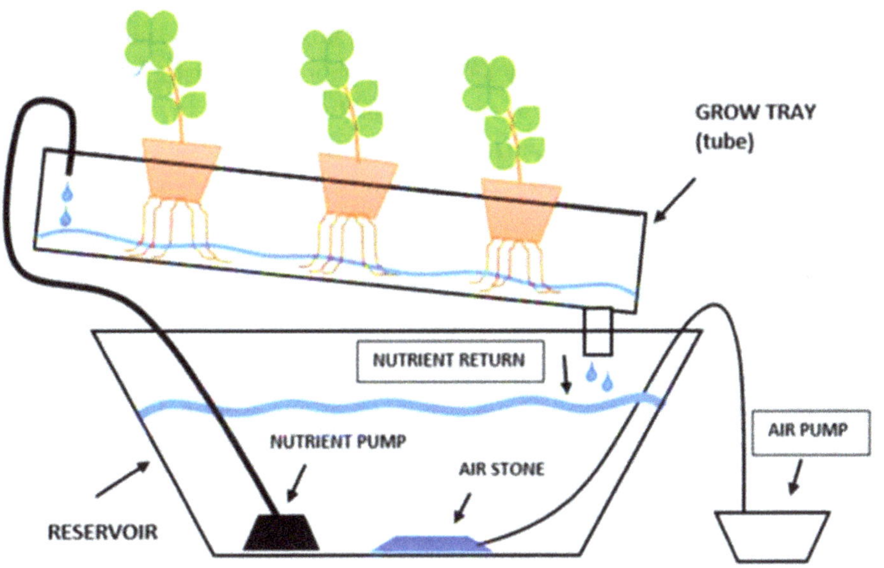

5V Relay: Relay modules are electrical switches that can be used to turn on or off circuits at voltages and currents that are far higher than those that a microcontroller is capable of handling. The microcontroller-controlled high-power circuit and the

Figure 3. Motor and Pump Controller

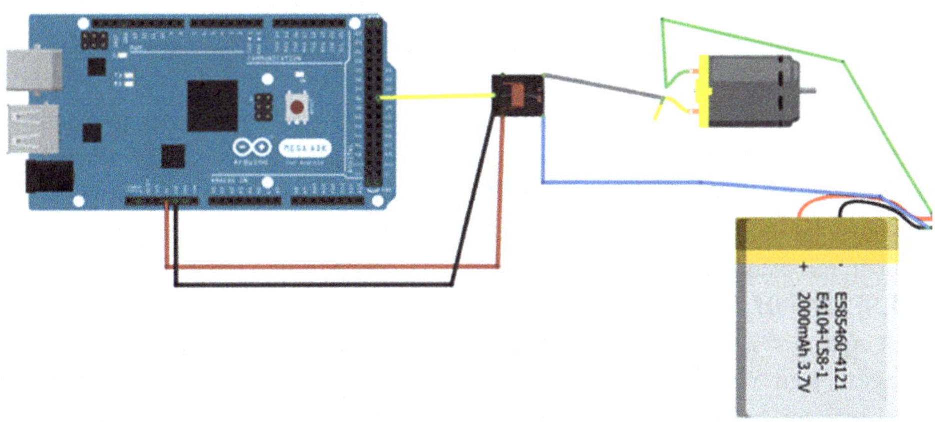

low-voltage circuit are not linked. Three connections—NC, COM, and NO—are available for each channel in the module.

A microcontroller that has the ability to regulate the voltages and currents of any system component. An electromagnet, which comprises of a magnet and a wire coil, is located inside a relay. magnetic switches that function at a low voltage level that the microcontroller is able to generate. A relay maintains an electromagnet when a modest voltage is introduced to it, allowing whatever current and voltage are connected to the relay to activate the desired component. Normally, these switches are open, preventing any current from flowing to the subsystem device. However, the electromagnets switch is activated when the relay pin is connected to ground. The user can convert the ground and VCC conditions on the microcontroller pins to AC current using a relay, which will turn the light on and off in a regular cycle. These relays allow for many different power conditions to be applied to components that cannot operate under 5 volt or low current settings (Goddek et al., 2019).

DS18B20 Temperature Sensor: The temperature of the device is indicated via the DS18B20 Digital Thermometer's 9 to 12-bit (adjustable) temperature readings. Without the use of an external power source, power can be obtained from the data line itself for reading, writing, and converting temperatures.

pH sensor: Most pH measurements are made with pH-sensitive glass electrodes, which frequently lack the necessary precision. This electrode has a PTFE diaphragm, a gel inside, and an integrated Pt100 temperature sensor for temperature adjustment. It is relatively stable against pressure fluctuations within the system. Figure 4 shows a pH sensor connected to an Arduino.

Calibration of pH sensor: According to Titrisol from Merck, the samples' pH was found to be 7.02, which can be attributed to SRM (NIST) and PTB (SRM, PTB).

EC Sensor: The conductivity of a solution used in hydroponic farming is measured by an EC sensor. An indoor gardener can use a mathematical method to calculate the amount of total dissolved solids (TDS) in the solution by taking an EC reading. Some manufacturers make combo meters, which combine the features of both types of metres into a single unit.

DHT22 Temperature and Humidity Sensors: A straightforward, inexpensive digital temperature and humidity sensor is the DHT22. It measures the ambient air using a thermistor and a capacitive humidity sensor. The only significant drawback of this sensor is that it can only provide you with fresh information every two seconds. Temperature and humidity sensors connected to the Arduino are shown in Figure 5.

ESP8266-01 Wi-Fi Module: The ESP8266 is an integrated TCP/IP protocol stack-equipped self-contained system on chip (SOC) that allows any microcontroller access to your Wi-Fi network. This indicates that you may easily connect this to your Arduino device and obtain roughly the same level of wi-fi functionality as a Wi-fi Shield. The ESP8266 is a very affordable board with a sizable and expanding

Figure 4. pH Sensor connected to Arduino

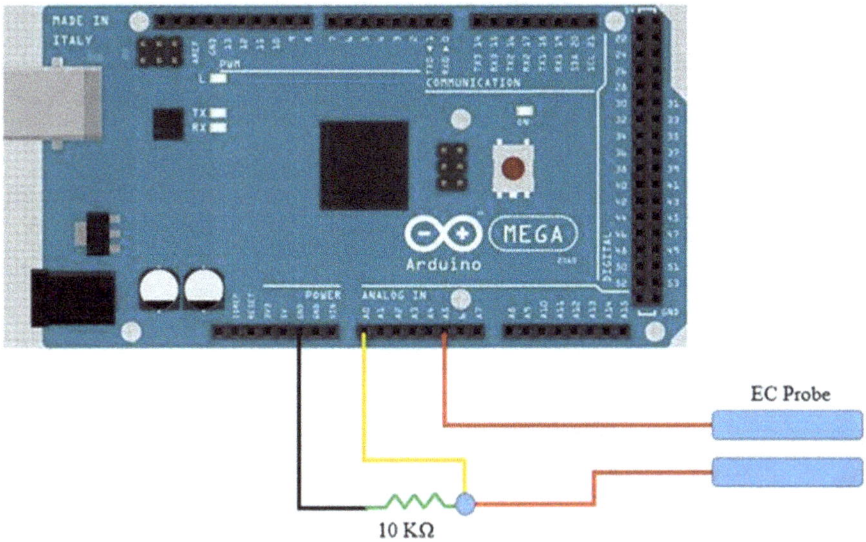

community. Through its GPIOs, this module may be combined with sensors and other application-specific devices thanks to its robust onboard processing and storage capacity. Both the initial development and the runtime loading are modest.

Figure 5. Temperature and Humidity Sensors connected to Arduino

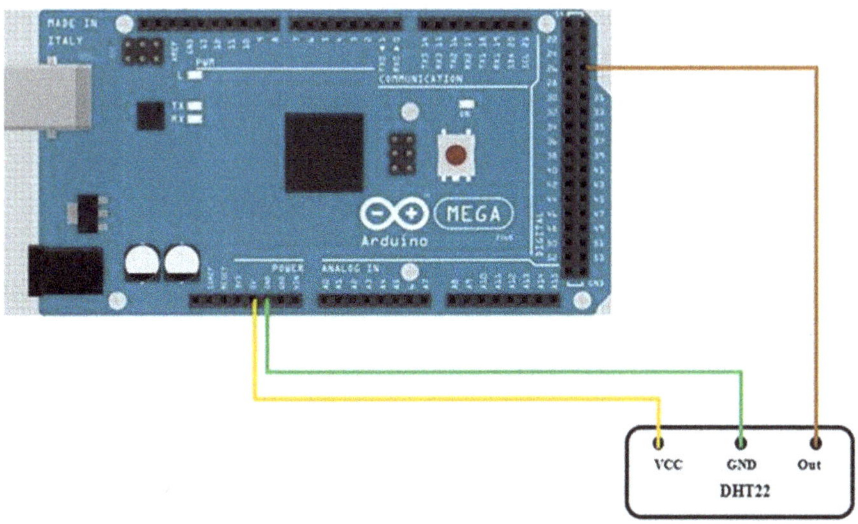

Figure 6. Connection of Sensors and Controllers for the Smart Hydroponic system

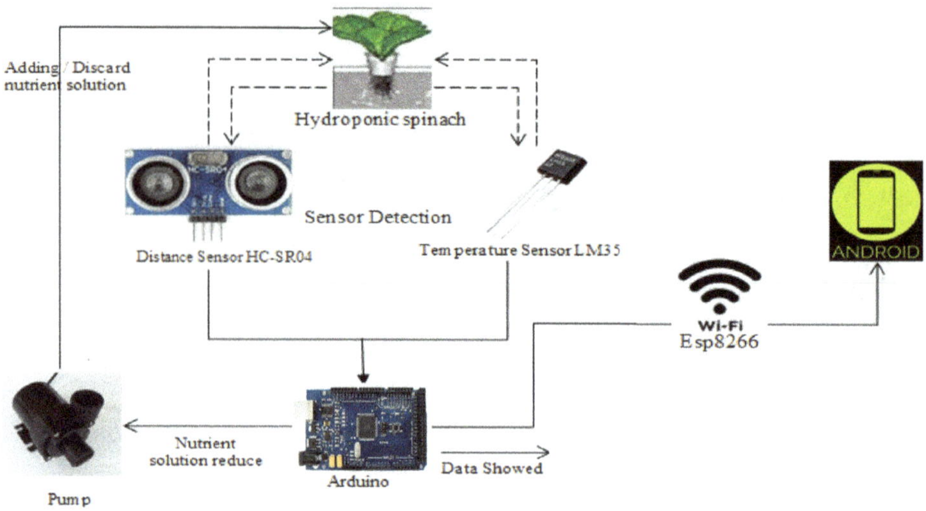

It features a self-calibrated RF that enables it to operate in various operational environments and supports APSD for VoIP applications and Bluetooth coexistence interfaces(Espinal & Matulić, 2019).

Hydroponic solution: A subset of hydroculture, or the process of growing plants without soil, is hydroponics. In hydroponics, mineral fertiliser solutions dissolved in water are used. Only the roots of terrestrial plants may be grown in the mineral solution, or the roots may be supported by an inert medium. For the hydroponic growth of plants, we employ two different types of nutrient solutions: organic and inorganic. In the lab at the University of South Wales in Swindon, we created the nutrition solution for our crops using the formula shown in the table below.

SMART HYDROPONIC SYSTEM: EXPERIMENTAL SETUP

The system is an automated hydroponic system that feeds weekly nutrients into water for tomato plants. When compared to a soil system, the system consumes less water and fertiliser. The nutrition pump, water pump, and relay are the main components of the automated system(Yanes, 2020). The connection of sensors and controllers for the smart hydroponic system is illustrated in Figure 6.

There is no need for soil, as critical minerals may be absorbed and consumed by plants in plain water, according to research. It has been discovered that soil is not even a vital element of a hydroponic system's plan for plants to develop. The quantity of crops in a plantation with soil is 4-10 times greater than in one without. Compared to growing in the ground, it helps generate more crops. Additionally, tomatoes planted inside a structure using the planting material show a 20–25% boost in production.

The system communicates the data over the internet using the IoT technology for this operation. It is impossible to constantly monitor every plant in the NFT. As a result, the sensor data are logged, and the user is alerted if a problem arises. This information comprises measurements of water level, water hardness, pH, humidity, temperature, and decreasing light intensity (in Lux). Any decrease or increase in these metrics will be reported to the user, and the system will take the appropriate action.

Temperature, humidity, pH, TDS, and alarm trip levels are all measured via sensors. The NFT system is connected to a water tank so that it can draw water from the reservoir to fill up in the event of a water shortage. It keeps track of temperature, humidity, and light intensity changes and alerts the user to rising environmental temperatures and water temperature changes. It sends the real-time data directly to the web by connecting to the Wi-Fi module, which is further connected to a router. It's crucial to select a microcontroller that can handle the task at hand because these sensors can occupy a lot of pins. With the help of this little open-source development board, you may expand the functionality of an Arduino microcontroller. Only 2 pins were used to transfer the sensor data to the microcontroller. It has a tiny USB port attached, which may be used to power the gadget and enable serial communication with the Arduino software.

The board contains 8 analogue inputs and 14 digital I/O pins. The board also has a 3.3V output pin, two active-low reset pins, and two reset pins. For use with an external hard drive, the board also has a reset button. For programming the Arduino Uno processor and uploading the data to the internet, an algorithm has been created. The algorithm is set up such that the complete NFT system functions flawlessly and optimally. The numbers acquired from the humidity and temperature sensors, the water temperature sensor, the light dependent resistor (LDRR), and the pH sensor are used as input to the algorithm.

When the values from the sensors are sent into the processor, the algorithm deals with efficiently and quickly turning on the motor drivers. The algorithm has set a threshold for these values, and if the condition is satisfied, the proper operations are carried out. The ability to turn on and off the motors in a matter of seconds is crucial. Using the Wi-Fi 802.11 b/g/n standard, the system must communicate with an online database and microcontroller sensors. The created algorithm helps to control the system's efficient operation. Every time a condition is satisfied, an alarm is also set off to let the user know. Low-cost Wi-Fi modules allow for two-way communication between the database and the system. They submit their data to a nearby router, which then sends it to a databank website online. The data will then be forwarded to a smartphone app, where it can be read or visually viewed.

The ESP826 Wi-Fi module from Adafruit was chosen due to its great performance at an affordable price. The Adafruit ESP8266 breakout board, soldered with pins for breadboard use, was the development board used for the initial programming and testing.

The National Fuel Tanker Federation (NFT) system's water pump and air pump are both managed by an Arduino controller. The water tank, which serves as a reservoir and fills the running water to the required level when there is a water shortage in the NFT. The system is connected to a router and a Wi-Fi module, which together transmit real-time data directly to the internet. It logs changes in temperature, humidity, and light intensity and alerts the user to rising temperatures nearby.

The system uses an online sheet that is accessible from anywhere in the world and can be used to upload data and control the system. After the necessary programming is completed, the processor is designed to perform the required functions. It uses an L293D motor driver IC. The hydroponic system is also connected to an alarm that is activated if there is an increase or fall in the sensor's value. The alarm can be in the form of various connected LED lights or speakers that emit a certain sound (a buzzer). For the job done in the developed system, an algorithm for the Arduino microcontroller has been designed. Coding is done for the Arduino board to interface with sensors based on this algorithm.

A separate algorithm is also prepared that is used for interfacing the system with the Wi-Fi module and is as follows: Connect to the router address, make a connection to the host (script.google.com), Receive the OK status, send a command to the Arduino UNO to send a value. Receive the values, reconnect to the server, Transmit the data, Get the response from the server, Wait for the next data from the Arduino UNO. Real time Smart Hydroponic system is depicted in Figure 7.

A built-in camera system has been installed on the Kivy plant monitoring gateway to keep an eye on the health of the controlled plants. The data is made available on the web interface and mobile application thanks to a local server operating on a Raspberry Pi that is connected to the gateway. An air pump is employed to add oxygen to the water for the plant to absorb. An electrical conductivity probe, a pH sensor, a water temperature sensor, and an air temperature and humidity sensor are all parts of the system.

Electrical conductivity probes measure the pH of the water that young plants are growing in. Measuring the pH allows you to adjust it before putting it into the system. Distilled water was used to have a base pH value close to 7 and was adjusted to be between 5.5 and 6 pH after adding nutrients. The pH of the water is then altered using acidic, basic, and carbon dioxide levels after it has been heated to a temperature of

Figure 7. Real time Smart Hydroponic system

25 C. In the UK, a sensor network was built using open-source software. As a result, nodes, gateways, and inter-node communication are made possible.

Hydroponic Mobile Application: On the Kivy platform, a mobile app has been created so that users can communicate wirelessly with the system using a web browser.

Accessing the camera: This code demonstrates using the mobile device's camera. The window with the Play button will be generated by the code, allowing the camera to be switched between the ON and OFF positions.

Creating code for Sensor Readings: In this code, a simple timer clock app is created to display the local temperature. This is similar to the clock that is in-built with Android, iOS, and other platforms such as Apple's iOS. The company also offers a mobile phone application that allows for more accessibility if you are unable to use the web interface but wish to control the lights remotely.

Code runs on a local server that is independent of all the electronics that are responsible for the operation of the system. The system also notifies the user if values of the parameters drift out of the bounds abnormally. This can be caused by not closing the system properly, sensor failure, or a node failure. The graphs of pH, moisture, water level, and temperature. These curves are analysed in specified time periods and can be monitored via a web browser.

The pseudocode of the smart hydroponic system:

```
Start
input Tn, TE, pH, humidity, EC
if Tn>30 then FAN ON
else FAN OFF
if pH < 5 then START pH Up Motor
else if pH > 7 then START pH Down Motor
if EC < 800 then START EC Up Motor
print Tn, TE, humidity, pH, EC
end
```

WORKING PRINCIPLE

The logs that the Kivy software kept were used for data analysis. Each log shows the values that were reached each day and is updated continuously. The current values are displayed on a single screen and include the humidity, pH, water temperature, and air temperature. For instance, the air temperature log in figure 5.4 depicts the temperature decline from 24 C to 20 C as the day gives way to night. Figure 5.5 displays how long the lights were on and when the condition was changed (from ON to OFF or from OFF to ON). Each log is crucial for ensuring that the controlling actions maintain all parameters at their ideal values. The pH of the water rises as the plant ingests more and more nutrients from it. This is due to the fact that plants in an environment like this require acidic nutrition. According to a study, plants grown inside of the food waste management system thrive more than plants grown outside of it. The colour, size, and length of the leaves and stems, among other things, are to blame for this. The temperature measurements of the water the plants are sitting in on a smartphone app. It demonstrates that the temperature is stable from 5 to 10 in the morning and subsequently rises from 5 to 10 in the morning. These values are taken every 24 hours at a frequency of 0.01°F (0.02°C) between the hours of 5am and 10pm. An automated hydroponic system enables the farmer to avoid onsite activities. Utilizing an Android application enables this. The user's smart phone receives data. The system is automatic since it consistently produces great crop yields even when the grower is not present.

Observation from Graphs: The monitoring of response parameters of the smart hydroponic system is shown in Figure 8. The solution temperature is measured at

Figure 8. Monitoring response parameters of Smart Hydroponic system

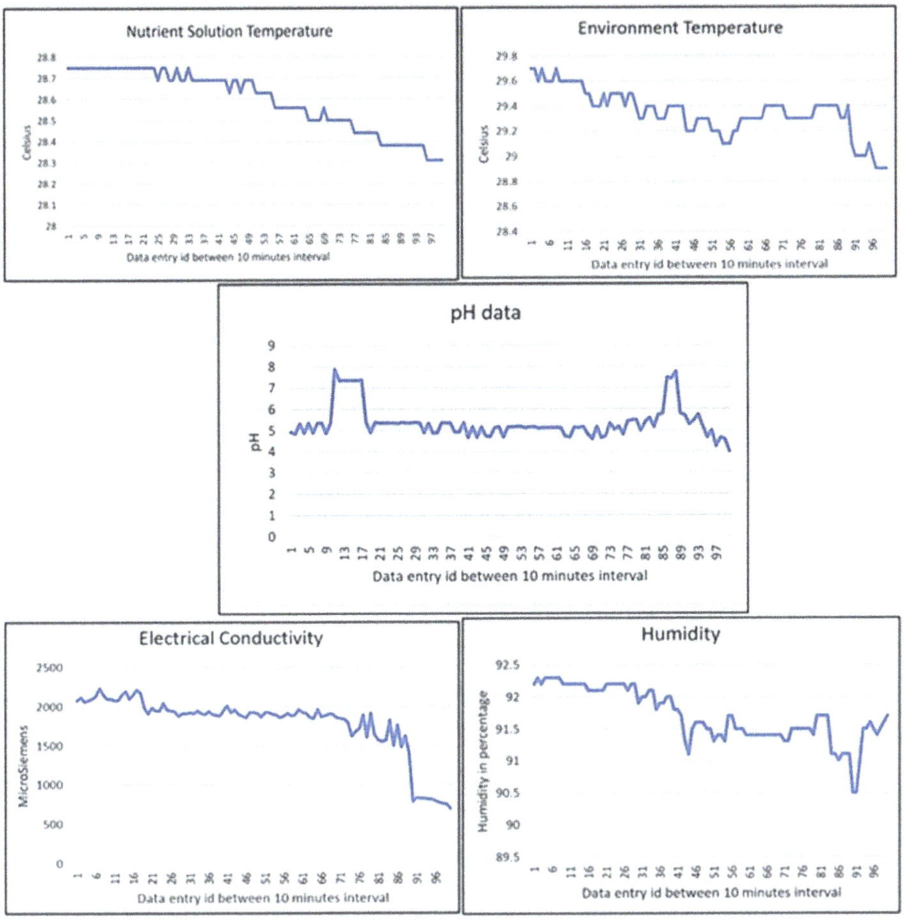

10-minute intervals. Here, we can observe that the values didn't abruptly change, indicating that the sensor readings are extremely reliable. spike with a greater pH value than other values. Following that, we can see if the ph value returns to its normal range. environment's temperature value, which is dropping as a result of the temperature management system. Plant nutrient uptake is causing a decrease in electrical conductivity. the system's humidity readings during the experiment.

Reliability of system: When a designed system is put through real-world testing and operates admirably, it is said to have been successful. Such information helps increase reliability in addition to helping identify problems. A device's performance, and consequently its reliability and failure, are influenced by a variety of factors. The hydroponic system's active status determines when it is operational. These data sources enable users to choose maintenance and inspection schedules. Trend analysis, which uses historical failure data from a system, is a popular statistical technique for determining the functioning and changes over time in a repairable system. A repairable system can be conceptualised as a counting failure process, where successive inter-arrival failure times typically increase for improving systems and decrease for degrading systems. System dependability trend tests seek to ascertain whether the pattern of failures has changed over time. Diseases may spread more quickly under conditions of relative humidity that exceed the limitations. High humidity contributes to the spread of powdery mildew and botrytis. Ventilating humid indoor air and exchanging it with drier outdoor air are two ways to reduce relative humidity. When compared to prior hydroponic systems, this results in a 40–50% reduction in water usage (Rouphael & Colla, 2005). Both the system and the soil-based system's growing times have been compared. As a result, a hydroponic system can be used to cultivate four different types of crops each year. An Internet of Things (IoT) network is also incorporated into the system for remote monitoring and control. The following are some of the designed system's key characteristics:

Total Yield: This strategy increased tomato plant output by over 50% when compared to conventional methods, yielding about 12 kg/m2/crop.

Parameters Controlled: A recirculating hydroponic system that seeks to keep the pH at ideal values of 5.5 to 6 moles/L will ensure that there are no nutritional issues with plants grown in this manner. The prolonged recycling of nutrient solutions has a significant impact on hydroponic growth and output.

Water conservation: In comparison to conventional hydroponics, it consumes less water and electricity to grow crops and generates greater magnitudes of about 12 kg/m2/crop.

Cost Efficient: A developed system costs around 20% less to operate and maintain than a system based on soil. No need for fertilisers, soil preparation for a subsequent crop, etc., can be met. A variety of crops can be grown using the system at the same price as using conventional techniques.

Labour saving: Although labour is needed in the soil-based system, after the germinated seeds have been planted there is no need for further human involvement because everything is managed remotely.

Time saving: A month and a half is saved since a single crop may be grown in a hydroponic system in 3 months as opposed to 4 months in a soil-based system. System may be tailored to meet the needs of different plants, and it is independent of its environment or surrounding atmosphere. It demonstrated its significance by being entirely automated, as well as by being compact and reasonably priced for consumer use.

FUTURE DEVELOPMENT ACTIVITIES

In order to provide water and nutrients to the roots of hydroponic plants, they are grown in a hydroponic environment using a nutrient-rich solution. Researchers are concentrating on how it might address upcoming food issues. The ability to grow a wider variety of food and provide a bio-regenerative life support system are two benefits of hydroponics in space. Future space programmes will benefit from hydroponic gardening.

Numerous space enterprises from many nations, including NASA, ISRO, and others, are conducting in-depth study in this field. An important step in the creation of a new scientific method could be made possible by a permutation-testing-based method for evaluating each profile's effectiveness. For instance, one user may choose certain upper and lower-bound pH values together with specific lighting and water cycles, but another user may specify entirely different settings. Limit the options available for each actuator to help solve this problem. The lack of easily accessible machine-readable data is one of the problems with agriculture. Misinformation abounds regarding the kinds of circumstances that yield the best outcomes.

CONCLUSION

- A hydroponic setup that is not dependent on the environment outside has been developed. The technology consumes less water and fertiliser as compared to a soil-based system. In order to boost crop output by 17% and maintain relative humidity between 55% and 65%, the study proposes a sophisticated configuration.
- Every crop has unique dietary, environmental, and cultivation requirements. Premium quality crops are grown in hydroponic greenhouses, where they fetch a premium price even for everyday veggies. Common veggies including lettuce, tomatoes, and cucumbers are grown hydroponically. The optimal temperature range for tomatoes is between 20- and 30-degrees Fahrenheit. Fancy Lettuce grows from seedling to harvest in 4 to 12 weeks.
- The majority of hydroponically grown cucumbers are European seedless types. Tomatoes and cucumbers may both be produced in the same greenhouse. In the world of subterranean hydroponics, carrots are grown.
- An Arduino Mega, a 5-volt relay, a 12-volt DC water pump, a humidity and temperature sensor (a DHT22), and a PH sensor are among the nine types of sensors and hardware that were used in this year's NFT System. The DS18B20 Digital Thermometer's 9 to 12-bit (adjustable) temperature readings can be obtained from the data line itself. The conductivity of a solution used in

hydroponic farming is measured by an EC sensor. Some manufacturers make combo meters which combine the features of both types of metres into a single unit. The ESP8266 is an integrated TCP/IP protocol stack-equipped self-contained system on chip (SOC) that allows any microcontroller access to your Wi-Fi network. It features a self-calibrated RF that enables it to operate in various operational environments and supports APSD for VoIP applications.

- A hydroponic system can be used to cultivate four different types of crops each year. An Internet of Things (IoT) network is also incorporated into the system for remote monitoring and control. This strategy increased tomato plant output by over 50% when compared to conventional methods, yielding about 12 kg/m2/crop. A hydroponic system is 20% less to operate and maintain than a system based on soil. It consumes less water and electricity to grow crops and generates greater magnitudes of about 12 kg/m^2/crop. System may be tailored to meet the needs of different plants, and it is independent of its environment or atmosphere.

REFERENCES

Calone, R., & Orsini, F. (2022). Aquaponics: A Promising Tool for Environmentally Friendly Farming. *Frontiers for Young Minds*, *10*, 707801. doi:10.3389/frym.2022.707801

Das, S., & Sahu, G. (2020). Soil-Less Farming- An Innovative Way Towards Sustainability. *Agricultural & Food: E-Newsletter*, *2*(10), 82–85.

Espinal, C. A., & Matulić, D. (2019). Recirculating Aquaculture Technologies. *Aquaponics Food Production Systems*, *01*, 35–76. doi:10.1007/978-3-030-15943-6_3

Ezzahoui, I., Abdelhouahid, R. A., Taji, K., Marzak, A., & Ghanimi, F. (2022). The Aquaponic Ecosystem Using IoT and IA Solutions. *International Journal of Web-Based Learning and Teaching Technologies*, *17*(5), 1–15. doi:10.4018/IJWLTT.20220901.oa2

Goddek, S., Joyce, A., Wuertz, S., Körner, O., Bläser, I., Reuter, M., & Keesman, K. J. (2019). Decoupled Aquaponics Systems. In Aquaponics Food Production Systems. doi:10.1007/978-3-030-15943-6_8

Kumar, P., Sampath, B., Kumar, S., Babu, B. H., & Ahalya, N. (2023). Hydroponics, Aeroponics, and Aquaponics Technologies in Modern Agricultural Cultivation. In Trends, Paradigms, and Advances in Mechatronics Engineering (pp. 223–241). IGI Global.

Rouphael, Y., & Colla, G. (2005). Growth, yield, fruit quality and nutrient uptake of hydroponically cultivated zucchini squash as affected by irrigation systems and growing seasons. *Scientia Horticulturae, 105*(2), 177–195. doi:10.1016/j.scienta.2005.01.025

Sampath, B., Sasikumar, C., & Myilsamy, S. (2023). Application of TOPSIS Optimization Technique in the Micro-Machining Process. In Trends, Paradigms, and Advances in Mechatronics Engineering (pp. 162–187). IGI Global.

Sathyanarayana, S. R., Gangadhar, W. V., Badrinath, M. G., Ravindra, R. M., & Shriramrao, A. U. (2022). Hydroponics: An Intensified Agriculture Practice to Improve Food Production. In Reviews in Agricultural Science (Vol. 10, pp. 101–114). doi:10.7831/ras.10.0_101

Taha, M. F., ElMasry, G., Gouda, M., Zhou, L., Liang, N., Abdalla, A., Rousseau, D., & Qiu, Z. (2022). Recent Advances of Smart Systems and Internet of Things (IoT) for Aquaponics Automation: A Comprehensive Overview. *Chemosensors (Basel, Switzerland), 10*(8), 303. doi:10.3390/chemosensors10080303

Yanes, A. R. (2020). Towards Aquaponics 4.0 : A framework for automated, IoT and smart implementations in indoor farming. *Information Processing in Agriculture, 9*(4).

Chapter 3
A Study on the Conversion of Ligninolytic Biomass to Biofuels

Lakshmana Kishore T.
University College of Engineering, Kakinada, India

Haribalaji V.
Narasu's Sarathy Institute of Technology, India

ABSTRACT

Due to increasing power demand and the utilisation of natural resources in growing countries, the use of biomass and biofuels is emerging. The role of thermophilic ligninolytic bacterial enzymes in the biomass-to-biofuel conversion process is discussed in this chapter. Various elements of biomass feedstocks, compositions, and viabilities of lignocellulosic biomass are illustrated. The lignocellulosic biomass pre-treatment methods, inhibitors during the pre-treatment process, hydrolysis methods, and bacteria production processes have been explained. The processes for isolation, screening, and maintenance of cellulolytic bacteria are exemplified through suitable schematics. The utilization of agro-waste in the ethanal and biofuel production processes is also exemplified.

INTRODUCTION

Industries, residences, services, and transportation use more energy overall on a daily basis. India is now the fourth-largest energy consumer in the world, thanks to growing primary energy consumption. Biomass-based fuels have the potential to produce

DOI: 10.4018/978-1-6684-4118-3.ch003

energy that is clean, carbon-neutral, and sustainable. Given its biodegradability, low toxicity, and ability to mix well with gasoline, biofuel is being looked at as a potential replacement for fossil fuels. The world's crude oil reserves are limited, and they are depleting considerably more quickly than in the past. A mesoscale biorefinery sector may be supported by the biofuel industry, which could also hold down the growth of agriculture and forestry and cut greenhouse gas emissions by an estimated 86 percent compared to gasoline.

Biofuels cover a wide range of feedstocks, conversion methods, and applications. About 51,000 million litres of ethanol have been produced worldwide, with the USA and Brazil ranking as the top two producers (Abualigah et al., 2022). The most plentiful bioresources are thought to be wood and agricultural waste, often known as lignocellulosic biomass, which are created by photosynthesis using solar energy and atmospheric CO_2. Crystalline celluloses make up 40–50% of lignocellulosic biomass, and obstinate lignin makes up 10–25% of it. It can be used as a powerful raw material for several biofuel generations and is generated on an annual basis in the range of 10-20 tonnes.

A fuel made from biomass is known as a "biofuel" and can be either solid, gaseous, or liquid. Examples of liquid biofuels include biodiesel, plant oils, methanol, and ethanol. It is mostly made up of timber, forestry products, animals, aquatic vegetation, and agricultural products. With 34 billion litres produced, the United States has maintained its position as the world's top producer of ethanol. Brazil's overall bioethanol production increased from 18 billion litres to 27 billion litres. Numerous other nations, like China, Thailand, Germany, France, Spain, and Sweden, are also carrying out sizable national biofuels' programmes. A key fuel source for long-term use is ethanol. In addition to the necessary herbicides and fertilisers for production, it requires extremely little water. Organic resources, such as plant and animal waste, are used to make biofuels. The most necessary source to obtain the end product as a fuel is living things that can use CO_2. Organic resources utilised in their natural state, such as wood chips, fuelwood, and pellets, are considered primary biofuels. Bioethanol, biodiesel, biobutanol, biogas, and biohydrogen are examples of secondary biofuels that are produced by the processing of biomass. They are divided into first, second, third, and fourth generation biofuel categories depending on the production process and type of feedstock used.

Fermentation is used to create bioethanol from source materials that include significant amounts of sugar and starch. The ethical dilemma of "food vs fuel" is linked to biofuels made from food crops.

Lignocellulosic biomass, non-food crops, agricultural waste, and energy crops are used to make second-generation biofuels. Algal biomass biofuels, also referred to as algae fuel, are categorised as third-generation biofuels. Crop leftovers are common feedstocks that are utilised all over the world to produce biofuels. Biofuel is a renewable energy source that reduces the need for fossil fuels. It is created using renewable plant-based energy sources like algae, hydrocarbons, and other alternative types of energy(Khan et al., 2022). When compared to traditional fuels, biofuels like biogas, bioethanol, and biodiesel can offer sources of fuel that are environmentally benign. Crops offset the CO_2 emitted by vehicles because plants take up CO_2 as they grow. They discuss issues including rural unemployment and energy-related problems.

Their growth and output have an impact on the policies that affect numerous industries involved in the nation's socioeconomic elements. Utilizing biofuels leads to better combustion, which reduces hydrocarbon emissions. New renewable energy sources are being developed by some nations, such as the manufacturing of cellulosic ethanol. Compared to other fuels, bioethanol has attracted increased attention as a clean and renewable fuel. Since fuel crops absorb carbon dioxide, using bioethanol will cut greenhouse gas emissions while also improving air quality. By blending bioethanol with gasoline, we may prevent over-dependence on countries that produce most of the world's oil while also extending the useful life of depleting non-renewable resources.

A fuel with a higher-octane number and clean burning properties is bioethanol. When compared to fossil fuels, it is less harmful and biodegradable. The absence of hydrocarbons makes ethanol burning completely and with few emissions possible. The most potential gasoline substitute is bioethanol, which is made from sugar and corn. By combining ethanol with gasoline, one can increase the volumetric efficiency of an engine and ensure that fuel is used as efficiently as possible, which helps to lower the price of petroleum. The moral concerns surrounding the use of food as fuel's primary raw material have sparked research to identify suitable substitute feedstocks (Jha et al., 2022). Due to its extensive large-scale availability, lignocellulosic biomass is a good source for the synthesis of bioethanol. In Asia, Europe, South America, and the United States, corn stover, rice and wheat straws, and sugarcane bagasse are among the most promising and plentiful feedstocks made from plant leftovers. The schematic arrangements for biomass to biofuel conversion is illustrated in Figure 1.

BIOMASS FEEDSTOCK

The various biomass feedstocks are illustrated in Figure 3. It was extracted from industrial waste, agro-waste, biofuel crops, and biological waste (Ha et al., 2020).

Sugar and starch-based sources: The use of lignocellulosic biomass for inexpensive ethanol production is now the focus of technological research and development. The key crops cassava, sugarcane, sorghum, and wheat may also be

Figure 1. Schematic arrangements for biomass to biofuel conversion

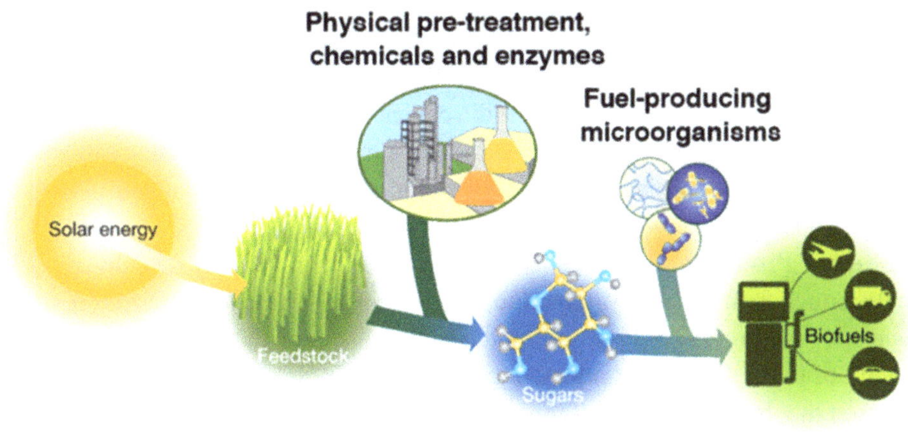

utilised as substrates. In the interim, maize, grains, dextrose, glucose syrup, food additives, etc. can be utilised for the production of heat and power as well as animal feed.

Lignocellulosic biomass: A strategy to cut back on both the use of fossil fuels and environmental damage is the generation of ethanol from lignocellulosic biomass (LCB). Because it can be found for less money and because it mostly consists of cellulose, hemicelluloses, and lignin content, LCB is regarded as a reliable source of energy generation.

Rice Straw: As one of the world's largest available biomass feedstocks, rice straw can be used to produce bioethanol. Due to its high silica, lignin, and limited digestibility, it is not a viable choice for animal feed. A potential feedstock for the manufacture of gasoline ethanol is rice straw. Its main components are cellulose (32–47% of its weight), hemicelluloses (19–27% of its weight), and lignin (5–24%). Due to their low bulk density and potential as pest and disease harbours, rice straws take up a lot of space. An effective method for turning rice straw into a value-added biofuel is bioconversion. It is economical to combine crude oil with gasoline given the current high costs.

Coconut Coir: Coconut coir pith is a by-product created when the coir fibre from the coconut husk is extracted. The pith of coir is extremely light, very compressible, and exceedingly hygroscopic. In India, particularly in the south, it is a significant agricultural waste product. Alpha cellulose makes up 39.31% of coconut husk, whereas hemicellulose makes up 16.15%, lignin makes up 29.79%, and extractives make up 28.48%. It is one of the renewable sources used in the synthesis of bio-ethane.

Coconut production: The southern Indian states of Kerala, Tamil Nadu, Karnataka, and Andhra Pradesh, as well as West Bengal, Odisha, Maharashtra, and Gujarat, account for about 90% of the country's total area dedicated to coconut cultivation. 20.82 lakh hectares produced 2395 crore worth of coconuts annually, yielding 11505 coconuts per hectare. India contributes 31% of the global production of coconuts, making it one of the top producers in the world. The major exporters of coconuts are Indonesia, the Philippines, India, Sri Lanka, Thailand, Vietnam, Mexico, and Malaysia, with India coming in third.

Ethanol Production: Farmers will be deterred from putting new areas under cane production by acute water scarcity and unusual weather patterns in major sugarcane growing zones. To accommodate its rising demand, India imports a sizable quantity of biofuel from leading ethanol producers. India's ethanol consumption is increased day-by day.

COMPOSITION AND A VIABILITY OF LIGNOCELLULOSIC BIOMASS

Cellulose: 35–50% of the entire biomass, measured in weight, is made up of cellulose. It is an extremely solid polymer made up of roughly 12,000 glucose monomer units connected to one another in a linear chain by a glyosidic link. Because of the way it is structured, cellulose cannot be broken down by microbes and is insoluble in the majority of solvents. The primary enzymes involved in the catalysis of cellulose polymer are called cellulases (glucoside bond, endoglucanase, and cellobiose).

Hemicellulose: Hemicelluloses are the second most prevalent element in the world after cellulose. By weight, it accounts for about 25 - 35 percent of the overall biomass. This bond is glyosidic. It is a relatively short heterogeneous polysaccharide, unlike cellulose, that contains 200 units or less of D-glucose, D-arabinoside, D-mannose, D-xylose, etc. When 1,4-xylanase, an enzyme, is used, it is easy to hydrolyse xylene, a crucial component of hemicelluloses.

Lignin: Lignin is a hydrophobic chemical molecule that accounts for between 10 and 25 percent of the weight of all biomass. It bridges the gap between cellulose and hemicelluloses, two other major constituents. A combination of the enzyme's manganese peroxidase and lignin peroxygenase may efficiently break down lignin.

Thermophilic Enzymes: Enzymes are replacing most chemical processes in a variety of sectors. Although these enzymes offer benefits, their use is limited because of their weak tolerance to extremely high temperatures and other environmental factors. Such enzymes are abundant in thermophilic bacteria and exhibit extreme stability at high temperatures. When converting lignocellulosic biomass into

Figure 2. Biomass Feedstocks

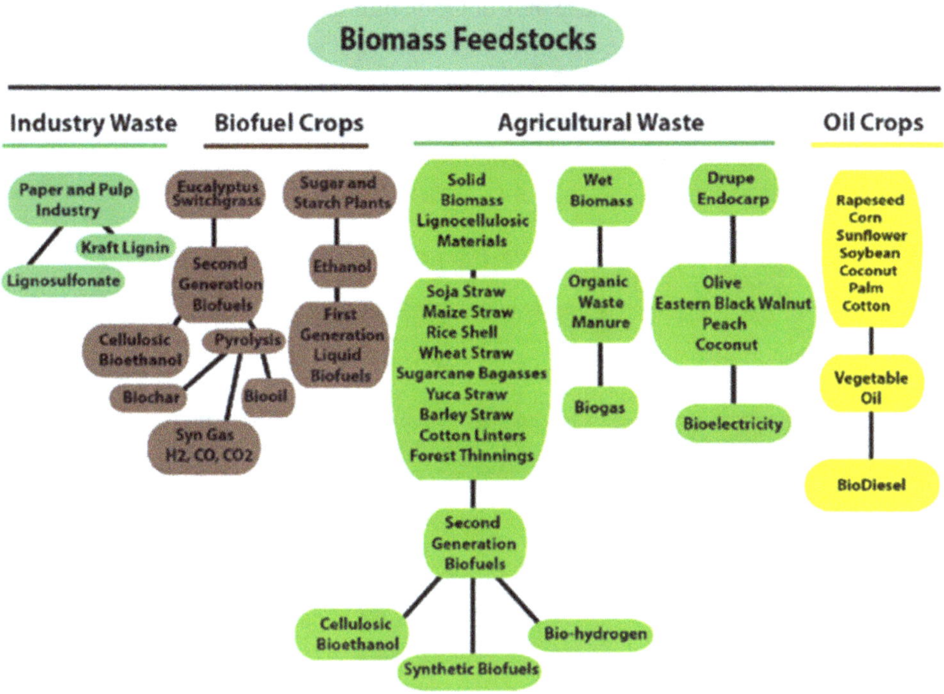

monosaccharides at high temperatures, cellulases with high activity and stability are typically preferred.

The majority of thermophilic cellulases function best between 80 and 90 degrees Celsius. Numerous studies have been conducted, and extensive use is being made in the laboratory of the creation of cellulolytic enzymes. The recovery of the product is thought to be faster and more stable in thermophilic procedures. Microbes are more capable than mesophilic species of acquiring high metabolic rates, stable enzymes, and better product yields. Due to the fact that the microorganisms isolated from these environmental circumstances are a reliable source of thermozymes, they are of great importance(Ha et al., 2020).

Ligninolytic enzymes: Three hydroxycinnamyl alcohol monomers, p-coumaryl alcohol, coniferyl alcohol, and sinapyl alcohol, are typically used in the formation of lignin in plant cells. Although the lignin polymer is extremely resistant to disturbance, it can be depolymerized using a variety of thermo-chemical techniques, which could have negative environmental implications. Laccases are multifunctional oxido-reductases that can break down a wide range of phenolic and non-phenolic substances by turning their oxygen molecule into water through four-electron reduction.

Four different sorts of living things, including bacteria, insects, higher plants, and fungi, create laccases. Because they degrade lignin and its analogues under mild conditions, microbes produce enzymes that are employed in various industries, in addition to ethanol production, for the degradation of aromatic hydrocarbons. Creating a low-cost growth medium for various microbial strains is crucial to comprehending the potential of various plant-based agricultural leftovers.

Figure 3. Classification of Ligninolytic-Biomass Pre-treatment

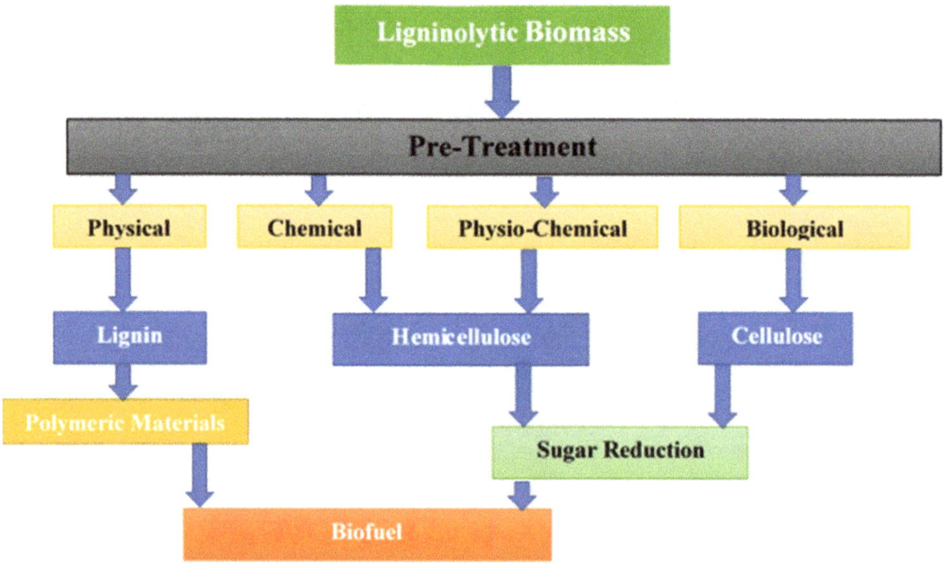

PRE-TREATMENT

The main obstacle to enzymatic hydrolysis when employing lignocellulosic material is recalcitrance. The pre-treatment of the complex polysaccharides is a crucial phase in the process because the main challenge with employing this material is breaking down the complex polysaccharides(Ha et al., 2020). A good lignocellulose pre-treatment should (1) minimise the cost of producing different inhibitors, (2) use less energy, and (3) use cheap chemicals. Some techniques have low ethanol and sugar yields despite being cost-effective, particularly when mild pre-treatment conditions are applied. An excellent pre-treatment procedure prevents particle size decrease, restricts the production of inhibitory chemicals, protects the hemicellulose fraction, and, most importantly, uses the least amount of money and energy. The

Figure 4. Schematics of pre-treatment process

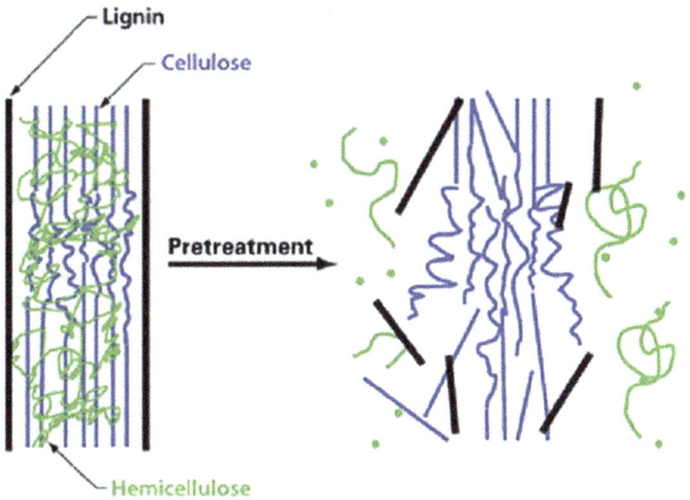

cost of pre-treatment and industrial cellulose production have a significant impact on ethanol production (Figure 4).

Mechanical Comminution: The size of biomass is decreased during mechanical comminution by a variety of techniques, including grinding, chipping, milling, etc. Vibratory ball milling has developed into a crucial technique for lowering the cellulose crystallinity and digestibility of LCB. The degree of polymerization can be reduced while the surface area can be increased using this technique(McKendry, 2002).

Pyrolysis: An innovative process for transforming the cellulose and hemicellulose fractions into fermentable sugars has been devised by researchers at the University of Bristol in the UK.

Physio-chemical pre-treatment (Steam treatment): One of the best physical-chemical pre-treatment techniques for various lignocellulose biomasses is steam treatment, explosion, or autohydrolysis. The biomass is heated swiftly and without the addition of any chemical or biological agents in a high-pressure environment. It is described as being free for personal use only.

Ammonia fib re explosion (AFEX) method: When lignocellulosic material comes into contact with liquefied ammonia at a high pressure and temperature, it quickly undergoes hydrolysis. The primary variables that can impact this process are temperature, time, ammonia and water loading, and the quantity of treatment cycles. Many LCBs, including switch grass, sugarcane bagasse, and rice straw, have been pre-treated using this technique.

Liquid hot water Method: Lignocellulosic materials are heated in hot water, which has long been helpful to the pulp industry. The biomass is then subjected to a set amount of time in hot water treatment, which results in the highest rates of recovery and the production of the least amount of inhibitor. But in this pre-treatment, there was 420 percent cellulose, 35 percent lignin, and practically all the hemicellulose.

supercritical fluid pre-treatment (SCF): A fluid with superior qualities, such as density similar to a liquid and transport characteristics similar to a gas, is known as a supercritical fluid (SCF). It has a penetration limit due to the material's crystallinity. At the lower temperature being used, sugars may be stabilised and biomass breakdown can be prevented.

Ozonolysis: Ozone gas is used in the ozonolysis method of pre-treatment to break down lignin and hemicellulose and speed up the biodegradation of cellulose. This pre-treatment typically takes place at room temperature and effectively removes lignins without producing any harmful by-products.

Alkaline pre-treatment: Any alkali reagent (such as NaOH, KOH, etc.) is used in the alkaline pre-treatment procedure to eliminate the lignin and other by-products that could make the enzyme more difficult to work with. This pre-treatment can be carried out under regular circumstances and takes hours as opposed to minutes and seconds. Lime ($Ca(OH)_2$) is less expensive and has fewer stringent safety standards than KOH and NaOH. Lime is an effective therapeutic agent, but only under the right pre-treatment conditions. It was discovered that lignocellulosic biomass is mixed with lime and water for a few hours to many weeks at a temperature of 313 to 423 K.

Acidic pre-treatment: After being treated with acids like hydrochloric acid, lignocellulosic biomass can produce a sizable amount of mono-sugars. Pre-treatment with diluted acid may improve the biomass's cellulose's ability to be broken down. After pre-treatment with 2% H_2SO_4, the cotton stalk's xylan content was reported to have been reduced by 95%.

Wet oxidation: Lignocellulosic biomass is milled and dried by mixing 6 grammes of biomass with one litre of water. Oxygen is necessary for this pre-treatment process in order to oxidise the chemicals that can dissolve in water. If wet oxidation is paired with an alkali solution, lignin is swiftly oxidised.

Organosolv pre-treatment: The quantity of hazardous by-products produced by the decomposition of lignins can be significantly reduced by the use of any potent inorganic acid in this process. The hydrolysate can be neutralised with fewer chemicals than with previous techniques. Higher carbon dioxide pressure and excellent lignin removal efficiency have created.

Biological pre-treatment: The optimal pre-treatment is said to be biological because there is no alteration to the structure or makeup. By using this technique, lignin and hemicelluloses are biologically degraded, making biomass more receptive

Figure 5. Production of Inhibitor during Biomass treatment

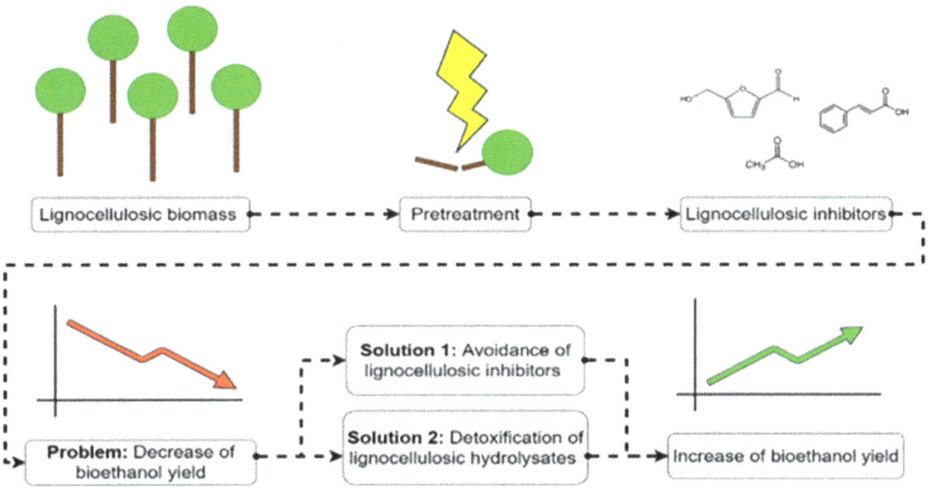

to enzyme digestion. Fungi, particularly white rot, are regarded as an efficient agent for biological pre-treatment among all biological agents.

PRODUCTION OF INHIBITOR DURING PRE-TREATMENT

Monomeric sugars will become progressively contaminated with aldehydes and organic acids if they are exposed to high temperatures and high acid concentrations. Additionally, this may result in yield loss and trouble with the fermentation and enzymatic hydrolysis processes. Furans, phenolics, and weak acids are three categories of the many inhibitors created during pre-treatment, particularly chemical pre-treatment. The breakdown products of C-5 and C-6 sugars are, respectively, furans. In terms of the acetyl groups found on the hemicellulose backbone, acetic acid is unique(Luo et al., 2014). The two most popular hydrolytic methods for converting cellulosic biomass into ethanol are used before fermentation. Hydrolysis is the process by which the hydrogen bonds in the cellulose and hemicellulose structures are broken down. other compounds like acetic acid, formic acid, and methanol. Production of Inhibitor during Biomass treatment is shown in Figure 5.

HYDROLYSIS

Acid Hydrolysis: This method calls for the conversion of complicated polymers into simple sugars when they come into contact with certain chemicals at a specific temperature and for a specific amount of time. You can do it with either diluted or concentrated acid. Acid hydrolysis is a method that uses the addition of sulfuric acid, which has a concentration of between 70 and 90 percent, to totally transform polymers into C-5 and C-6 sugars at a rapid rate from cellulose to glucose.

Enzymatic hydrolysis: The proportions of hemicellulose and cellulose are broken down using a variety of enzymes. Cellobiohydrolase, endoglucanase, exoglucanase, and glucosidases are used to break down the cellulose component. This technique produces monosaccharides from processed biomass in an economical and highly efficient manner that is also environmentally benign. The endoglucanase enzyme generates short polymer chains by randomly attacking cellulose. Cellobiose and other complex polymers are catalysed by glucose sidases to generate glyco. The entire process is very dependent on variables including pH, temperature, time, enzyme concentration, and others.

Fermentation: The final product we receive after enzymatic hydrolysis is often primarily made up of certain C–6 sugars. The theoretical yield of ethanol from these sugars is found to be nearly 0.51 g/g glucose. The fermentation of all the sugars generated from lignocellulose into ethanol requires a large number of microorganisms.

C6H12O6 2H5OH + 2CO2 + ATP: Fed-batch reactors outperform batch cultures because they have more benefits than those offered by other processes or modes. Some of the main advantages include increasing the concentration of viable cells and building up goods to the highest level. The formation of various acids and phenolic compounds is the main challenge to increasing ethanol yield. A different strategy to reduce the price of enzymes is to produce cellulose-degrading enzymes using SSF (simultaneous saccharification and fermentation). The main drawback of SSF is that cellulose activity is hindered by glucose, combining hydrolysis and fermentation in one container. In terms of producing ethanol, SSF is a little better than SHF, particularly when it comes to rice(Akia et al., 2014). Enzymatic hydrolysis and simultaneous fermentation are coupled in SSF to produce ethanol from carbohydrates. Although the stages in SSF are almost identical to those in SHF, it has lower capital expenditures and higher process efficiency. Lynd proposed the term "consensual bioprocessing" (CBP) in the beginning. This CBP, which combines the phases of enzyme synthesis, hydrolysis, and fermentation into one, can take the place of the conventional microbial conversion process.

ETHANOL PRODUCTION

A complex combination of hydrolysis, sugar reduction, product separation, and purification is used to turn biomass into ethanol. Different mesophilic bacteria, such as Escherichia coli, Saccharomyces cerevisiae, Zymomonas mobilis, etc., carry out the fermentation process. Several enzymes are required to completely degrade cellulose and separate it into its component parts. The action of thermophilic bacteria or their enzymes occurs between 50 and 80 C. Thermophiles are preferred over mesophiles because of their stability and propensity for higher-temperature activity. Utilizing artificial metabolic pathways in vitro has become popular in recent years(Osman et al., 2021). Because of its well-known characteristics, suitability for many industrial applications, and availability of genetic tools, Saccharomyces cerevisiae has drawn the attention of cellulosic ethanol producers. Due to this fermenting strain's inability to use xylose sucrose, the success of the transition from first generation to second generation biofuel production has been limited. Enhancement of xylose metabolism in S. cerevisiae has been achieved through genetic modification of a common CYC8 co-repressor. Economical and effective co-fermentation has also benefited from the mutant transporter's improved half-life.

BACTERIA PRODUCTION

For the manufacture of diverse biofuels, including ethanol and biodiesel, numerous strains of bacteria found in nature have been employed. Role of bacteria in biomass to biofuel production is illustrated in Figure 6.

***Zymomonas mobilis*:** Compared to yeast, Z. mobilis converts carbohydrates into ethanol much more effectively (Ajit et al., S. cerevisiae). It only has one drawback: despite the availability of pentose sugars, which are inexpensive, it cannot use them. But lately, a recombinant strain with several enhanced characteristics was created, and it was demonstrated to produce ethanol with a yield of 136 g/L.

Bacillus subtilis: By inactivating, the strain BS37 was produced, and it demonstrated long-term ethanol output of up to 8.9g/L.

Escherichia Coli : Due to the scientific community's familiarity with the molecular biology of the bacteria E. coli, it has been frequently employed to produce ethanol. It was altered to become KO11, an unique strain with a high ethanol tolerance. Z. mobilis genes that code for pyruvate decarboxylase were present in KO11.

Figure 6. Role of bacteria in biomass to biofuel production

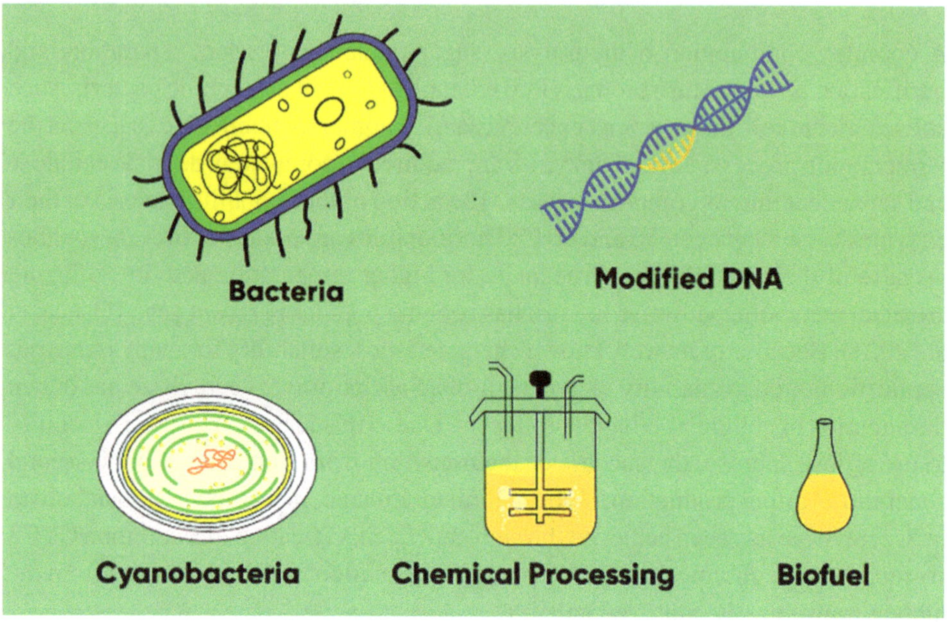

THERMOPHILIC BACTERIA

The production of a significant amount of bioethanol in an efficient and cost-effective manner is now a requirement for thermophilic microorganisms. Because they catalyse the formation of complex carbohydrate polymers, thermophiles are preferred over mesophiles. Mesophiles are rapidly infected with undesirable bacteria, rendering them useless for future use. Processes that can use lignocellulosic biomass (LCB) for ethanol generation are required. It is necessary to pre-treat LCB in order to obtain a significant quantity of fermentable sugars(Ong et al., 2020). Thermophiles that can hydrolyze the LCB and convert reducing sugars into ethanol at the same time are favoured.

Clostridium thermocellum: The bacterium clostridium thermo-cellum is most frequently employed to make ethanol from cellulose, although it is also capable of producing a variety of other products, including acetate, lactate, formate, and CO_2, without generating butanol.

Geobacillus: Researchers from the University of Bristol in the UK have identified a novel class of bacteria that can convert carbohydrates into ethanol, acetate, lactate, and formate in soil and water at temperatures between 40 and 70°C. Recently, the efficient halogen bacteria Geobacillus thermo-glucosidasius was created for ethanol production from lignocellulosic biomass. By eliminating the lactate dehydrogenase from this bacterium, the fermentation process was improved. The genetically modified strain was discovered to exhibit high ethanol yield at 70°C, inhibitor resistance, and substrate tolerance.

GLASSWARE, CHEMICAL REAGENTS AND INSTRUMENTS

Glasswares: A 2000 ml beaker, a 1000 ml conical flask, a 1000 ml volumetric flask, a 500 ml volumetric flask, a measuring cylinder, a 500 ml titration flask, sieves, a Gooch Crucible, and other equipment like a china dish and a Kjeldahl flask.

Chemical reagents: The chemical formula for ethylene glycol monoethyl ether includes acetrimide, sulphuric acid, potassium permanganate, silver sulphate, distilled water, ferric nitrate, silver nitrate, and hydrochloric acid, among other substances (EDTA).

Instruments: Weighing scales, heated stirrers, muffle furnaces, autoclaves, hot air ovens, desiccators, shaker cum incubators, water baths, FTIR, SEM, and other lab equipment are available.

ISOLATION, SCREENING AND MAINTENANCE OF CELLULOLIGNOLYTIC BACTERIA

Site Selection & collection of samples: Hydrocarbon-contaminated waste water samples have been gathered from numerous locations across the nation. In sterile sample bottles, samples were gathered from various locations and brought right away to the lab. Furthermore, soil samples were transported to the lab and stored there for later analysis at 4°C in the refrigerator.

Media preparation: For secondary screening, K_2HPO_4 (7.0 g/L), KH_2PO_4 (3.0 f m), $CaCl_2.2H_2O$ (0.10 g/l), 10.0 ml of trace elements, and Nutrient Agar (10.0 mg/L) were added to Minimal Salt Medium (MSM).

Isolation: Peptone, beef extract, yeast extract, and sodium chloride were added to samples of soil and waste water to create the nutrient broth medium. A 500 ml conical flask containing 100 ml of broth medium was added to each sample in order to enrich it for the isolation of bacterial strains from soil and water samples. Media- and sample-containing beverages were cultured in flasks for 72 hours in Minimum Salt Medium (MSM) before being shaken in an incubator for one week. Following a 7-day incubation period, the incubated broth medium was serially diluted (up to 10-8) by adding 1 ml of each sample to 9 ml of sterile water. New NA plates were streaked with bacterium stains a further 1-2 times, and then the bacteria were cultivated into colonies and fed on nutrient medium.

SCREENING OF THE ISOLATES FOR LIGNOLYTIC, CELLULOLYTIC AND XYLANOLYTIC ACTIVITY

Screening for Lignolytic activity: Methylene blue decolorization was used to identify isolates having lignolytic activity. The MB-containing broth medium was infected with the isolates, which were then centrifuged for five minutes at a speed of 10,000 rpm while being stored in an incubator. The resulting supernatant was placed in a cuvette so that its optical density could be measured at 520 nm.

Screening for cellulolytic activity: After being streaked on plates, isolates were put in an incubator with a shaker for 7 days at 50 °C and 120 rpm. Agar containing CMC was coloured dark, and regions where CMC had been broken down by bacteria were transparent (the "halo zone"). Halo measurements are used to verify the bacteria's cellulase activity.

Screening for Xylanolytic activity: The experiments used the same media as the cellulase activity assays, with the exception that Birchwood xylan was used as the carbon source. On agar plates, bacteria colonies were streaked and cultured for 7 days at 50°C with 120 rpm shaking. In contrast to the surface's reddish-purple tint, a yellowish opaque region will be visible around the colonial spot. Flood the plate with iodine solution after incubation, let it sit for 5 minutes, then pour off the stain and wash the surface with distilled water(Alonso et al., 2010).

Identification of bacterial isolate: Isolated strains of Eubacteria were screened for lignolytic potential using 16S rDNA gene sequencing, followed by homology using an rDNA database, in this study.

(a) **DNA Extraction:** Lysis/homogenization: Cells grown in monolayers should be lysed by suspending them in 1–3 centrifuge tubes and lysing the cells by repeated pipetting. To minimise shearing of the DNA molecules, mix DNA solutions by inversion. Centrifuge the tubes for 20 minutes at 14. Following centrifugation, transfer the resultant microcentrifuge tube without disturbing the pellet. Transfer the spin column to a sterile 1.5-ml microcentrifuge tube containing the DNA, run it through a centrifuge at 14,000 rpm for 2 minutes, discard the flow-through, and then turn off the centrifuge for 5 minutes.

(b) **Polymerase Chain Reaction (PCR):** On a single-stranded DNA template, DNA polymerase is the enzyme that controls the synthesis of DNA from deoxynucleotide substrates. When a specially created oligonucleotide is annealed to a longer template DNA, DNA polymerase inserts nucleotides at the 3' end of the oligonucleotide.

(c) **Annealing:** After that, the mixture is heated until there is enough oxygen present for the primers to bind (anneal) to the DNA's complementary sequence.

(d) **Extension:** Using the target DNA as a template, a succession of nucleotides are added one after another to create DNA primers. In order for DNA polymerase to effectively work on its target DNA strand, the reaction is then heated to 72 °C.

(e) **Purification of PCR Production:** A single-pass sequencing kit (ABI PRISM® BigDyeTM Terminator Cycle Sequencing Kits with AmpliTaq® DNA polymerase (FS enzyme)) was used to prepare DNA and protein sequences (Applied Biosystems). Using an ethanol precipitation technique, the fluorescently tagged fragments were separated from the unincorporated terminators.

Bioinformatics protocol: The NCBI blast similarity search tool was used to blast the 16S rDNA sequence. Sequence alignments were performed numerous times using MUSCLE 3.7. PhyML was demonstrated to be an order of magnitude faster and at least as accurate as other phylogeny tools using simulated data.

PRE-TREATMENT OF LIGNOCELLULOSIC BIOMASS

Different pre-treatments were applied to the biomass made from rice straw and coconut coir. Samples were neutralised with 1N HCl and NaOH after being filtered through muslin cloth. Then, using GC-MS, they were examined for inhibitory substances including acetic acid, formic acid, HMF, and furfural.

Physical pre-treatment: The goal of the study was to decrease the particle size of biomass in order to enhance its surface area. This decrease makes it easier for cellulases to reach the biomass surface, which increases cellulose conversion. Rice straw and coconut coir, two types of lignocellulosic biomass, were ground through sieves with pore sizes of 1 mm and 2 mm.

Chemical pre-treatment: To get rid of dust, the material was washed under running water and sieved through 1 mm and 2 mm mesh. Before being treated with acid and alkali, it was first sun-dried to get rid of all the moisture.

(a) **Acid pre-treatment:** Here, lignocellulosic material underwent an acidic preparation using hydrochloric acid (HCl). Estimating the post-pre-treatment concentrations of cellulose, hemicellulose, and lignin allowed researchers to examine the impact of pre-treatment on wheat straw and coconut coir. After being neutralised with deionized water, the recovered residue was dried at 50 °C.

(b) Alkaline pre-treatment: Wheat straw and coconut coir were used as lignocellulosic biomass, and they were processed in an autoclave with sodium hydroxide at a concentration of 2% (v/v) as a wetting agent for 60 minutes at 121°C. After pre-treatment, residues were collected and thoroughly washed to pH neutrality.

Physico-Chemical pre-treatment: Only microwave alkali treatment was used as a physical-chemical preparation of biomass (Biofuels) in a study published in the Journal of Bioassays and Biofuels. Biosciences are composed of biomass, which has been subjected to a variety of circumstances and substances that alter its physical and chemical characteristics.

Microwave alkali Pre-treatment: In this experiment, we employed lignocellulosic biomass (wheat straw and coconut coir) prepared in a microwave oven for 10 minutes at a solid loading of 10%. The cakes were submerged in a 2% sodium hydroxide alkali solution. The residues were collected after pre-treatment and thoroughly rinsed with deionized water until pH neutral.

ENZYME PRODUCTION FROM LIGNOCELLULOSIC BIOMASS

Using 250 ml Erlenmeyer flasks watered with BLM to reach an initial moisture content of 70% at pH 7, a small-scale reactor was created and autoclaved for sterilisation. Flasks were incubated at 50 °C and 65 °F for 10 days with a temperature and humidity regulated inoculum of 1 ml of newly made bacterial broth suspension.

Optimization of enzymes production: Batch tests in 250 ml conical flasks were conducted at a temperature of 50 °C and an agitation rate of 120 rpm. 100 ml of BLM for bacterial culture were used in all optimization trials. The reducing sugar concentration was determined using the DNS method, and inoculation was carried out using 1 ml/100 ml of culture.

Effect of pH: Using HCl and NaOH, 100ml flasks of medium were given pH adjustments ranging from 4 to 9. Flasks were then incubated for 7 days at 500C and 120 rpm. Each flask had samples taken at intervals of 24 hours, and the decreasing sugar level was calculated using the DNS method.

Effect of contact time: First, flasks containing media with an ideal pH were autoclaved for 15 minutes at 15 pressure (121°C). After being autoclaved, flasks were transferred to laminar air flow and inoculated with 1 ml of inoculum. Flasks were then incubated for 10 days at 50 °C and 120 rpm.

Effect of inoculums dose: First, flasks containing fluid with an ideal pH were autoclaved for 15 minutes at 15 psi (121 °C). The flasks were placed in laminar air flow after being autoclaved, where an optimal dose of vaccination was administered. Flasks were then incubated for the optimum contact period at 50 °C and 120 rpm.

Enzyme production at bioreactor level: In this work, 100 g of each biomass—coconut coir and rice straw—was extracted and mixed in separate sets with the medium in a bioreactor. For the purpose of producing enzymes, a bioreactor was maintained under ideal conditions for five days. Samples were obtained and examined for various enzyme activities after 5 days of incubation.

Measurement of enzyme activities: By analysing the concentration of hydrocarbons contained in the samples, the enzyme activity—which is expressed in units per gramme of dry substrate—is discovered.

ENZYMATIC HYDROLYSIS AND SACCHARIFICATION OF AGRICULTURAL WASTES

Figure 7. Biofuel from Agro-waste utilization

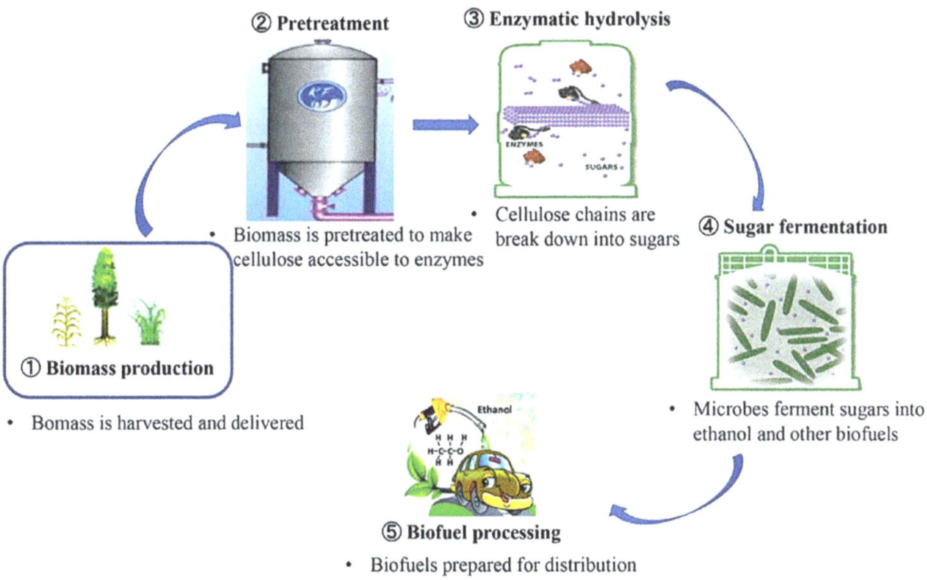

The process involved using 1% w/v of dried and alkali-pre-treated coconuts and 1% of alkali-treated rice straw, as well as 1% ethylene oxide. The experiment was carried out in a 250-ml Erlenmeyer flask carrying a castration incubator and cumulus shaker at 120 rpm. The substrate was soaked in citrate buffer for about 3 hours before being treated with 0.005% sodium zide (Kumar et al., 2023; Sampath et al., 2023). The processes for the extraction of biofuel from agrowaste are illustrated in Figure 7.

Enzymatic Saccharification of Lignocellulosic Biomass in Bioreactor

Enzymatic saccharification of alkali pre-treated coconut coir and rice straw (50 g each) was carried out in a 3 L capacity bioreactor. The lignocellulosic biomass was soaked in citrate buffer for 2 hours before pouring a crude enzyme cocktail. Sodium zide was also added to prevent microbial growth during the enzymatic hydrolysis (equation (1)).

Saccharification (%) = reducing sugars released x 0.9 x 100/grams cellulose in substrate (1)

Fermentation (conversion of sugars into ethanol): Bioethanol production from alkali pre-treated coconut coir was carried out by separate hydrolysis and fermentation (HGE) processes, as described below.

(a) **Substrate:** As substrates for ethanol production, alkali-treated coconut coir and rice straw were used.
(b) **Crude Enzyme Cocktail:** The crude enzyme cocktail produced from lignocellulosic biomass during the enzymatic saccharification step was used in the fermentation step.
(c) **Fermenting microorganism:** In the experiment, Sacchromyces cerevisiae was used for fermentation.
(d) **Growth and maintenance of microorganism**: The above-mentioned yeast inoculums were prepared by growing them for 24 hours in a 100-ml growth medium in a 250-ml Erlenmeyer flask in an incubator on a cumulus orbital shaker.
(e) **Composition of Growth medium:** Yeast 3.0 g, Peptone 10 g, Dextrose 20 g and 1 liter Distilled water.

Optimization of ethanol production: Cerevisiae and coconut coir were used as substrates for the production of ethanol. Conditions like pH, temperature, agitation speed, etc. were studied by changing one variable (factor) at a time while keeping others fixed. The supernatant was then analysed for ethanol concentration using a UV-Vis spectrophotometer.

Separate Hydrolysis and Fermentation in bioreactor: It was carried out under optimised conditions in order to evaluate the effect of pre-treatment on reducing sugar production and ethanol concentration.

Procedure: In this experiment, coconut coir, rice straw, and yeast extract were added to a flask of NaOH and Urea and left at room temperature for 2 hours. The pH was adjusted to 6.0, and then 0.5 percent yeast extract and peptone were added.

Detoxification: The pH of the medium is increased to 10 due to the addition of $Ca(OH)_2$, and it is thought that at this high pH, the inhibitors are precipitated by giving calcium hydroxide. The mixture was then centrifuged at 15,000 rpm for 30 minutes to remove the precipitates before being used for fermentation.

Ethanol estimation: In this experiment, 34 g of potassium dichromate were dissolved in 500 ml of distilled water and 375 ml of concentrated sulfuric acid. The final volume was made at 1000 ml by adding distilled water. The ethanol concentration was estimated using the Caputi method. A standard curve was prepared by using a standard ethanol solution containing 2–12% (v/v) ethanol in deionized water. Distillation was set at 60 °C, and tap water was turned on. The ethanol concentration of each sample was estimated through graph coordinates. Total bioethanol was estimated in terms of g/l of fermented liquid and g/g of biomass on a dry weight basis. Fermentation efficiency was calculated using the following equations (2), (3), and (4) (Jha et al., 2022; Luo et al., 2014; McKendry, 2002).

Fermentation Efficiency = ethanol produced (g/g) x 100/theoretical yield of ethanol (2)

in contrast, theoretical yield was defined as a standard value of 0.511 g/g sugars.

Ethanol yield (g/g) = ethanol concentration (g/L) divided by reducing sugars (g) (3)

Ethanol productivity (g/L/h) equals ethanol concentration (g/L) divided by time (h) (4)

SUMMARY

- The various biomass feedstocks were extracted from industrial waste, agro-waste, biofuel crops, and biological waste. Coconut coir pith is a by-product created when the coir fibre from the coconut husk is extracted. The key crops cassava, sugarcane, sorghum, and wheat may also be utilised as substrates. India contributes 31% of the global production of coconuts, making it one of the top producers in the world. Farmers will be deterred from putting new areas under cane production by acute water scarcity and unusual weather patterns.

- India imports a sizable quantity of biofuel from leading ethanol producers. Microbes are more capable than mesophilic species of acquiring high metabolic rates, stable enzymes, and better product yields. Laccases are multifunctional oxido-reductases that can break down a wide range of phenolic and non-phenolic substances by turning their oxygen molecule into water through four-electron reduction.
- Pre-treatment and industrial cellulose production have a significant impact on ethanol production. Some techniques have low ethanol and sugar yields despite being cost-effective. An excellent pre-treatment procedure prevents particle size decrease, restricts the production of inhibitory chemicals, protects the hemicellulose fraction, and uses the least amount of money and energy. Lime (Ca (OH)2) is less expensive and has fewer stringent safety standards than KOH and NaOH. Lime is an effective therapeutic agent, but only under the right pre-treatment conditions.
- Lignocellulosic biomass is milled and dried by mixing 6 grammes of biomass with one litre of water. Monomeric sugars will become progressively contaminated with aldehydes and organic acids if they are exposed to high temperatures and high acid concentrations. Furans, phenolics, and weak acids are three categories of the many inhibitors created during pre-treatment. The breakdown products of C-5 and C-6 sugars are, respectively, furans and phenolics. Acid hydrolysis is a method that uses sulfuric acid to totally transform polymers into C-5 and C-6 sugars.
- Fermentation of all the sugars generated from lignocellulose into ethanol requires a large number of microorganisms. Fed-batch reactors outperform batch cultures because they have more benefits.
- Coconut coir, rice straw and lignocellulosic biomass were used to extract biofuel from agrowaste. The process involved using 1% w/v of dried and alkali pre-treated coconuts and 1% of alkali-treated rice straw as well as 1% ethylene oxide. In this experiment, 34 g of potassium dichromate were dissolved in 500 ml of distilled water and 375 ml of concentrated sulfuric acid. The ethanol concentration was estimated using the Caputi method. Total bioethanol was estimated in terms of g/l of fermented liquid and g/g of biomass on a dry weight basis.

REFERENCES

Abualigah, L., Zitar, R. A., Almotairi, K. H., Hussein, A. M., Elaziz, M. A., Nikoo, M. R., & Gandomi, A. H. (2022). Wind, Solar, and Photovoltaic Renewable Energy Systems with and without Energy Storage Optimization: A Survey of Advanced Machine Learning and Deep Learning Techniques. *Energies*, *15*(2), 578. doi:10.3390/en15020578

Akia, M., Yazdani, F., Motaee, E., Han, D., & Arandiyan, H. (2014). A review on conversion of biomass to biofuel by nanocatalysts. *Biofuel Research Journal*, *1*(1), 16–25. doi:10.18331/BRJ2015.1.1.5

Alonso, D. M., Bond, J. Q., & Dumesic, J. A. (2010). Catalytic conversion of biomass to biofuels. *Green Chemistry*, *12*(9), 1493–1513. doi:10.1039/c004654j

Ha, G. S., El-Dalatony, M. M., Kurade, M. B., Salama, E. S., Basak, B., Kang, D., Roh, H. S., Lim, H., & Jeon, B. H. (2020). Energy-efficient pretreatments for the enhanced conversion of microalgal biomass to biofuels. *Bioresource Technology*, *309*(April), 123333. doi:10.1016/j.biortech.2020.123333 PMID:32305842

Jha, S., Nanda, S., Acharya, B., & Dalai, A. K. (2022). A Review of Thermochemical Conversion of Waste Biomass to Biofuels. *Energies*, *15*(17), 1–23. doi:10.3390/en15176352

Khan, A. A., Minai, A. F., Pachauri, R. K., & Malik, H. (2022). Optimal Sizing, Control, and Management Strategies for Hybrid Renewable Energy Systems: A Comprehensive Review. *Energies*, *15*(17), 6249. doi:10.3390/en15176249

Kumar, P., Sampath, B., Kumar, S., Babu, B. H., & Ahalya, N. (2023). Hydroponics, Aeroponics, and Aquaponics Technologies in Modern Agricultural Cultivation. In Trends, Paradigms, and Advances in Mechatronics Engineering (pp. 223–241). IGI Global.

Luo, J., Fang, Z., & Smith, R. L. Jr. (2014). Ultrasound-enhanced conversion of biomass to biofuels. *Progress in Energy and Combustion Science*, *41*(1), 56–93. doi:10.1016/j.pecs.2013.11.001

McKendry, P. (2002). Energy production from biomass (part 1): Overview of biomass. *Bioresource Technology*, *83*(1), 37–46. doi:10.1016/S0960-8524(01)00118-3 PMID:12058829

Ong, H. C., Chen, W. H., Singh, Y., Gan, Y. Y., Chen, C. Y., & Show, P. L. (2020). A state-of-the-art review on thermochemical conversion of biomass for biofuel production: A TG-FTIR approach. *Energy Conversion and Management, 209*. Advance online publication. doi:10.1016/j.enconman.2020.112634

Osman, A. I., Mehta, N., Elgarahy, A. M., Al-Hinai, A., Al-Muhtaseb, A. H., & Rooney, D. W. (2021). Conversion of biomass to biofuels and life cycle assessment: a review. In Environmental Chemistry Letters (Vol. 19, Issue 6). Springer International Publishing. doi:10.100710311-021-01273-0

Sampath, B., Sasikumar, C., & Myilsamy, S. (2023). Application of TOPSIS Optimization Technique in the Micro-Machining Process. In Trends, Paradigms, and Advances in Mechatronics Engineering (pp. 162–187). IGI Global.

Chapter 4
Sustainable Developments of Modern Soil-Less Agro-Cultivation Systems:
Aquaponic Culture

Sampath Boopathi
https://orcid.org/0000-0002-2065-6539
Muthayammal Engineering College, India

Praveen Kumar Siva Kumar
Lagaram Engineering Solutions, India

Radhey Shyam Meena
Indian Institute of Technology, Kanpur, India

Samson Isaac J.
Karunya Institute of Technology and Sciences, India

Suresh Kumar P.
RVR & JC College of Engineering, India

M. Sudhakar
Sri Sai Ram Engineering College, India

ABSTRACT

In this chapter, the various terminologies of soilless Agro-cultivation systems have been explained. Various factors have been used to explain the classification of soilless cultivation systems. Different fabrication elements of soilless, aquaponic systems are illustrated with the utilization of bio-fish and plant waste components. The essential components for the design and fabrication of small and large aquaponic systems have been explained. The implementation of the internet of things (IoT) in the aquaponic system has been illustrated using microcontroller devices.

DOI: 10.4018/978-1-6684-4118-3.ch004

INTRODUCTION

Agriculture is a sacred occupation because it provides food for living things. All plants grow by taking in nutrients from water and growing in relation to time. As the world's population grows, so do the demands for food and other resources. As a result, large quantities of high-quality healthy vegetable foods must be produced using both traditional and modern farming methods. In the traditional farming process, the land is watered, and healthy fertilizers are poured into the soil to reach the roots of the plants. In the modern cultivation process, soilless new technologies are applied to grow the plants anywhere in the world or even in outer space (Lakhiar et al., 2020). In the conventional farming methods, the harvests are grownup in the air atmosphere, using irrigation and the use of artificial fertilizers. The big land with large quantity of water is required due to supplying water into the land itself (Sathyanarayana et al., 2022).

The soil is an important medium for growing plants in the earth. Water moves micronutrients and macronutrients from one location to another in the soil. The soil's nature and properties are important for the plant's growth and development. However, due to the world's growing population, soilless agriculture research is also critical for growing plants and vegetables in space vehicles or other land-free environments. Hydroponics, aquaponics, and aeroponics are important technologies to promote soilless agricultural activities. These technologies promote the growing of useful vegetables in soil-free house places such as roofs, balconies, apartments, and in the various house holding places. Due to increasing vegetable demands, the high production of quality food products should be encouraged. In this new technology, the soluble nutrients are mixed into the water and fertigated using various types of innovative technologies. Plants, fruits, leafy vegetables, and ornamental plants are grown by these methods. In this approach, the required mineral nutrient solutions have been supplied to such plants without using soil supports. The organic and inorganic components in the water solutions are artificially supplied to physically supported plants. The special hoses, pumps, and control values are also used to control and timely supply such solutions to the plant without the presence of soil. One of the advantages of the proposed technologies is that the plants grow under controlled and optimized atmospheric conditions. The plant growth supporting elements are: quality of nutrients, pH level of water, quantity of nutrients, atmospheric temperature, air and humidity ratios, and light intensity, colour, and light intensification—the air mixing components are carefully monitored and controlled.

TERMINOLOGIES OF SOILLESS CULTIVATION SYSTEMS

Soilless Cultivation Key-Elements

- **Nutrient solution:** Nutrient solutions have been employed an important role in soilless cultivation. N2, B, Zn, Cu, Mo, Mn, Mg, P, Fe, K, and Ca are presented in the nutrient solution (Pickering et al., 2002). The quantity of each element can be varied with respect to plant growth.
- **pH Level:** The pH level of the nutrient solution should be 5 to7 based on the vegetable types. critical for the plant's growth and can be artificially controlled by new technology.
- **Electrical Conductivity: In the** soilless methods, 1.5-2.5 dS/m ranges of electroconductivity of nutrient solutions have been used for the growing the better growth (Rouphael & Colla, 2005).

Substrates

The substrates are made up of organic, inorganic, and synthetic materials. Coir and straw can be used to create organic origins. The inorganic components are made from agricultural byproducts, plants, and urban waste. Synthesis processes yield the industrial origin (e.g., polystyrene). (Rouphael & Colla, 2005). The organic and inorganic components of the porous subtract are illustrated in Figure 1.

- **Peat:** Bryophytes, cyperaceous plants, and other residues are used to make peat. The peat has good structural stability, low nutrient availability, and an acidic pH.
- **Coconut Fiber:** Coco coir is another name for coconut fiber. The fibrous husks of coconuts are used in the extraction. To make peat, its pH level must be at its highest. components are preprocessed before being used in the form(Fornes et al., 2003).
- **Volcanic Tuffs:** Sand particles range in size from 0.05 to 2.0 mm. Fine sands are better suited for hydroponic systems. The coarse sands are used to improve the drainage capacity of the substrate. (Bunt, 1988).
- **Sand:** The silica sand can be used to make the inorganic material. (Maucieri et al., 2019). Small sands with a particle size of 0.1-2.0 mm are ideal for constructing a hydroponics system.
- **Vermiculite:** Vermiculite is collected by Al, Mg, and Fe components.
- **Perlite:** Perlite is made of more silicon oxide and aluminium oxide with various sizes of particles(Verdonck et al., 1984).

Figure 1. Porous Subtracts: Organic and inorganic components

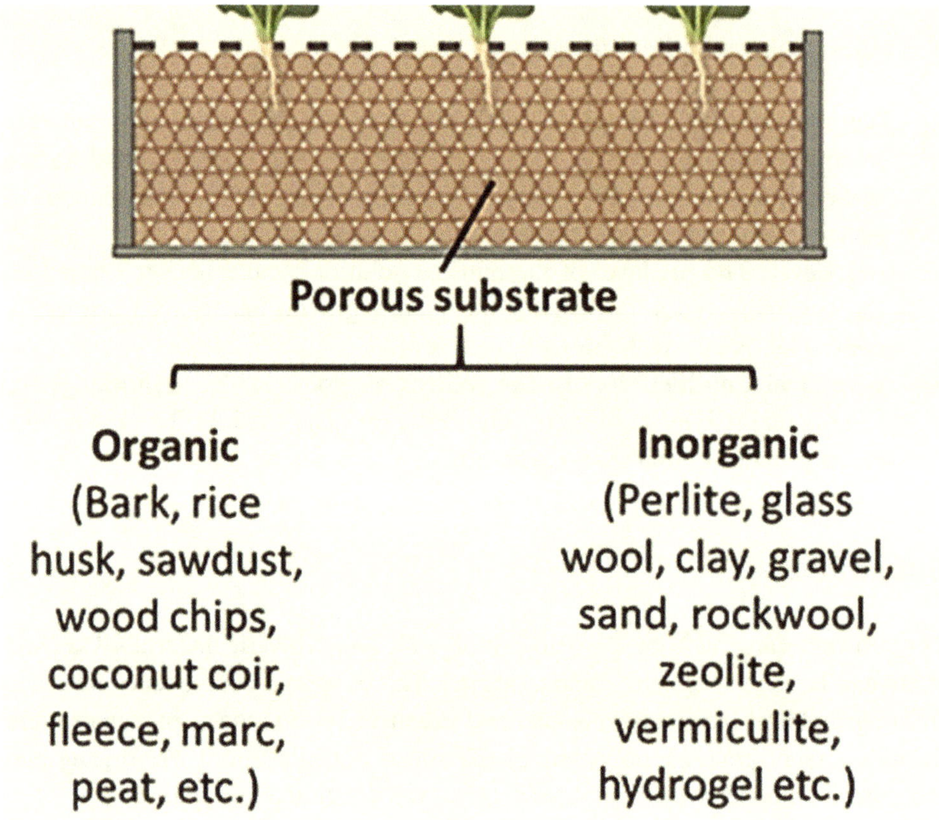

- **Expanded Clay:** 40 percent of the clay powder is used to improve the pH level of the form, which is then heated to 850 degrees centigrade.(Rahmad et al., 2019).
    - **Stone Wool:** At a temperature of 2000 °C, magnesium, calcium, coal, and coke combine to form stone wool.(Abad et al., 2002).
    - **Zeolites:** The zeolite is composed of aluminium silicate with micro- and macro-elements containing K+, NH4+, Cl-, Na+, and Fe ions.
- **Expanded Polystyrene:** The porosity and drainage can be improved by adding 3–9 mm of expanded polystyrene.
- **Polyurethane Foam:** Because polyurethane foam has a higher water content than soil, it is used in soilless agriculture (Espinal & Matulić, 2019).

The factors for soilless cultivation system (Lakhiar et al., 2020)

- A uniform distribution of root systems to supply and deliver nutrient solutions.
- The root system must have a low density and a high porosity.
- The particle size of the substance should support the air and water systems.
- pH levels in supply water were adjusted to be between 5-7.
- The salt content of the water must be kept to a minimum.
- The chemical composition of the solution should be uniform, not disturbing the inorganic contents.
- The solution should be toxic-free.

Advantages Soil-less cultivation process

- No soil is required.
- Artificially controlled cultivation system
- Compact enough.
- Run-off water and nutrient utilization.
- Nutrients are directly given to roots, no wastage occurs.
- Natural, balanced-protein foods could be produced.
- Weeding, tilling, and free lands are not required.

Disadvantages Soil-less cultivation process

§ Ecological issues may arise.
§ Investment costs are high.
§ We need skilled people.
§ Running cost also high.
§ Plant diseases are an issue.

CLASSIFICATION OF SOILLESS CULTIVATION SYSTEMS

The nutrient supply systems are classified as open and closed drain and supply systems. The drain system has been made from inorganic (sand, grit, nuggets, perlite, rock fabric, and vermiculite), organic (sawdust, rice hulls, coconut coir mulch, and straw), and synthetic (foam vessels, sponges, moisture-permeable flexible fiber) materials.

Hydroponics: The roots of the plants are placed on perlite, gravel, or coco peat. The liquid state of unused vegetables, fish flushes, and liquid wastes are used as energy solutions. In this method, no soil contents have been used. Only 10% of water has been utilized when compared to the soil system.

Hydroponics =Hydro' è Water + 'Ponics'è Labour.

Figure 2. Schematic of soil-less Hydroponics system

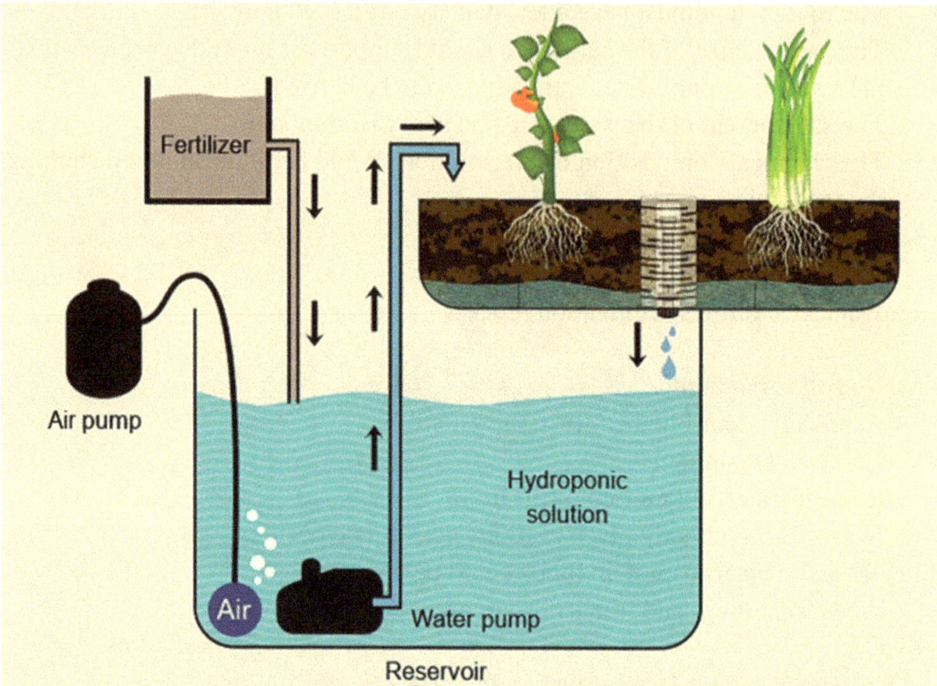

The schematic of a soilless hydroponics system is illustrated in Figure 2. Water can be used to stimulate plant growth. In this system, some subtractions are used to make the root-bottom supports that hold the oxygen and water for the plants. The air and fertilizers are separately supplied to the water to make the optimal nutrient solution for the particular plants. An automatic system can control the amount of mixing air and fertilizers to make a large amount for production. Pump is used to circulate the solution in the container. The excess water from the form has been drained through the bottom droplets.

Aquaponics: The waste fish and plants are combined with water to prepare the nutrient solutions. Ammonia has been created for plant growth by naturally nitrifying bacteria found in fish waste. The aquaponic system is shown in Figure 3.

Aeroponics: This is one of the advanced soilless farming methods in which air is used to supply the nutrient solution to the root systems of plants. The oxygen in the air is used to promote the faster growth of the plant. The tomato, herbs like chives, mint, oregano, basil, sage, and rosemary, and leafy greens like lettuce and kale are grown by this method. The aeroponic system is illustrated in Figure 4.

Nutrient Film Technique: The submersible pump, reservoir, and stones, as well as some short, saturated plants, have been used to fabricate this system. The drained

Figure 3. Schematic of soil-less Aquaponics system

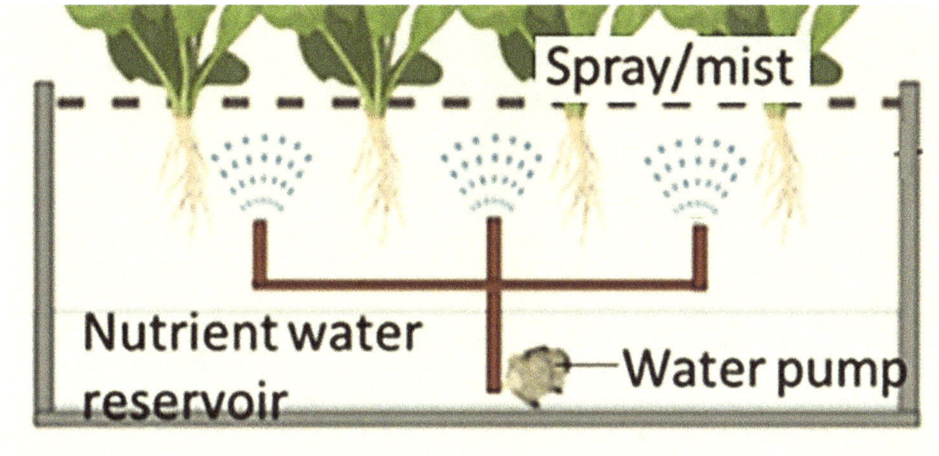

water is again recycled by this method. The solutions are uniformly supplied to the root without any timer system.

Ebb and Flow (Flood and drain): The nutrient solution is frequently supplied to the roots every two hours using a control system and pump. The plants are supported by thin materials in order to provide a high amount of oxygen while preventing fruit damage.

Drip system: This system is made of coco peat and rock wool to retain maximum water content for the plants. The control system is used to turn the pump on and off. The solutions are transferred from the reservoir to the system through small

Figure 4. Schematic of soil-less Aeroponics system

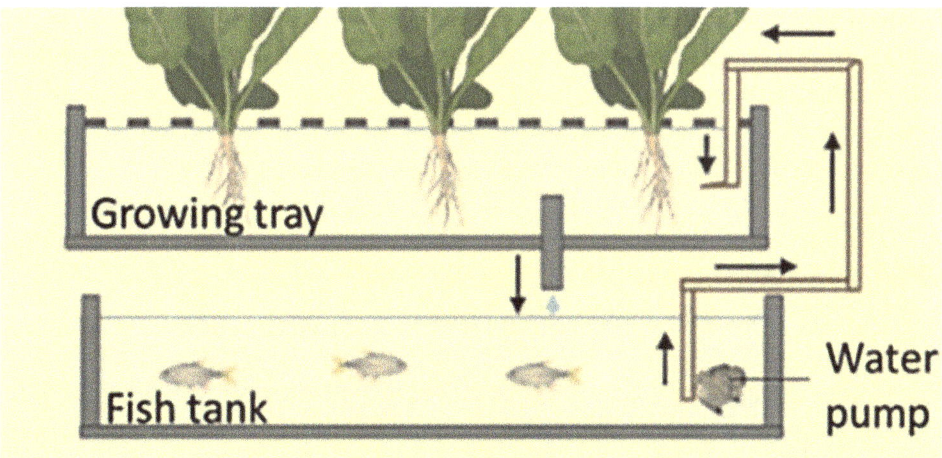

Figure 5. Deep water soilless culture

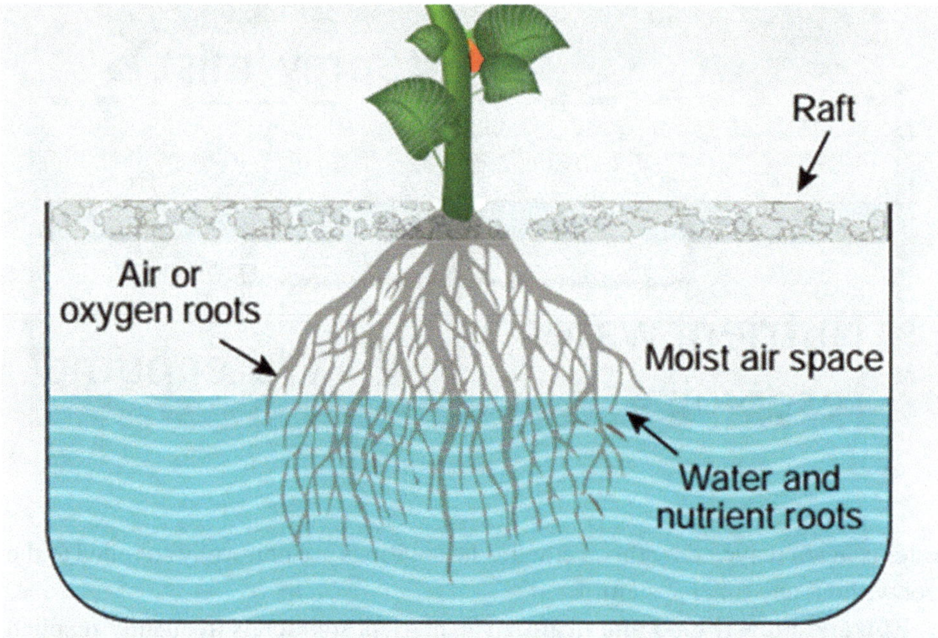

drip hoses. It is faster than other soilless cultivation methods and is widely used due to its simplicity. The synthetic fibres are also used to minimize the cost and reuse them for a long time.

Wicking system: The wicks are used to suction the nutrient solution from the solution chamber to the plant root systems. The capillary actions are carried out to supply the system with the next set of nutrient solution without any dynamic parts on it.

Deep water culture: Floats made of Styrofoam insulation boards are submerged in the nutrient solution, and an air pump delivers air to the air stones, which are then delivered to the plant roots. The plants are placed on the floats on which the water is recirculated. The air and water are sprayed on the roots. There is no plant support required. The deep water soilless culture is illustrated in Figure 5.

VARIOUS FABRICATION ELEMENTS OF SOILLESS AQUAPONICS SYSTEMS

The various components of a soilless aquaponics system are shown in Figure 6. In soilless aquaponics systems, recirculating aquaculture and hydroponics techniques

Figure 6. Various Components of Soilless Aquaponics system

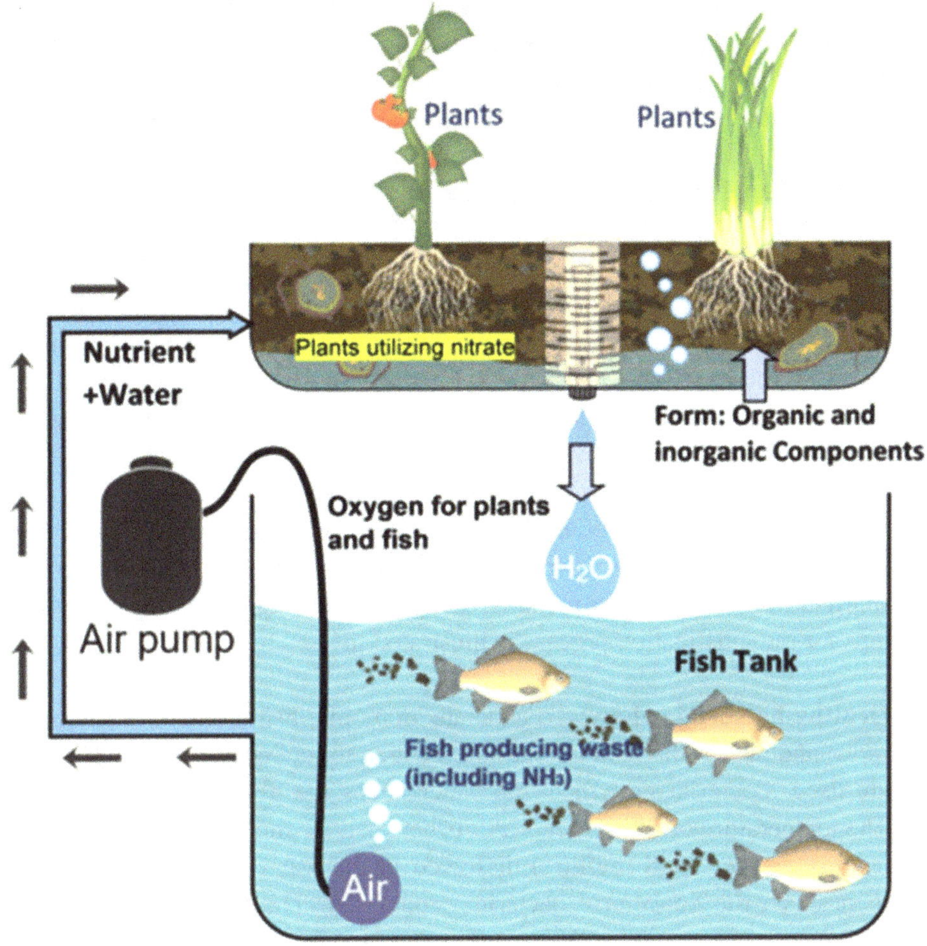

are integrated for food production. Water is circulated from the fish tank through filters to the roots of the plants. After utilization, it will be recirculated into a fish tank. Two types of filters are utilized: mechanical and biofilters. The unwanted solid contaminants are removed from the water by mechanical filters. The nitrification process will be done by the biofilter using toxic waste from fish integrated with bacteria to produce ammonia. Circulating water to the plants' grow beds contains nitrate and nutrients. The purified water is returned to the fish tank after the plants have used it (Kumar et al., 2023; Sampath et al., 2023).

The three organisms—plants, fish, and bacteria—are integrated to make an environmentally friendly soilless cultivation system. It is important to make sustainable developments in the modern agricultural system to enhance a cost-effective and chemical-free system. These two products, such as fish and vegetables, could be obtained by this method. Water resources are effectively used to make organic food without using fertilizers or pesticides. The waste produced by this technology is very minimal when compared to the conventional method.

Components of Aquaponics systems

Nitrogen fixation: Ammonia (NH_3) and nitrate (NO_{3-}) components will be used by the plants. These are produced in the biofilter, in which bacteria are used to process the wastes drawn out from the fish. During this process, the nitrogen was consumed from the atmospheric air. The waste products of the living organisms are well utilized by the bacteria to make useful components for the plants. The Nitrogen cycle for nitrogen fixation process is show in in Figure 7.

Nitrification process: During this process, ammonia (NH_3), nitrates (NO3-), and nitrite (NO2-) gases are produced by nitrification processes. In the nitrification processes, ammonia- and nitrite-oxidizing bacteria are employed to convert the ammonia to nitrite and the nitrite to nitrate components through oxidizing processes. The steps for the nitrification process is illustrated in

- **Healthy bacterial colony:** To generate ammonia, fish biomass, waste plant roots, and tank walls with a large surface area for collecting healthy bacteria are used. Bacterial growth in the water keeps the pH between 6 and 8.5. The temperature of the source water is between 17 and 34 °C. The optimum level of oxygen (4–8 mg/litre) dissolved in the water should be important to produce the concentric bacterial production. The bacteria colony will be improved by passing UV light on the water for 3 to 4 days because the bacteria are photosensitive organisms.
- **Nitrate balance:** It can be determined when the amount of fish waste in the aquaculture system equals the amount of plant production.
- **Feed rate ratio:** It consists of three important elements: quantity of fish feed and plant type, bacteria, and the plant's growing space. This ratio is also called the signal ratio.
- **Health check of fish and plants:** When defective fish and plants are introduced into the system, the system will not be balanced, resulting in nutrient deficiencies.
- **Nitrogen testing:** Fish stocking density, types of plants, planting capacity, nitrite requirement, and biofilter sizes have been examined to improve the overall efficiency.

Figure 7. Nitrogen cycle for nitrogen fixation process

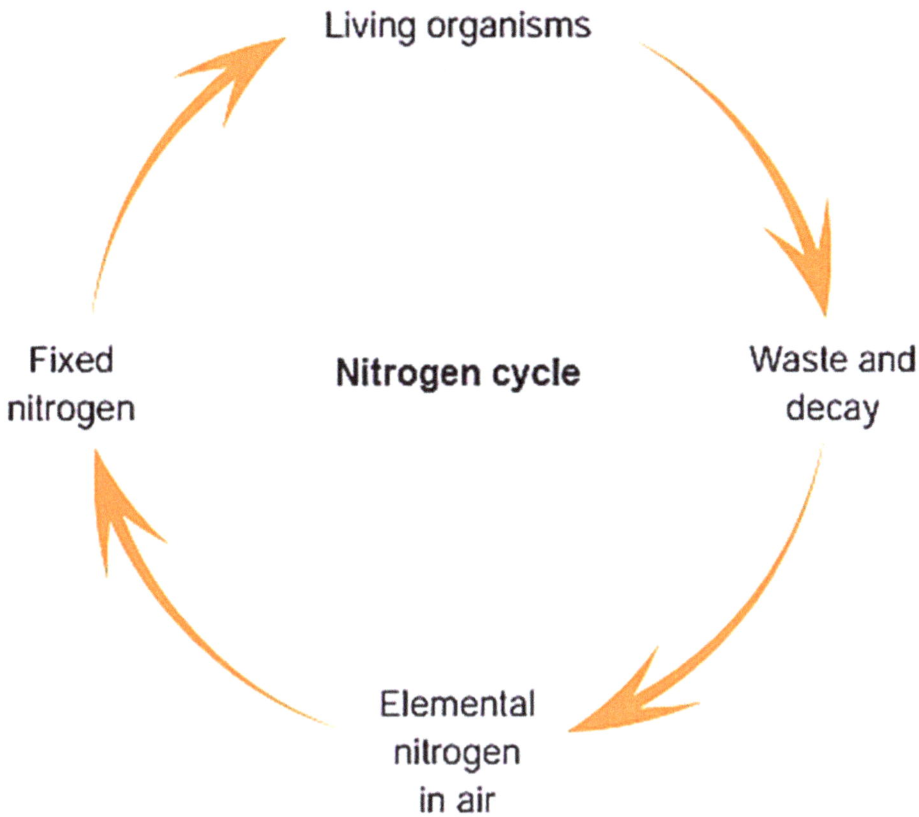

Quality of water controlling parameters (Somerville et al., 2014)

- Oxygen Level: 4–8 mg/litre.
- Water pH Level: 6-8.5.
- Water Temperature: 17 and 34 °C.

Figure 8. Flow chart for Nitrification process

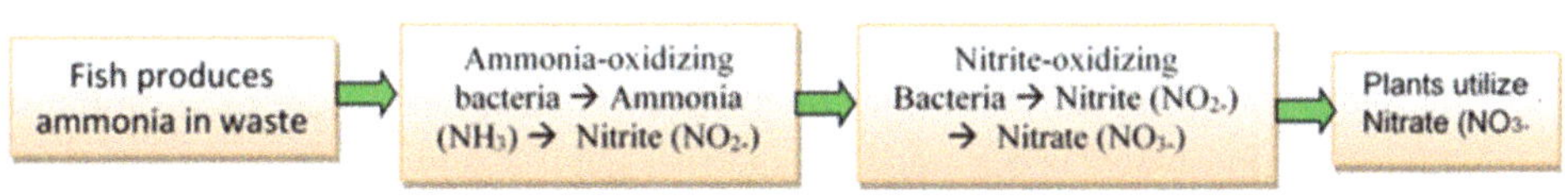

- Water nitrogen components: ammonia, nitrite, and nitrate.
- Water Hardness: 60–140 mg/litre.

Design of Soilless Aquaponic Systems

Selection of Suitable Location

The straight and stable surface is suitable for the start of the aquaponic plant system. With any green house and oblique net covers, entire plants should be placed in direct sunlight. The sunlight is important to produce the bacteria for the plant. The entire system is placed in a wind, rain, and snow protected zone to prevent diluting the nutrient solutions. The entire system should be placed on concrete or steel tables to support the weight of the fish tank and plant support systems.

Selection of Essential components

- **Fish tank:** Most of the time, the transparent flat bottom shape of the round-walled, square-shaped fish tank can be used. The water and air pumps are used to circulate the water and provide oxygen to the fish. A bottom dwell is provided to remove water from the tank while cleaning it.
- **Filtration:** Two types of filtration methods are used: (1) mechanical filtering and (2) biofiltration.

Mechanical filtering used to remove and separate the solid waste from the fish to avoid the formation of harmful gases in the fish tank. It prevents the fish's health problems. The simple screen is located at the bottom of the grow bed. 60% of the waste can be removed by a mechanical filter. The outlet of the filter or separators is connected to the biofilter unit.

Biofilter: Biofiltration is utilized to generate the ammonia, nitrite, and nitrate that are suitable for plant growth using living bacteria. The biofilter system for the aquaponic plant is shown in Figure 9.

- **Mineralization:** It is used to metabolize and process the fish solid waste into nutrients for plants by using bacteria. The solid waste is coming from mechanical filters, which are used as a source for this system.
- **Water movements:** The water is moving from the fish tank through mechanical separators, biofilters and nutrient solutions are sent to the root of the plant through plastic hoses. The nutrient is dissolved in the plant pot, and the remaining water is drained into the fish tank. The water is purified by the

filter in the drain system. The impeller-type submersible water pump is used to circulate the water throughout the entire system.

- **Piping System:** The holed PVC pipes and connectors are used to make the portable aquaponic system. Hydroponic media consists of volcanic gravel, and rockwool. The water is drained at the bottom of the pipe. The piping system for the small aquaponic system is shown in Figure 10. The piping system for a large aquaponic plant is illustrated in Figure 11.

Figure 9. Biofilter unit

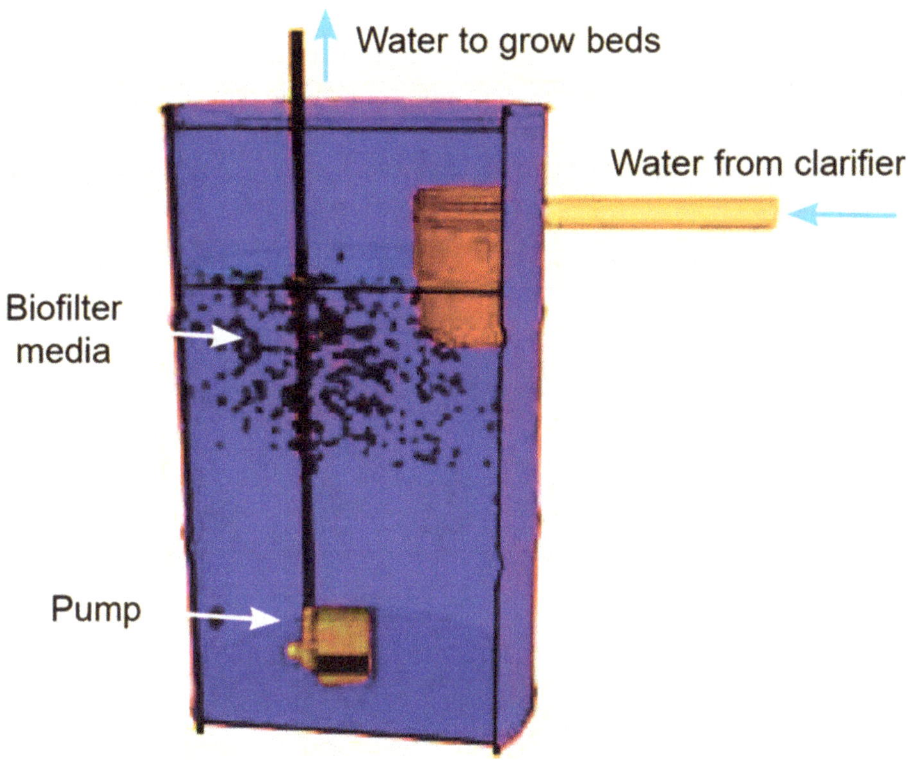

- **Three zones of media beds:** The three zones and drain-filter components of the plant media bed systems are shown in Figure 12. The dry, dry/wet, and wet zones are arranged in the top, middle, and bottom of the plant media, respectively (Das & Sahu, 2020).

Figure 10. Piping system for the small aquaponic system

Dry zone: The dry zone is used to provide a light barrier, reduce light intensity, prevent nutrient solution dilution, and allow for algal growth. The thickness of the layer is between two and five centimeters at the top of the bed.

Combined dry and wet zone: The moisture and gases are exchanged, and the nutrient solutions are directly connected to plant roots. It is connected to the hoses from the water flow system and drain system. The layer thickness of the zone is between ten and twenty centimeters.

Figure 11. Piping system for Large aquaponic system

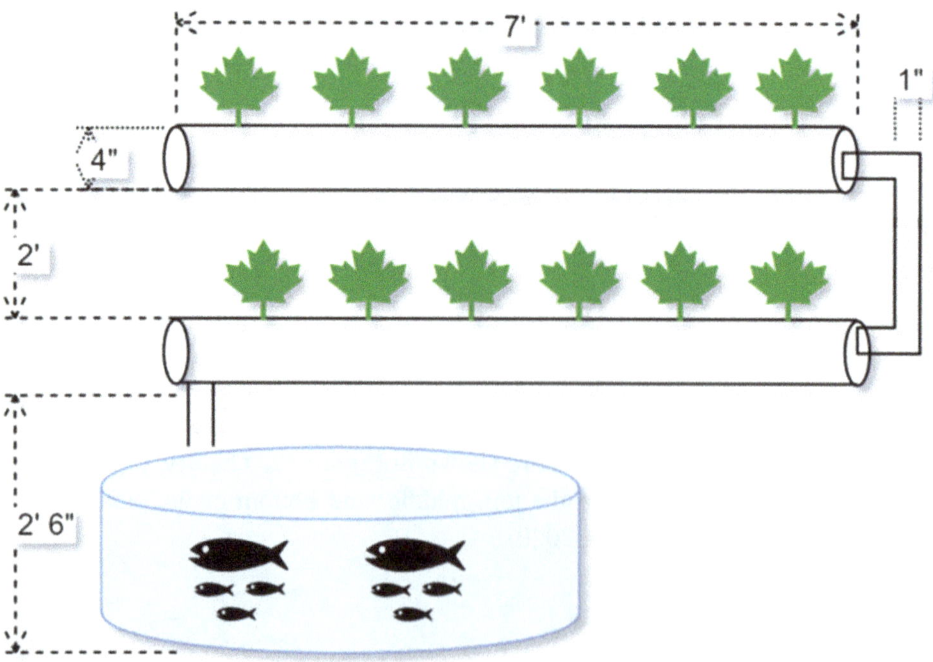

Figure 12. Three zone of plant media beds of the aquaponic system

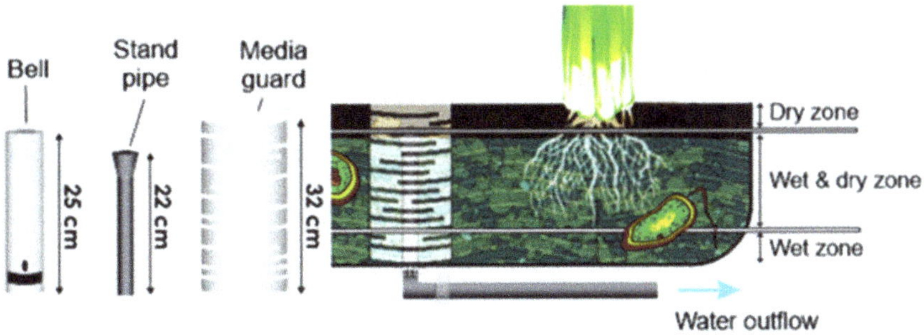

Pure wet zone: It is a permanent wet zone that consists of solid waste and mineralized components. It includes many microorganisms and heterotrophic bacteria. It is connected to the root of the plant to transfer the nitrate components to the plant's growth.

Timer mechanism: It is applied to the pump control system to switch on and off the pump and drain systems based on the types and water levels in the plant.

INTERNET OF THINGS (IoT) ENABLED SOILLESS AQUAPONICS SYSTEM

The internet of things technology is applied to controlling the physical activities of an aquaponics system using various sensors and microcontrollers. The sensors and actuators are employed to control the entire system with minimal human interpretation. Machine learning and artificial intelligence are also applied to predict the quantity of nutrient solutions based on the various processes of the system. In this system, pH sensors, humidity sensors, TDS sensors, temperature sensors, and RGB intensity sensors are used to control the system activities as illustrated in Figure 13.

Power to system: The system's power is generated through solar panels and a battery storage system.

Sensor: All the sensor signals are integrated by the Arduino pregame. The programme is executed to monitor and activate the various devices, such as water flow and a balanced system (Sharma et al., 2018).

- **pH Sensor:** Measure the pH level of water entering the system.
- **Air Temperature/ Humidity Sensor:** Measure and monitor the air temperature and humidity.

Figure 13. Schematic Arrangements of IoT for aquaponic system

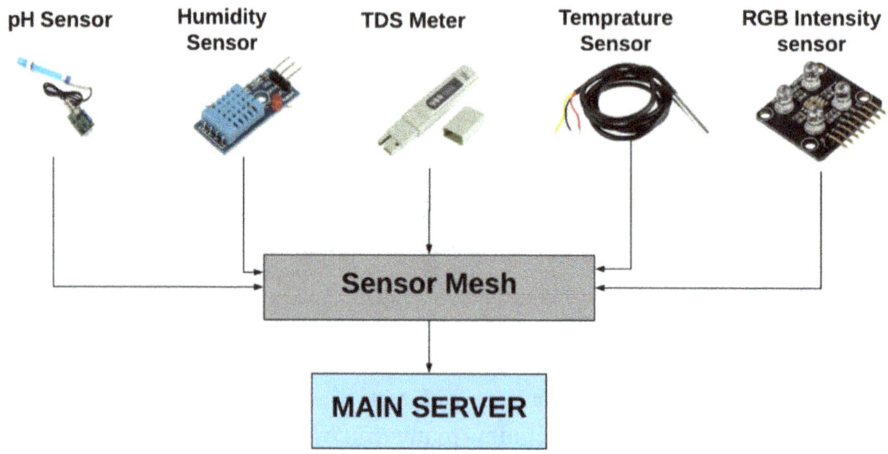

- **Electrical Conductivity sensor:** Monitor and measure the electrical conductivity of water and total dissolved solids.
- **Water Temperature sensor:** Monitor the water temperature.
- **RGB Intensity sensor:** Monitor and measure the light intensity on the system.
- **Gas sensor:** measure the oxygen level and quantity of supply to the fish tank.

Microcontroller: Arduino is the best microcontroller to integrate the sensor signals and various actuators of the soilless system. The analogue and digital signals can be manipulated using sensors, signals, and actuators. A logical programme can be performed to control the quantity of water, air, concentration of nutrient solutions, plant growth, and humidity of the roots of the plants.

Actuators: Water and air pumps are mainly controlled by signals from the microcontroller.

Server: It is used to store the entire data from the sensors, output, and signal information to the actuators. The timing of programme executions and the switching times of actuators are also recorded on the server.

Cloud Storage: All the server information is synchronized by the cloud storage to communicate the information to the operator or housekeepers in time.

Mobile data / WIFI connections: The control system is fully automated through internet connectivity. The information can be conveyed through internal signals.

CONCLUSION

- As the world's population grows, so do the demands for food and other resources. High-quality healthy vegetable foods must be produced using both traditional and modern farming methods. Soilless new technologies are applied to grow the plants anywhere in the world or even in outer space. Hydroponics, aquaponics, and aeroponics are important technologies to promote soilless agricultural activities. These technologies promote the growing of useful vegetables in soil-free house places such as roofs, balconies, apartments, and in the various house holding places.
- N2, B, Zn, Cu, Mo, Mn, Mg, P, Fe, K, and Ca are presented in the nutrient solution. The organic and inorganic components porous Subtract is illustrated. Sand is a naturally occurring inorganic material formed by various weathering materials. Fine sands are better suited for hydroponic systems. Perlite is ideal for horticulture because it retains water and creates an airy compost for seedlings.
- Clay powder is heated to 700 degrees Celsius to produce expanded clay. Stone wool is formed by the fusion of magnesium, silicate, aluminum, carbon coke, and calcium at temperatures ranging from 1500o C to 2000o C.
- Open and closed nutrient supply systems are classified as drain and supply systems. Drain system made from inorganic (sand, grit, nuggets, perlite, rock fabric, and vermiculite) and organic (sawdust, rice hulls, coconut coir mulch, and straw) materials. Only 10% of water has been utilized when compared to the soil system. Aeroponics is one of the advanced soilless farming methods in which air is used to supply the nutrient solution to the root systems of plants. The tomato, herbs like chives, mint, oregano, basil, sage, and rosemary, and leafy greens are grown by this method.
- Synthetic fibres are used to provide a high amount of oxygen while preventing fruit damage. Floats made of Styrofoam insulation boards are submerged in the nutrient solution. An air pump delivers air to the air stones, which are then delivered to the plant roots. In soilless aquaponics systems, recirculating aquaculture and hydroponics techniques are integrated for food production. Water is circulated from the fish tank through filters to the roots of the plants. The waste produced by this technology is very minimal when compared to the conventional method. The waste products of the living organisms are well utilized by the bacteria to make useful components for the plants. During this process, ammonia (NH_3), nitrates (NO3-), and nitrite (NO2-) gases are produced.
- The Internet of Things technology is applied to controlling the physical activities of an aquaponics system using various sensors and microcontrollers.

The sensors and actuators are employed to control the entire system with minimal human interpretation. Machine learning and artificial intelligence are also applied to predict the quantity of nutrient solutions.

REFERENCES

Abad, M., Noguera, P., Puchades, R., Maquieira, A., & Noguera, V. (2002). Physico-chemical and chemical properties of some coconut coir dusts for use as a peat substitute for containerised ornamental plants. *Bioresource Technology*, *82*(3), 241–245. doi:10.1016/S0960-8524(01)00189-4 PMID:11991072

Bunt, A. C. (1988). Media and Mixes for Container-Grown Plants. In *Media and Mixes for Container-Grown Plants*. British Library Cataloguing in Publication Data Bunt. doi:10.1007/978-94-011-7904-1

Das, S., & Sahu, G. (2020). Soil-Less Farming- An Innovative Way Towards Sustainability. *Agricultural & Food: E-Newsletter*, *2*(10), 82–85.

Espinal, C. A., & Matulić, D. (2019). Recirculating Aquaculture Technologies. *Aquaponics Food Production Systems*, *01*, 35–76. doi:10.1007/978-3-030-15943-6_3

Fornes, F., Belda, R. M., Abad, M., Noguera, P., Puchades, R., Maquieira, A., & Noguera, V. (2003). The microstructure of coconut coir dusts for use as alternatives to peat in soilless growing media. *Australian Journal of Experimental Agriculture*, *43*(9), 1171–1179. doi:10.1071/EA02128

Kumar, P., Sampath, B., Kumar, S., Babu, B. H., & Ahalya, N. (2023). Hydroponics, Aeroponics, and Aquaponics Technologies in Modern Agricultural Cultivation. In Trends, Paradigms, and Advances in Mechatronics Engineering (pp. 223–241). IGI Global.

Lakhiar, I. A., Gao, J., Syed, T. N., Ali Chandio, F., Tunio, M. H., Ahmad, F., & Ali Solangi, K. (2020). Overview of the aeroponic agriculture – An emerging technology for global food security. *International Journal of Agricultural and Biological Engineering*, *13*(1), 1–10. doi:10.25165/j.ijabe.20201301.5156

Maucieri, C., Nicoletto, C., van Os, E., Anseeuw, D., Van Havermaet, R., & Junge, R. (2019). Hydroponic Technologies. In Aquaponics Food Production Systems. doi:10.1007/978-3-030-15943-6_4

Pickering, H. W., Menzies, N. W., & Hunter, M. N. (2002). Zeolite/rock phosphate - A novel slow release phosphorus fertiliser for potted plant production. *Scientia Horticulturae*, *94*(3–4), 333–343. doi:10.1016/S0304-4238(02)00006-7

Rahmad, I. F., Nababan, E. B., Tanti, L., Triandi, B., Ekadiansyah, E., & Fragastia, V. A. (2019). Application of the Alcohol Sensor MQ-303A to Detect Alcohol Levels on Car Driver. *2019 7th International Conference on Cyber and IT Service Management, CITSM 2019*, (pp. 247–255). IEEE. doi:10.1109/CITSM47753.2019.8965395

Rouphael, Y., & Colla, G. (2005). Growth, yield, fruit quality and nutrient uptake of hydroponically cultivated zucchini squash as affected by irrigation systems and growing seasons. *Scientia Horticulturae*, *105*(2), 177–195. doi:10.1016/j.scienta.2005.01.025

Sampath, B., Sasikumar, C., & Myilsamy, S. (2023). Application of TOPSIS Optimization Technique in the Micro-Machining Process. In Trends, Paradigms, and Advances in Mechatronics Engineering (pp. 162–187). IGI Global.

Sathyanarayana, S. R., Gangadhar, W. V., Badrinath, M. G., Ravindra, R. M., & Shriramrao, A. U. (2022). Hydroponics: An Intensified Agriculture Practice to Improve Food Production. In Reviews in Agricultural Science (Vol. 10, pp. 101–114). doi:10.7831/ras.10.0_101

Sharma, N., Acharya, S., Kumar, K., Singh, N., & Chaurasia, O. P. (2018). Hydroponics as an advanced technique for vegetable production: An overview. *Journal of Soil and Water Conservation*, *17*(4), 364. doi:10.5958/2455-7145.2018.00056.5

Somerville, C., Cohen, M., Pantanella, E., Stankus, A., & Lovatelli, A. (2014). Small-scale aquaponic food production. In *FAO Fisheries and Aquaculture*. https://www.fao.org/3/contents/1dea3c92-1faa-47bb-a374-0cf4d9874544/i4021e00.htm

Verdonck, O., Penninck, R., & De Boodt, M. (1984). The Physical Properties of Different Horticultural Substrates. In Acta Horticulturae (Issue 150, pp. 155–160). doi:10.17660/actahortic.1984.150.16

Chapter 5
IoT-Based Smart Farming

Pawan Kumar Singh
Honeywell Technology Solutions Labs, India

ABSTRACT

Internet of things (IOT) is a rising technology, having become very popular with the rise of wearable devices. IOT based infrastructure provides connectivity to everything, and not just humans. With affordable wireless high speed internet connectivity, IOT is playing measurable roles in various industries and changing the way it has been traditionally. Agriculture is also not untouched by IOT enabled solutions, which automates several process and captures more accurate data about the climate, soil, crop, livestock, and inventory with the help of connected sensors, drones, etc., to help farmers. With more data, farmers predict more accurately about adverse climate changes and can make better decisions in real time to save losses and optimize farm yield. IOT based solutions in agriculture enable scientists and farmers to understand and innovate the farming techniques for growing food demand for a rapidly growing human population.

INTRODUCTION

In the early 70s, computer scientists were trying to connect computer nodes to share a simple message and today after 50 years technology has grown so far that not just computer nodes are connected but the whole ecosystem is connected over the Internet. Vinton Cerf and Robert Kahn developed Transmission Control Protocol and Internet Protocol, or TCP/IP, for data transmission between multiple networks in 1974. They may not have even visualized while developing TCP/IP that Internet will not be just used for message exchange between computers but virtually any

DOI: 10.4018/978-1-6684-4118-3.ch005

electronic device can connect to the Internet and exchange data and automate several processes with the help of connected systems.

Today, with advancements in technology especially in Internet connectivity using fast fiber optic cable networks, enhancement in wireless communication (4G/5G), and cloud platforms, many more devices are connected to the network. These connected devices can share the data across the network with other devices and control centers to control, monitor or analysis. For example, various sensors can be deployed in a smart home automation system to control or monitor the air conditioning system, lighting, music system, fan, refrigerator, garage door, door lock, geyser, etc. Similarly, in a factory, several systems can be controlled or monitored using various sensor-based control units. The complete airport terminal could be controlled using well-connected IoT-enabled systems like an HVAC system, fire safety equipment, flight information system, lights, etc. All equipment in an operation theatre or ICU in a hospital could be connected with the help of several sensors monitoring and controlling the comfort of the room for better patient care. Corporate office buildings, Shopping complexes, power distribution systems in a power grid, every industry has some room for IoT-enabled solutions.

IoT includes everything from a fitness band, mobile phones, smartwatches, coffee makers, smart lighting, smart refrigerator, washing machines, headphones, music systems, to the jet engine of an airplane or the drill of an oil rig—and almost anything else you can think of. IoT applications are gaining popularity due to demand and changes in user behavior with changing technological advancement.

Bluetooth Low Energy, Zigbee, LoRaWAN, Z-Wave, etc. enabled devices can operate for months and sometimes a few years on single battery power. Using such low power consumption technology and highly optimized hardware to send data up to a few hundred meters, several new applications in the IoT space are evolving fast and changing the overall efficiency and user experience.

DEFINITION OF IOT

Internet of Things could be defined as a system or infrastructure (or can be viewed as a global infrastructure) for information sharing by interrelated, interoperable connected devices (physical things) or virtual data points which can collect, generate and share data over a network (Internet). These devices (a piece of hardware) mandatorily have an identification (physical serial number, MAC Id, etc.) and capabilities of communication to share data in the network with optional capabilities of data collection, storage, processing, sensing, and actuation. A typical IoT infrastructure consists of a central/distributed data collection and processing unit(s).

Figure 1. Internet of Things

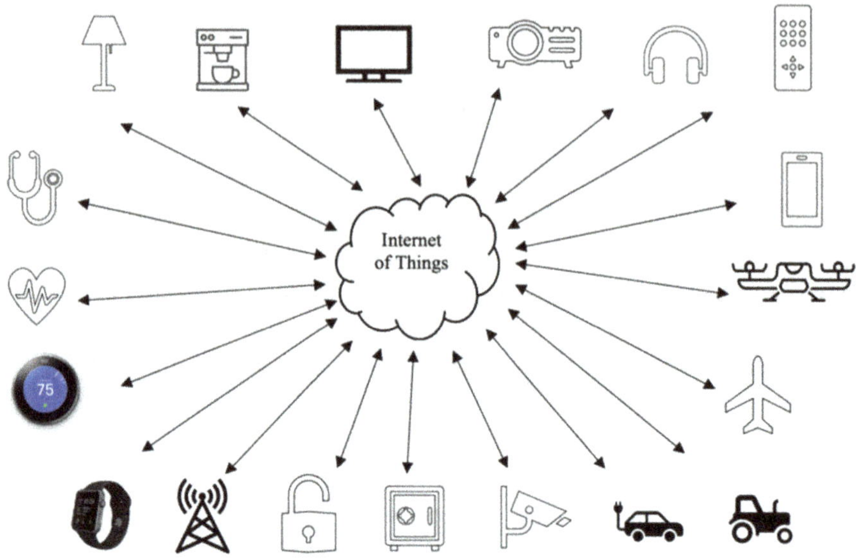

With growing, IoT applications and services cyber security requirements such as data confidentiality, integrity, authorization, privacy, availability, etc. are becoming an essential part of the definition and they must be ensured in an IoT application to provide uninterrupted IoT-based services.

EXAMPLES OF IOT APPLICATIONS

Home And Building Automation

There are 10s – 1000s of sensors reading data and sharing it with an automation system running to control and monitor the building's ambiance such as lighting, heating, ventilation, air conditioning, and fire and safety system. A facility manager sitting in a room can control and monitor their facility using some supervisor software running on a central computer or service hosted on the cloud using a portal. The central service computer is collecting data from all the sub-systems in an HVAC system directly or indirectly.

Healthcare Industry

A smartwatch, or fitness band connected to the human body to track health parameters using various sensors for measuring heartbeat, temperature, blood pressure, etc., and sending the data to a smartphone. The application running on a smartphone will monitor and raise an alarm when parameters are outside of normal range could be another example of IoT for personal healthcare. In fact, pacemakers put up in a heart patient's body is nowadays smart to monitor several parameters such as blood temperature, sinus node rate, breathing, and other vitals, and these data could control heart rate, slowing or speeding it depending on a patient's current activity level. These next-gen pacemaker devices are IoT devices. Healthcare professionals can monitor the data of their patients and provide more personalized care.

In hospitals IoT devices tagged with sensors are used for tracking the real-time location of medical equipment like wheelchairs, defibrillators, nebulizers, oxygen pumps, and other monitoring equipment. Deployment of the medical staff at different locations can also be analysed in real-time.

Aerospace Industry

In the production facility of aerospace, there are several data points that are monitored to keep quality and compliance in check such as emission norms, inventory readiness of individual parts, and manufacturing details that could be easily tracked and maintained using IoT devices connected to them. Ground crew and engineers do a regular check for the flying condition of an airplane using various IoT devices reporting the reading of sensors monitoring the sub-systems. This helps them to quickly diagnose, optimize the cost of maintenance, avoid human errors and increase the efficiency and safety of flying machines.

Automobile Industry

In the next generation of cars, there are more sensors and connected digital components than mechanical equipment driving or controlling the car. For example, an autonomous vehicle is connected over the internet to receive the latest software updates, map navigation data, GPS-enabled components to track location and route, LIDAR system to identify the object, its distance, speed, size, and shape. In another example, if there is any problem detected by a sensor in the engine then a component called the diagnostic bus would start collecting the data from various sensors and will be passed to the gateway in the car. This gateway or central control unit will send the most relevant information to the manufacturer's platform to get an appointment for fixing the faulty part or taking some care if a false alarm is raised.

This data could also be used to speed up the car diagnostic and procuring inventory by the nearby service center.

SMART AGRICULTURE USING IOT (RAVINDRA, 2020; SHALIMOV, 2020)

The world is expected to touch the 9.7 Bn human population mark by 2050 and 11Bn by 2100, so the increase in food demand is obvious. Also, the middle-class population in developing countries like India, and China is rising faster than last century so the demand for quality Agri-products needs to be dealt with the help of technology. Since the agricultural land is not going to grow hence more yield from the same land mass is being tried to achieve by adopting smart and innovative ways using IoT-enabled agricultural/farming techniques. This includes monitoring of climate, optimal usage of water, fertilizers, greenhouse, remote monitoring the livestock, etc.

Smart agriculture (IoT enabled smart farm, n.d.) can be visualized as farming using IoT-enabled infra-structure where connected sensors are deployed for monitoring crops, soil, humidity, fields, livestock, and storage facilities. and acting smartly to control the resources for better quality farm products. Smart agriculture vehicles, drones, autonomous robots, and actuators are used to optimize the fertilizer and pesticide spray cover crop, and it helps reduce manual labor. Connected agriculture spaces such as smart greenhouses, hydroponics, and aeroponics are some examples of alternative farming techniques with the help of IoT technology. With the use of IoT techniques massive data is captured which provides transparency in agriculture processes and farmers can get precious insights into the performance of their fields, greenhouses, barns, etc.

APPLICATION OF IOT IN AGRICULTURE

Climate Condition Awareness (Climate-Smart Agriculture, n.d.)

Agriculture is the oldest profession and with a growing population, food demand has grown. Over time farmers have improved their farming techniques and moved from rain as the only means of irrigation to today's advanced irrigation techniques such as sprinkler irrigation, drip irrigation, etc. In farming, climate and weather condition plays a very critical role. Droughts are very common in India due to a lack of rain and water shortage in rivers & canals. Sometimes this leads to farmers' suicide and price rise for food essentials like onions, tomatoes, green vegetables, etc.

Having improper knowledge about climate conditions impacts the quality and quantity of crop production. IoT solutions enable farmers to have real-time information about weather conditions using the sensors placed on farms (inside or outside). These systems can be connected with the meteorological department's computer system to access the climate predictions for few weeks in advance so that irrigation and other farming activities could be carried out effectively. These IOT-enabled sensors can send alerts about unfavourable weather conditions.

The data collected from the sensors about the environment is used to choose the right crops which can grow and sustain in particular climatic conditions. IoT ecosystem is made up of sensors that can detect accurately in real-time the weather conditions like humidity, rainfall, temperature, etc. Low-cost, energy-efficient durable sensors are available that help monitors the condition of the crop and weather surrounding them, customized as per the farming requirements. These systems eliminate the need for physical presence during disturbing climatic conditions.

Conventional farming will not be effective for densely populated cities due to less space for crops to flourish. So IoT-enabled climate controller vertical farming could provide a healthy sustainable small food supply chain. These indoor farms don't need soil or natural light to thrive. IOT-controlled indoor farming setup can offer prosperous harvests year-round and it will eventually reduce carbon footprint due to farm product transportation in conventional.

Precision Farming (Crop Health Management, n.d.; Crop Irrigation Management, n.d.)

Precision farming or precision agriculture is a term that is used for controlled and accurate farming practices when it comes to growing crops, raising livestock, inventory management, field observation, etc. This is about improving the key decisions farmer make about planting, fertilization, crop protection, and irrigation on the inputs as well as yield and quality outputs at harvest. Traditionally, farmers relied on experience and historical knowledge to undertake farmer practices according to the schedule for each season. IOT-enabled technologies, sensors, actuators, robotics, drones, driverless farm machinery, etc. make precision farming a reality. The goal of precision farming techniques is to monitor every individual tree, plant, and livestock (bird, animal, fish) for its growth, health, disease, and other climatic condition with the help of data collected from sensors and take intelligent countermeasures.

With the availability of affordable and reliable high-speed wireless internet services, low-cost satellites (for imagery and mapping), GPS, and low-energy sensors, the precision agriculture trend is steadily growing. The agricultural drones can be equipped with cameras to capture images of a field that can be processed to create maps for fields capturing several metrics such as elevation and vegetative index.

These maps can be used to optimize crop inputs such as water, fertilizer, herbicides, and other growth regulator chemicals through variable rate applications.

Applying the optimum level of irrigation is very important. Water is the most valuable natural resource, applying excess irrigation is an unnecessary cost, leeches nitrogen, and can even lower yields. On the other hand, applying too little irrigation limits yield potential. Various sensors are used to understand and accurately identify each field's variability factors such as soil, topography, yield, organic matter content, moisture levels, nitrogen levels, pH, EC, Mg, K, etc. This data is used in conjunction with satellite imagery by variable rate technology (VRT) including seeders, sprayers, etc. to optimally distribute resources.

Agricultural Drones (Winn, 2020)

A drone is definitely a very good tool to use in agriculture for various purposes. A drone can be used to spray pesticides, spray fertilizers, irrigation, field analysis, crop health monitoring & assessment, planting, soil analysis, etc.

Using integrated GIS mapping with drones, data-rich maps of farms and fields could be created which helps to make more informed crop management decisions and better understand farm product performance.

Drones fitted with sensors and cameras for imaging, mapping, and surveying agricultural land. Farmers can perform in-flight monitoring and observations by entering parameters like what field to survey at which altitude or ground resolution in the drone control dashboard.

With the use of drones with a mounted camera and GPS sensor, farmers can now collect multispectral, thermal, and visual imagery on drones which could be processed with satellite maps to create a data-rich map of the field having various metrics such as vegetative index. With the advancement of technology data captured by drones could be sent back to the control station while the drone is flying and its flight path could be controlled with a joystick to get a better image of the farm. Data captured using a drone could help draw an insightful conclusion regarding plant health indices, plant counting and yield prediction, plant height measurement, canopy cover mapping, field water ponding mapping, scouting reports, stockpile measuring, chlorophyll measurement, nitrogen content in wheat, drainage mapping, weed pressure mapping, and so on.

Using drones aerial survey of the farm becomes very simple and less labour intensive and helps the farmer to make a better decision. The use of drones has shown a positive impact on the environment by severely reducing harmful chemicals such as pesticides, and herbicides, reaching underground.

Figure 2. Agricultural Drones

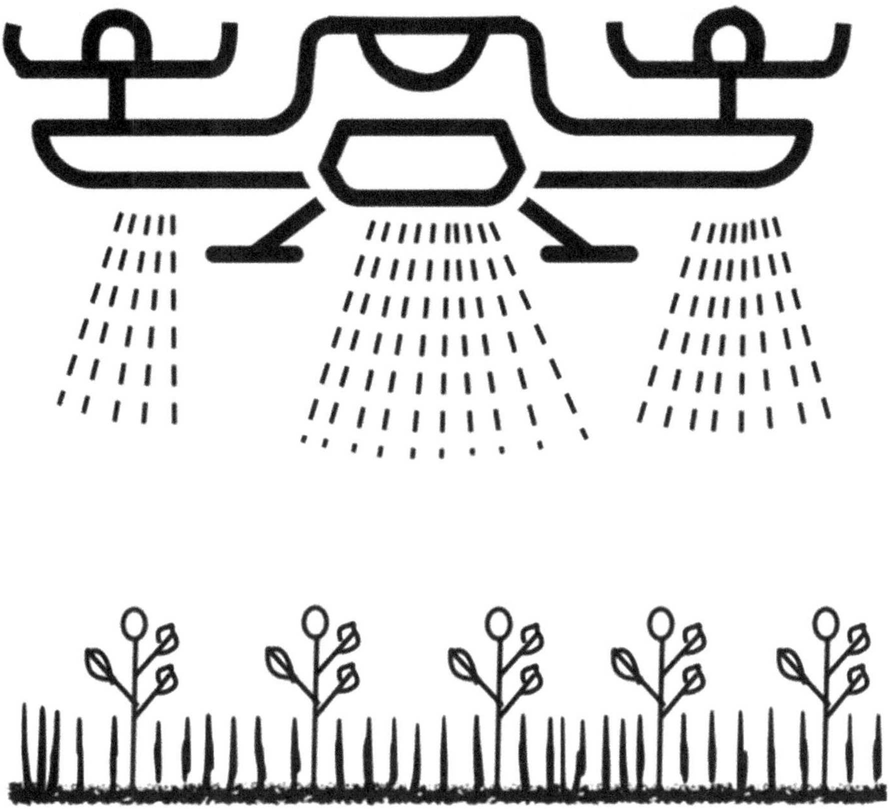

Livestock Monitoring (Ravindra, 2020)

IoT-enabled sensors can help farmers monitor the health, physical activity, and physical location of cattle, birds, and other domestic animals. This can help the farm owners to identify unhealthy animals, birds, etc, and separate them proactively to avoid the spread of disease. It lowers the labour cost as locating the livestock is easier with the help of location data being collected by IOT-enabled connected sensors.

Collar tags are an example of such IoT sensor-enabled device which is used by farmers to record information such as body temperature, health, activity, etc about the cow. Dairy farm owners use such devices to observe the health parameters of pregnant cows and can track their delivery due dates and take precautions to avoid any uncalled emergency. Some sensors even enable farmers to get notifications when the animal is about to deliver. The battery-powered sensors send information to the herd manager or the rancher when the water breaks right before the delivery of any animal. This gives precocious time for farmers to stay focused and spent time with the heifers that are giving birth.

Autonomous Greenhouse

Greenhouse farming is a climate-controlled farming technique within an enclosed structure having walls and roofs made chiefly of transparent material, such as glass called a greenhouse or glasshouse. The atmosphere inside the greenhouse structure when exposed to sunlight, observes heat and keeps it warmer than the outside atmosphere, protecting its contents in colder climate conditions. The Netherlands has around 4,000 greenhouse enterprises that operate over 9,000 hectares of greenhouses producing €7.2 billion worth of vegetables, fruit, plants, and flowers, some 80% of which is exported.

The closed environment of a greenhouse needs different farming techniques than outdoor production. Controlling the environment artificially to the best environment needed for different plants to grow and produce quality products through manual intervention could be a daunting task and costly. It could lead to production loss, energy, and resource wastage. With the help of IoT-enabled environments having various sensors and actuators, intelligently design IT solutions could control the whole ecosystem very effectively with very less or no manual interventions. For example, sensors that measure soil moisture level could be used to control irrigation automatically than controlling the drip-irrigation system manually. An array of sensors measuring CO2 level, chlorophylls, nitrogen, temperature, pressure, humidity, and light level could help automated actuators to control the opening/closing of windows, turn on/off lights, control heaters, and turn on/off fans. All these climate controls can be done automatically with these sensors measuring and sharing data over WiFi signal to the control centre and the control centre sending the data to the actuators/regulators connected to windows, heaters, fans, and other equipment. This eliminated the need for constant manual monitoring and provide a cost-effective and optimal solution for farmers with minimal manual interaction.

Cloud-connected IOT-enabled greenhouses could provide remote monitoring and controlling capability. Data collected from a greenhouse IoT ecosystem could further be analyzed using big data to get a better insight. Along with machine learning techniques, the IOT-enabled greenhouse could be further made smarter to act automatically in changing environments.

Considering the unbeatable benefits, environmental friendliness, and agility of autonomous greenhouses, they could become the solution to the challenges farming confronts, and thus build the future of IoT in agriculture.

BENEFITS OF USING IOT IN AGRICULTURE

There are several benefits of using technology in the field of agriculture such as using a tractor to reduce the effort to plough, till, and plant fields in addition to routine lawn care, landscape maintenance, moving or spreading fertilizer, and clearing bushes. Similarly using IOT-enabled solutions in modern times, the agriculture industry is reaping benefits and helping farmers optimize resource wastage, and monitor the farming process better. This helps farmers solve the problem of high quality and high farm product yield with fewer resources and a greener way.

Excelled Efficiency

With the growing population, receding farmland, natural resources, and severe weather conditions, the human race is facing a hard time fulfilling food demand. Farmers are looking for innovative but cost-effective solutions to fight all odds and produce more food products. Smart farming using IOT is helping farmers to better monitor the product and surrounding conditions in real-time. It also helps optimize the use of farming resources such as water, chemicals, etc in real-time to improve the quality of farm products, reduce cost, and better yield. Now farmers can have more information to make better decisions and accurate predictions about upcoming challenges than in conventional farming where farmers relied on past experience and historical knowledge. Constantly improving technological space is making advancement faster than our thoughts and more and more agricultural equipment will are expected to have built-in sensors in the future to help farmers significantly improve the quality of farm yield with lesser investment.

Expansion

With the rise in population and faster urbanization, it is expected that 70% of the population will live in urban areas which mean fewer farmers and far-situated farmlands. Urban population will have more demand for quality food products and IOT-enabled small greenhouses, hydroponic, and aeroponic systems will play a major role to enable a short food supply chain and should be able to feed these people with fresh fruits and veggies. IoT-enabled smart micro-farming will allow growing food almost everywhere like on walls and rooftops of skyscrapers, hotels, supermarkets, balconies of an apartment in high-rise buildings, small houses in suburban areas, university campuses, etc.

Cleaner Process

IOT-based precision farming help farmers save water, energy, pesticides, fertilizer, and other resources thus it makes farming greener. This approach helps with cleaner and more organic final products compared to conventional farming methods. Organic farming requires more labor to tackle weeds and crop rotation to manage soil nutrients without the use of fertilizer, unlike conventional farming where the use of herbicides, insecticides, and fertilizer saves manual labour and contaminates the soil and environment too. The IoT-enabled solution helps farmers manage the environment by saving resources and consumers' health by serving organic farm products.

Agility

Using IOT-enabled infrastructure in agriculture, reaction response time for any change in weather, humidity, air quality, the health of each crop, pest attack, soil health in the field, etc has become quicker with the help of real-time monitoring and prediction systems that IoT offers. This has increased the agility of the process and helped reduce damages from unfavourable conditions.

An IoT-enabled solution helps really well with dairy farming, fishery, horticulture, poultry, etc to monitor the livestock in real-time and act immediately. The farmers with this precise information can act quickly in event of extreme weather changes and save the crops and livestock better.

Improved Product Quality

Data-driven agriculture helps both grow more and better products. Using soil and crop sensors, aerial drone monitoring, and farm mapping, farmers better understand detailed dependencies between the conditions and the quality of the crops. Using connected systems, they can recreate the best conditions and increase the nutritional value of the products.

Customer Satisfaction

Customers are more satisfied when they can access more trustworthy information about any product. In conventional farming, there is hardly any information available for the customer. With IoT-enabled farming more data about every stage of farming could be captured in more detail and made available for them to read using some smartphone app with a QR code attached to the product. Customer care about various details these days such as climate conditions, pesticide usage, geographical location, quality of the crop, age of the product, quality control approval from some trustworthy agency, post-harvesting storage, processing and shipping to market, etc. Products that are approved by agencies such as USFDA, USDA, etc give the customer more satisfaction before purchasing any food product. IoT-enabled solutions help farmers get these certifications and approvals easily with the data available to them which were captured during the whole process of farming using various sensors and stored in a data center.

Agriculture Research

IoT devices generate a massive amount of data which needs months if generated using manual efforts. There are chances that some data might be corrupted due to some faulty sensor but given the fact that there would be multiple similar sensors/ devices capturing similar data the accuracy of these data can be higher than the manual process. This correct information when analysed with the help of a machine learning algorithm, Agriculture scientists can arrive at a better result. They can try out some experiments with the help of farmers more reliably in a larger area to optimize resource usage to get more farm yields of superior quality.

CHALLENGES OF USING IOT IN AGRICULTURE

Connectivity

In order to work with IoT devices in agriculture the basic requirement is power and network connectivity. With the advancement in technology low powered devices can work for months and years without any need for battery replacement. In devices are solar-powered and can work on solar energy. So power problem can be solved but network connectivity could be sometimes complex due to proprietary and non-standard technology for data exchange in various IoT devices. In remote areas Internet connectivity especially 4G may not be present or signal strength might be poor for uninterrupted data communication. However, it is expected with the introduction of 5G connectivity network problems would be resolved and IoT devices can work without many interconnectivity issues.

Design and Durability

An IoT-enabled device used in agriculture will have its own drawbacks and strength. Not a single solution can work universally. Sensors or devices might work well in certain weather conditions such as hot or cold, rainy or dry, indoor or outdoor, and may not work properly in other conditions. So careful design of such systems is required and a professional technician can help on these aspects. To increase the durability and reliability of the IoT system regular maintenance would be needed. Devices like Drones, sensors, and weather monitoring stations might have simple functional designs and robustness to work on the farm but these devices will have their own complexity.

Limited Resource and Time

With the fast-growing population and reduced farmland and constantly worsening climate conditions due to global warming, the farmers are racing against time and resources. With technology integration in agriculture, the farmers are expecting to improve the production of farm products but they can't yield more by multiplying the number of automation equipment, unlike other industries. With natural pollinators dying due to extreme weather conditions, it is very critical to create an environment that sustains and scales with time to provide sufficient food. IOT-enabled farming can't solve all the problems but definitely can improve techniques of how farming should be done in the future.

Cost

Technology has its own cost, it improves the ease and efficiency of the process and sometime it may reduce the cost in the long run but usually, it comes with initial capital investment. For example drones, tractors, temperature sensors, humidity sensors, soil condition sensors, monitoring devices, internet infrastructure, and data centres all need to be bought and installed. Installation and maintenance will be recurring costs and it might need IT engineers and experts to support the whole infrastructure. The IoT-based solution has many components and all of them feed data to a controlling computer system which could be in premise or cloud. These integrated software solutions are not cheap and they require software professionals to maintain and training to the farmers. IOT-enabled hardware also needs specialized maintenance by professionals as they are not traditional equipment that can be fixed without technical understanding.

Data Security and Privacy

Data security and privacy is an age-old problem with any system which collects and transmits data. IoT systems collect enormous amounts of data through various sensors and data points in real-time. If data is not properly secured and transmitted then the possibility of data leakage, and stealing by cyber-criminal increases. If some cybercriminals get access to the IoT infrastructure then they can infect the live-stock monitoring system or can disturb the drones which are spraying pesticides, and fertilizer or they can disturb the supply chain of farms. Although the nature of these attacks may or may not be very critical still security and privacy is critical area of concern that needs careful design in an IoT solution.

Data Overload and Accuracy (Remote food and plant growth monitoring with IoT technology, n.d.)

As mentioned earlier, IOT-enabled sensors, farming and agriculture equipment, etc transmit enormous amounts of data using various protocols (proprietary or open standards) causing a data overload. Although more data available is better for analysis and getting better insight into crop growth, weather conditions, livestock's health, livestock's location, etc, it is becoming extremely difficult for farmers to pick a smaller relevant set of data points to arrive at a conclusive decision quickly. This needs sophisticated software to reduce redundancy and pick relevant points. Storing these data and keeping them safe is another area of concern.

FUTURE OF AGRICULTURE USING IOT

The future of every industry is constantly changing with fast technological advancement and innovation. For example, when the industrial revolution began, the textile industry changed forever, and instead of traditional handloom big industrial machines changed the way textiles were produced. Similarly, with the invention of the car, the transportation industry changed forever and now autonomous vehicles are expected to change the way we drive our cars today and travel to our workplace. The agricultural industry is also consistently under innovation with a new breed of cattle being produced with the help of technology or meteorological data helping farmers taking better decisions about sowing, irrigation, harvesting, etc. Similarly, IOT enabled solution is going to change the way farming was done conventionally to support increasing food demand for a steadily growing population.

Smart Climate-Aware Farming (Climate-Smart Agriculture, n.d.)

Since farming depends on favorable climate conditions otherwise the farm yield would be poor. So, using IoT-enabled climate monitoring devices can help farmers optimize the usage of resources based on climate such as controlling water usage, harvesting the farm product, store farm products after harvesting. Livestock can be raised better under climate-aware facilities to produce better quality dairy products, eggs, meat, fur, wool, leather, etc. AI and big data technology will further enhance the capability to revolutionize the automation of climate-aware farming by taking several steps automatically to reduce crop damage due to harsh climate conditions. Robotics and autonomous vehicle connected to the IoT solution could further strengthen the farming process by removing weeds, reaping, and farm harvesting farm products.

Micro-Farming (Schipani, 2019)

Fresh farm products were always in demand and with the growing urban population and increasing city limits it is becoming a challenge to provide fresh farm products such as vegetables, fruits, milk, etc. to market. Although, refrigerated containers are used to keep vegetables and others farm products stay fresh but the cost of transportation from farm to market is rising. Micro-farming is a small-scale, high-yield farming technique that is helping urban/sub-urban populations to grow fresh vegetables and fruits in their courtyards/terraces, etc. Micro-farming with the help of IoT enables techniques that can help control the farming conditions/climate and help the urban population to get high yields. Since micro-farming is ideally carried out in a small area for farming which is usually an artificially created farm so IoT-enabled environment can control the overall farming condition automatically such as humidity, temperature, lighting condition, etc.

Data-Driven Organic Farming

Organic farm products are in greater demand and customers are looking for genuine organic products due to health benefits. Organic farming is not cheaper due to its production cost is higher and there is a need to optimize the cost of farming, and processing and improve the supply chain. Customers trust the product if they read more information about the farming, processing, and storage techniques used while transported from farm to market. With the help of sensors in farm/agricultural fields, several data can read such as humidity, moisture, light condition, temperature, atmospheric factors, etc could be captured, and accordingly, measures can be taken manually or automatically to improve the yield. Additionally, these data might be stored in a blockchain for customers to see the details of farm products. Data collected at various stages can be analyzed to improve production, storage efficiency, and supply chain.

CONCLUSION

IoT-enabled farming has several benefits such as saving resource wastage (water, fertilizers, electricity, manual labour), collecting a large amount of data regarding crop condition, fertilizer usage, pesticide usage, etc, and automating several processes based on climate data. Although it comes with some investment to procure IoT-enabled equipment and data centers, internet connectivity and maintenance of these systems benefits can significantly enable the future of farming. With more usage of IoT-based agriculture, farmers could meet the increasing demand for food grains for a growing population. Data collected from IoT devices will help farmers and scientists understand several conditions that could cause lower yield and improve the overall yield with limited farmland and less wastage of resources. IoT-enabled micro-farming, vertical farming techniques can change the way the current urban population consumes farm products, they can have their own small farm on a balcony of a high-rise building apartment or in the backyard of a house. These techniques will definitely help the urban population to have fresh and healthier food in their daily life and with ease, as IOT-enabled techniques will help them focus on their day-to-day job, and farming can be automated and monitored/controlled from a remote place like the workplace.

REFERENCES

Agriculture IoT Solutions and Cloud Connectivity. (2019). https://www.soracom.io/blog/agriculture-iot-solutions-and-cl
oud-connectivity/

Climate-Smart Agriculture. (n.d.). https://www.fao.org/climate-smart-agriculture/knowledge/en/

Crop Health Management. (n.d.). https://www.iotforall.com/use-case/crop-health-management/

Crop Irrigation Management. (n.d.). https://www.iotforall.com/use-case/crop-irrigation-managemen
t/

Internet of Things in Agriculture. What is IoT and how is it implemented in agriculture? (n.d.). https://www.cropin.com/iot-internet-of-things-applications-a
griculture/

IoT enabled smart farm. (n.d.). https://www.soracom.io/customers/agriculture/iot-enabled-smart-cows/

Ravindra, S. (2020). *IoT Applications in Agriculture.* https://www.iotforall.com/iot-applications-in-agriculture/

Remote food and plant growth monitoring with IoT technology. (n.d.). https://www.soracom.io/customers/agriculture/food-plant-growth-monitoring-iot/

Schipani, S. (2019). *What Is Micro Farming?* https://hellohomestead.com/what-is-micro-farming/

Shalimov, A. (2020). *IoT in agriculture. 8 technology use cases for smart farming (and challenges to consider).* https://easternpeak.com/blog/iot-in-agriculture-technology-use-cases-for-smart-farming-and-challenges-to-consider/

Winn, M. (2020). *How Drones Are Keeping the Lights On.* https://www.iotforall.com/drones-keeping-lights-on/

KEY TERMS AND DEFINITIONS

Drones: A small unmanned flying vehicles used for monitoring, air surveillance, geo-mapping, spraying pesticides/herbicide, etc.

Greenhouse: Is an eco-system of controlled farming in an enclosure made of glass and advanced technology to control the climate within the enclosure.

Internet of Things: Is the global infrastructure to exchange data for any computing device over the Internet from any place and anytime.

Micro Farming: Is a farming technique to grow farm products in smaller areas with specialized technology for the urban population.

NB-IoT: NB-IoT stands for NarrowBand Internet of things. It is a standards-based low power wide area (LPWA) technology. It provides improved indoor and outdoor coverage and supports huge numbers of IoT devices having low throughput, low cost, and low power consumption. It works along mobile networks on 2G, 3G, and 4G (GSM/GPRS/LTE).

Smart Farming: Is a concept where farming techniques evolves by adopting the latest and advanced technology to get high-quality farm product and more yield.

Vertical Farming: Is a soilless farming technique for growing crops in vertically stacked layers in a controlled environment.

Chapter 6
Sustainable Development in Modern Aquaponics Cultivation Systems Using IoT Technologies

Ravi Samikannu
https://orcid.org/0000-0002-6945-6562
Botswana International University of Science and Technology, Botswana

Ashok Kumar Koshariya
https://orcid.org/0000-0002-7427-4835
School of Agriculture, Lovely Professional University, India

E. Poornima
Gokkaraju Rangaraju Institute of Engineering and Technology, India

S. Ramesh
Saveetha School of Engineering, India

Ashok Kumar
Banasthali Vidyapith, India

S. Boopathi
https://orcid.org/0000-0002-2065-6539
Muthayammal Engineering College, India

ABSTRACT

The main problem in developing reliable and economically feasible solutions for aquaponics is combining aquaculture (the farming of fish) and hydroponics (the growing of plants without soil). The purpose of this chapter is to aid in the development of a workable commercial aquaponics system. Additionally, it looks at the internet of things (IoT) and smart systems in the literature. For the purpose of real-time monitoring and physical system control in an aquaponics system, an application that uses digital twinning is created. A tool that models the relationship between crop size and fresh weight, and then estimates crop size as it grows, using image

DOI: 10.4018/978-1-6684-4118-3.ch006

segmentation is suggested. The various parameters of the aquaponics system and the sensors used in the IoT and digital twin framework system are demonstrated. The controlling strategies, various smart technologies, and implementation procedures are also illustrated.

INTRODUCTION

Due to its ability to conserve resources, as well as its high efficiency and low consumption, aquaponics has gained increasing attention. Symbiotically cultivating aquatic organisms and plants is known as aquaponics. Aquatic creatures emit waste, which bacteria subsequently break down into nutrients that plants can absorb to develop. The aquaponics system has successfully addressed the issue of environmental pollution by realising the change from waste to nutrients. Future low-carbon production methods will include aquaponics because it is circular, efficient, and intensive. It has less of an influence on the environment and mimics natural systems, where water efficiency is greatly boosted. In 2018, acute hunger affected more than 113 million people in 53 countries, necessitating immediate food, nutrition, and livelihood assistance (IPC/CH Phase 3 or above). The world's current food needs cannot be met by using more natural resources or exploiting the land. Instead, we must seek out sustainable answers that support the production and consumption of food (Kumar et al., 2023).

At least 43 countries throughout the world already practise aquaponics; however, 84 percent of those who do so do so as a hobby. Aquaponics' effective development could ensure a significant portion of a more sustainable world food supply. If it spreads widely as a commercial alternative, it will be successful in assisting the food issue and global sustainability. An aquaponic system's design and administration are challenging tasks. Because it is a greenhouse and a symbiotic habitat, a variety of characteristics and elements (such as light, temperature, pH, moisture, etc.) must be managed. Due to their many parts and requirements, such as disease prevention and water purity, these systems are highly complex. A new window has opened up for the advancement of these aquaponics systems with the introduction of automation, smart farming practises, and networking in the agricultural sector. Smart automation is predicted to reduce the amount of manual labour required significantly and improve process management. It appears appropriate to continue working on the creation of tools and frameworks for precision farming as aquaponics gains popularity as a farming technique. The ideas listed above (sensing, automation, etc.)

This analysis is based on a review of the literature on aquaponic monitoring, smart systems, and IoT systems. The goal of this study is to compile the existing understanding and methods for aquaponic farming monitoring systems. Few publications have been made in the literature, so certain hydroponic systems were

Figure 1. Aquaponics cultivation system

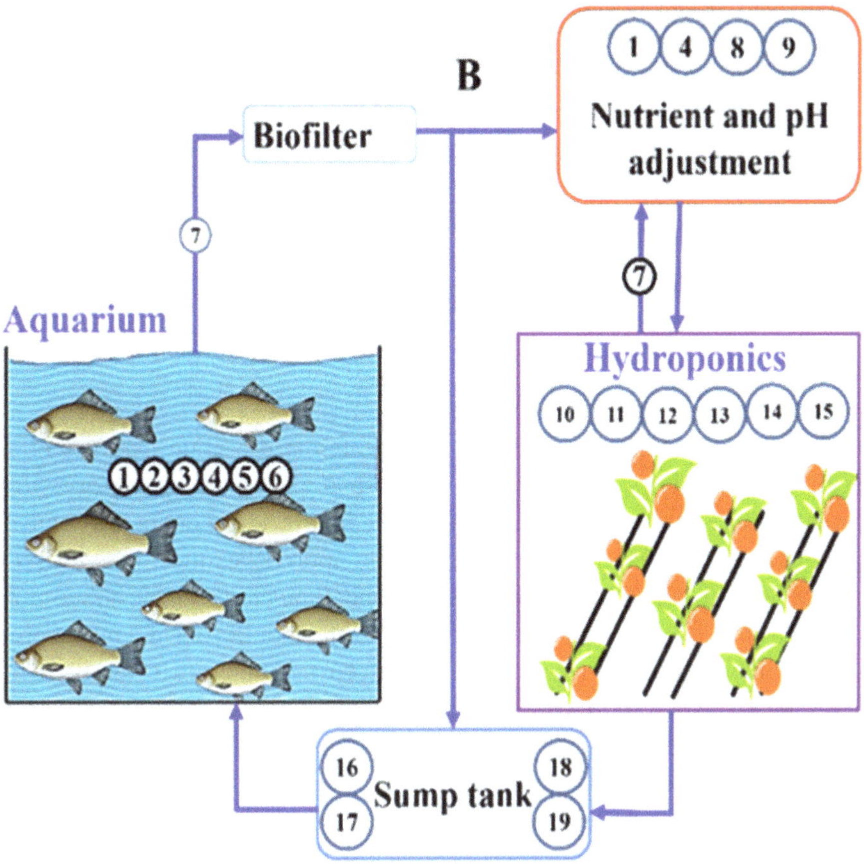

analysed to supplement the data. The authors of a new book on aquaponics hope to lay the groundwork for the commercial introduction of automation in the industry (Sathyanarayana et al., 2022). Frameworks and methods for smart cities and IoT are offered. We also talk about some potential applications for aquaculture and hydroponics in the future. The schematic arrangements of the aquaponic system are shown in Figure 1.

MONITORING PARAMETERS

Depending on the system, a different method may be used to measure the repeating parameters. A pH sensing system, for example, could include pH test strips, freestanding sensors with an LCD panel, or analogue sensors that can transfer data wirelessly or not to a controller (PLC, microcontroller, etc.). Automated sensing approaches need to be assessed with the goal of building a dependable, sustainable, and financially viable system. The stability of the aquaponics system depends on the environment and machinery being monitored and controlled by intelligent technologies. Aquaponics systems of many kinds, particularly coupled and decoupled systems, have been designed, used, and described over time. For the aforementioned aquaponics systems, Figure 2 summarises the connections between the sensors and parameters. Each parameter is described in detail along with its function in aquaponic systems in the subsections that follow. Following that, a description of how writers measured, employed, and controlled them is provided (Calone & Orsini, 2022).

Water

Systems can differ in their methods for measuring the repeating parameters. For instance, the pH sensing system can use pH test strips, freestanding sensors with an LCD panel, or analogue sensors that can transfer data wirelessly or not to a controller (PLC, microcontroller, etc.). Automated sensing approaches must be assessed in order to develop a dependable, long-lasting, and financially viable system. The stability of the aquaponics system is predicated on the monitoring and management of the environment and machinery through intelligent technologies. Over the years, a number of aquaponics system types, particularly linked and decoupled systems, have been suggested, accepted, and explained. Maintaining the ideal levels of nutrients, pH, temperature, dissolved oxygen, and salt in water, as well as keeping an eye on other factors including dissolved solubility, pH, and dissolved oxygen levels, is crucial for fish survival in ponds (Das & Sahu, 2020).

Nitrification

The most crucial mineral for plants is nitrogen. Ammonia and ammonium are two different forms of it. The pH, temperature, and salinity all affect how much of each is present in the water solution. the overall nitrogen-ammonia concentration. Ammonia's oxidation and nitrification are combined to become TAN. Ammonia oxidising bacteria convert TAN to nitrite. Nitrite oxidising bacteria then convert nitrite NO_2 into nitrates NO_{3-} (NOB). An aquaculture component, a biofilter for the nitrification process, and a hydroponic component make up a conventional aquaponic system.

Figure 2. Various Sensors and Parameters used in IoT Aquaponic System

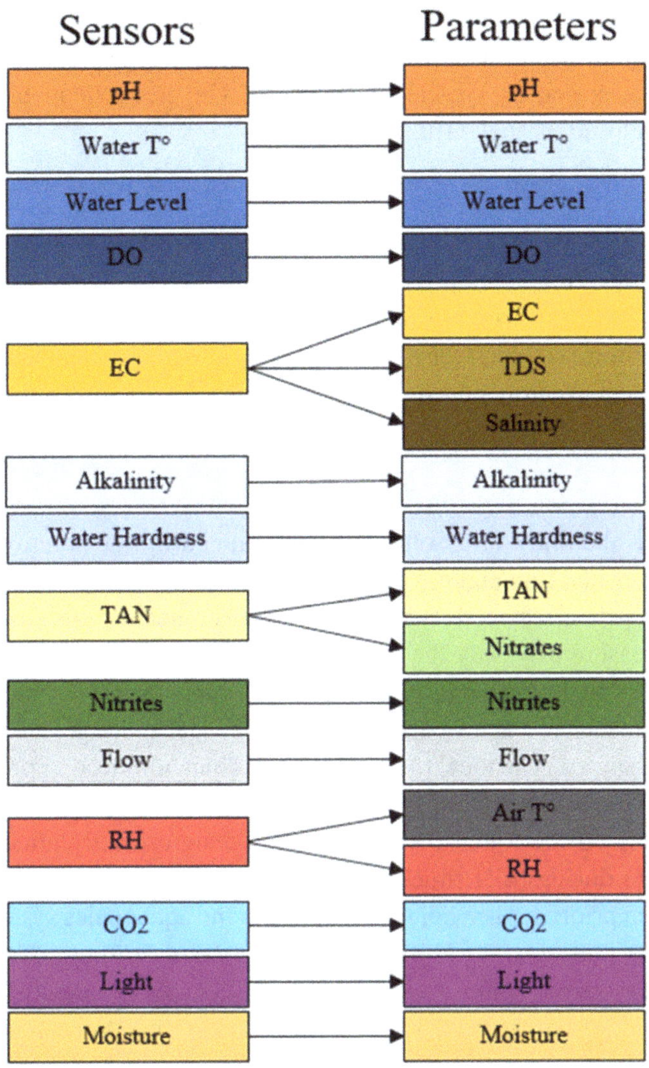

TAN

If fish or other species are exposed to excessive levels of pollution, they may emit ammonia, a type of nitrogen, into the atmosphere. The US Department of Agriculture estimates that 10% of the protein in fish feed will be converted to ammonia in the system water.Small levels of ammonia are particularly hazardous to fish, and when the water solution is strongly acidic or alkaline, it becomes predominant (relative proportion increases). The ideal range for fish containing TAN is 0–2 mg/L, and the ideal value for bacteria, specifically AOB and NOB, is 3 mg/L. The amount of ammonia in water solutions is determined by ammonia sensors. Sensing ammonia is the best way to detect its existence because it is odourless and colourless in small doses. These sensors typically consist of a wire electrode in an ion-selective membrane-enclosed, bespoke fill solution.

Nitrite

Another type of nitrogenous waste that is fatal to aquatic life is nitrite. For fish, plants, and microbes to thrive in water, nitrite concentrations must be between 0 and 1 mg/l. Nitrite-ionized electrodes and a plastic membrane-based sensing element are typically used in nitrite concentration sensors.

pH

The acidity or alkalinity of a solution is determined by the solution's pH level. Plant nutrient availability and nitrification rate are influenced by water pH. Three methods—test strips, manual electronic probes, and automatic probes in controllers—are used to measure pH. Only the third method will be covered in this essay. The ideal pH for the hydroponics part is around 6.0. Fe or Mn will precipitate at pH values higher than 7.0, and root damage and nutritional shortages may result at pH values lower than 4.5. The ideal pH range for the aquaculture component's water is 6.5–9.5, while the acceptable range is 5.5–10. Systems for aquaponics are sensitive to pH variations in the water. Fish health can be significantly affected by small variations (0.3) during brief times (18–24 h). Bases like potassium and calcium should be given priority when adjusting the pH in the aquaponics system since they act as bases for nutrients. A pH balance-focused smart farming IoT platform was developed to ensure maximum yield. They employed a pH metre with a temperature range of 0 to 13 and a maximum of 80 °C. They claimed that their technology, which used an ISFET (ion-sensitive field-effect transistor), performed nearly as well as a digital pH metre.

Temperature

Most of the other water-related characteristics in the aquaponics system are related to the water's temperature. The ideal temperature range for nitrification is between 17 and 34 °C. Water that is too hot can limit how much calcium can be absorbed by plants. The most popular and extensively used tool for determining the temperature of water is a thermometer. Different species require different temperatures. This sensor's temperature range is -55 °C to 125 °C (Kumar et al., 2023).

Level

Natural water losses occur in every aquaponics system for four primary reasons: 1) the removal of fish sludge, 2) evaporation, 3) plant evapotranspiration, and 4) fish splashing during feeding. It has been found that it is simpler in larger systems to mitigate changes in the water quality parameters. The water level in tanks can be checked manually using a sight glass or floating objects, although there are several automatic measuring systems on the market. Ultrasonic, radar, or laser-based sensors are the most sophisticated types of fluid level sensors. Depending on the hydroponic/ fish tank ratio, water usage ranges from 0.1% to 3%. The third most extensively studied parameter in the literature at this time is water level. Omron's K8AK-LS1 water level controller with a 50 °C maximum temperature tolerance was utilised. An analogue output water level sensor coupled with an Arduino controller was also used.

Dissolved Oxygen

The amount of oxygen that is dissolved in the water and available to aquatic life is measured by the term "dissolved oxygen" (DO). Any aquaponics system must carefully monitor the dissolved oxygen level because it can change significantly in a short amount of time. Hydroponically grown plants often need high DO concentrations of more than 3 mg/L, but the majority of fish species in aquaculture need more than 5 mg/L. Fish swimming near the surface or with red zones around the eyes are signs that DO levels are low. Temperature, pressure, and salinity have an impact on DO measurements, and some adjustments might be required. Electromechanical and optical technologies are both available to measure the concentration of DO. Due to interactions between oxygen molecules and the dye when DO is present, the returning wavelengths are constrained or changed. TCP/IP was used to transmit data from a DO sensor attached to a Modbus device to a web server.

Electro-Conductivity

The electro-conductivity (EC), a measurement of a medium's ability to conduct electric current, is closely related to salinity in aquaponics systems. High EC values suggest that the water is contaminated, which could lead to the extinction of the fish population. However, a minimum salt level is preferred to aid fish in maintaining their osmotic balance. Although a greater range has been proven to be appropriate, the ideal range for fish is 100–2000 S/cm. The amount of nutrients in the water can be estimated by measuring the electroconductivity of the hydroponic component. Water EC measurements may provide information on nutrient absorption on a daily basis.It might assure maximum nutrient use without overfertilizing and provide consistency with each crop cycle (Kumar et al., 2023).

Total Dissolved Solids

Water contains dissolved solids (TDS) by nature. TDS levels show how much organic matter, inorganic salts, and other dissolved components are present. The majority of fish species can become hazardous due to high TDS levels in the water. TDS can be measured with TDS metres, which are comparable to EC metres.

Salinity

The salinity (SL), which measures the amount of salt in the water, is a key element that influences the abundance and growth of fish. Similar to TDS, salinity is frequently determined by the electroconductivity technique. SL's desired range varies depending on the type of fish.

Water Hardness

While low levels can cause plant death, high levels of water hardness can stress fish. Fish growth is affected by the metabolism of calcium and magnesium, which are necessary for the creation of scales and bones in fish. TDS or EC measurements are used to qualitatively determine the hardness of the water. The ideal range for water hardness is 50–150 mg/L, but most species can tolerate 10 mg/L. When lower values are anticipated, some colorimeters or spectrophotometers are utilised.

Alkalinity

In aquaponics systems, the concentration of bases, mainly carbonate and bicarbonate, is measured by the alkalinity. It gauges the water's capacity to neutralise acids or withstand pH fluctuations. Ammonia that is not poisonous becomes hazardous at high concentrations.

Flow

It is highly advised that the flow in the system be kept consistent to prevent stress in the fish and nutrient deficiencies in the plants. Water flow through the aquaponics system is vitally necessary to estimate the capacity of filtration (solids) and biofiltration (nitrification). Water flow for NFT should not be greater than 1-2 L/min. A syphon is employed in the media-based approach to filter the water through the media. Channel size and water capacity are used to determine the ideal water flow. A flowmeter is a tool that can gauge how much water is moving through a pipe. Water flow throughout the system can be ensured by flowmeters, which can also help identify serious issues with piping systems.

Environment

In the hydroponic system, it is required to monitor and regulate some environmental parameters in order to achieve and maintain balance and secure optimal crops. Specifically, the air temperature, where there are variations among plant types. The plant's stage of growth affects the light intensity.

Air Temperature

It is required to monitor and regulate some environmental parameters in order to achieve and maintain balance and secure optimal harvests in the hydroponic system. Specifically, the air temperature, which varies depending on the variety of plants. depending on the plant's stage of growth and the amount of light.

Relative Humidity

The phrase for atmospheric moisture is "relative humidity." The majority of the plants cultivated in aquaponic systems require humid air to thrive. The ability of warm air to hold moisture is greater than that of cooler air. Another way to quantify air humidity is as a percentage of air-water saturation or relative humidity (RH). Depending on the type of crop and the stage of plant growth, different RH levels are ideal. The most typical concerns range from 50% to 80%, however, and they are temperature dependent. The plants' ability to transpire would be hampered by an excess of RH in the air, which would also prevent roots and stems from getting enough water.

Carbon Dioxide

In order for plants to sustain their photosynthesis, carbon dioxide is a crucial component. Plants may utilise all of the CO_2 present in the air in large-scale indoor systems. Therefore, artificial CO_2 addition and level control are required. To allow for optimum fish growth, recommended CO_2 levels must be less than 5 mg/L. Aquatic life is hazardous at levels higher than those specified. Chemical gas sensors or infrared gas sensors (NDIR) are commercially available sensors that monitor CO_2.

Media Moisture

The moisture in the media base's soil is referred to as "media moisture." Only when using the media-base type in the hydroponic component is this measurement required. Installing a moisture-soil sensor is an excellent idea to ensure that the medium contains the right amount of water for the plants. By examining the media base and crops being used, it is possible to identify the ideal range for soil moisture, which ranges from 0-100 centibars (0-100 km/h). The soil moisture sensor uses an FC-28 to determine the dielectric permittivity of the water in the medium.

Light Intensity

Sunlight is essential for plants, yet it is scarce or nonexistent in indoor settings. Aquaponics systems use artificial illumination as a replacement for natural lighting. High levels of PAR can cause some light-sensitive crops, like lettuce, to bolt, seed, and turn bitter and disagreeable. When bacterial colonies first form, photosensitive species like algae and nitrifying bacteria are present. In new aquaponics systems, it is recommended to shield the area from direct sunlight for the first three to five days. Water does not require direct light radiation like plants do; hence, it is crucial to isolate any water system. A balance between the appropriate PAR, which is typically expressed as photosynthetic photon flux density (PPFD), and the appropriate light intensity (lux or lumens), must be achieved in efficient lighting systems. Plants require different amounts of PPFD depending on their stage of growth and their type, but an average ideal range is needed.

Smart Systems

In the academic community, "intelligent" or "smart" systems are a broad notion. Precision farming is becoming more popular as smart systems are adopted in farming. Smart is a concept that is more closely tied to Industry 4.0 than to traditional logic. It encompasses complicated logical processes and algorithms and is not just restricted to elementary logical operators. The incorporation of clever approaches into aquaponics is assisting in lowering production times, requiring less manpower, and improving product quality (Yanes, 2020). These systems are portrayed as layers of cooperation between extended production system analysis, self-configuration (machines), and collaboration between different actors such as clients, farmers, etc. The smart system for aquaponics cultivation is illustrated in Figure 3.

Figure 3. Smart system for Aquaponics cultivation

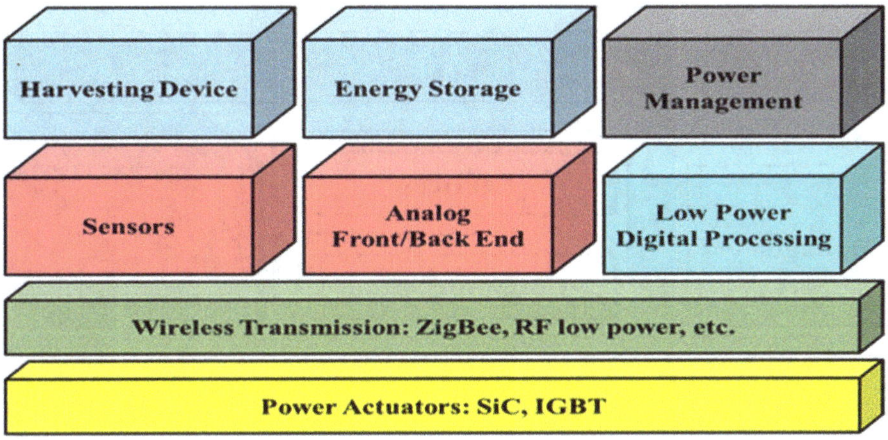

PARAMETERS PREDICTION

In order to forecast future values for certain of the sensed parameters (nitrate and pH), Kumar et al. created a wireless system that employs machine learning. A deep neural network was trained in hydroponics to forecast pH, humidity, light intensity, temperature, and water level. The parameters were monitored and controlled using the error between projected values and sensor outputs.

QUALITY AND GROWTH RATE

In the research on aquaponics, image processing and prediction approaches based on photographs are not frequently seen. Lu et al. used deep convolutional neural networks to monitor the growth rate of lettuce in a hydroponics system. In the future, the aquaculture component's properties, such as the health of the fish, can be monitored using this picture prediction model.

IOT SYSTEMS

The Internet of Things (IoT) intends to enable intercommunication amongst industrial gear and to offer a framework for autonomous, data-driven decision-making. The viability of IoT implementations rises with the network capacity improvements brought on by 4G and 5G technology (Ezzahoui et al., 2022). Three categories—

Figure 4. Block diagram for IoT

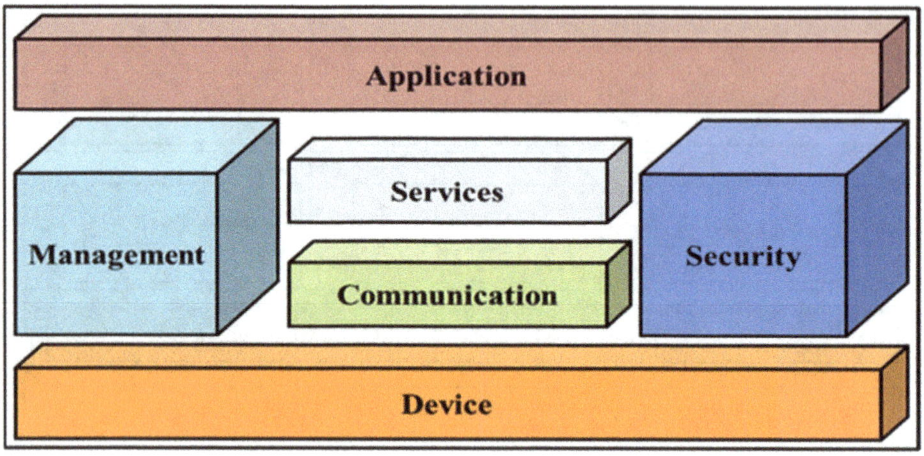

monitoring interfaces, remote applications, and wireless technologies—are used to group the findings. Combinations of these categories were discovered in some instances. The block diagram of IoT system is shown in Figure 4.

Remote Monitoring Interfaces

Monitoring interfaces is frequently a setting where a user or stakeholder can view some of the interesting parameters. These monitoring interfaces can display values across wireless networks thanks to IoT technology. A web application created a dashboard linked to a microcontroller to track particular aquaponics parameters. An iOS application that makes it possible to continuously monitor the system environment by obtaining data straight from the systems' microcontrollers. These partnerships are moving in the direction of real-time dependability and mobility (not only web based but also applications for mobile devices).

Remote Control Applications

Applications for remote control go beyond the system monitoring described in the preceding subsection. Operators, for instance, can switch water pumps or lights on or off as needed. A GSM Arduino-based monitoring and control system was put into place. This system may send operators alert messages when measurements fall outside of predetermined ranges. Operators can communicate with the receptor to control the water supply or temperature in real time. The system currently lacks adequate parameter display, which is a requirement for a better system (Taha et al., 2022).

Figure 5. Wireless Technology in Aquaponics system

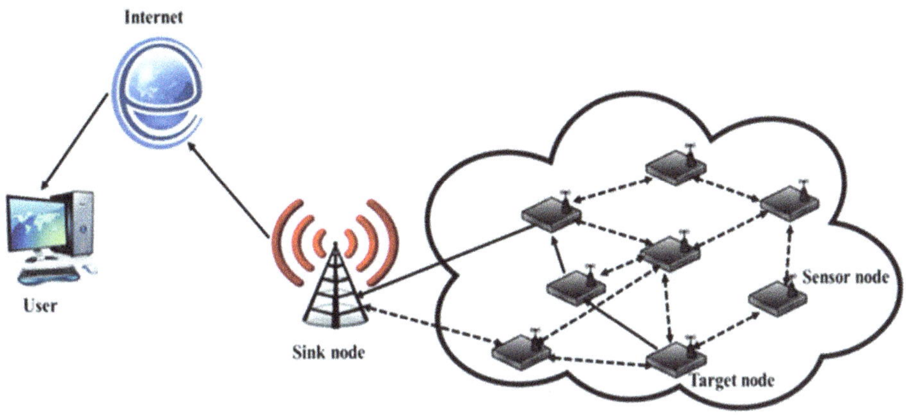

Wireless Technologies

Using wireless technologies, some wireless technologies were developed and applied to aquaponics. GSM technology is used to transmit alerts when pH and temperature readings deviate from normal. A web-based monitoring system was developed. For internet access, Jacob utilised a Raspberry Pi and a Wi-Fi adapter (Figure 5).

Control Strategies

Microcontrollers, such as the Raspberry Pi and Arduino, are commonly used in aquaponic systems. Murad et al. kept the automation and deployment local. Several sensors were used and controlled with an Arduino. The system was connected to a GSM (Global System for Mobile Communication) interface. An Arduino and a WRTnod were used by Wang et al. to monitor and manage an aquaponic system built from components that can be connected wirelessly via the internet. Temperature, humidity, light, water level, and DO sensors were built into the sensor acquisition module.

The data was sent wirelessly to a control and management center. Users were then able to analyse the information and make decisions with regard to the state of the air and water pumps, lights, and so forth. In their system, wireless sensor network devices (temperature, pH, and nitrate sensors) are connected to each other and monitored wirelessly via the Internet of Things (IoT) in a controlled environment. Complex control strategies found in the literature aim at achieving autonomous systems through a variety of smart techniques that go from linear regression to more complex prediction models such as neural networks. An IBM platform and cloud data storage system were used to monitor nitrate and pH in an aquaponic system. No discrete or event-triggered control strategies were found during the analysis, as authors tend more to apply distributed control systems (DCS). These missing strategies are key to the implementation of PLCs in such environments.

CURRENT LIMITATIONS

Some components of an aquaponics system, such as pH and water temperature, are being overlooked by the automation community. Research is necessary to provide practitioners with a thorough grasp of the consequences of each of the aforementioned criteria. The prosperity of numerous industries, including the automotive and aerospace sectors, has been impacted by the deployment of IoT technologies. Sensors that can offer real-time data on variables like temperature, pH, electroconductivity (EC), and relative humidity (RH) are used in aquaponic farming equipment to improve the productivity and quality of crops in greenhouses.

Commercially available monitoring options exist for linked aquaculture systems. These are capable of monitoring things like temperature, dissolved oxygen (DO), and pH, with the same limitations as before. Due to the high level of complexity and interconnectedness of the recommended systems, it is imperative to have a control system that is flexible enough to monitor the parameters in those systems. Water is the most important and challenging element of an aquaponic system. Aquaculture alone has been one of the fastest-growing areas of the food production industry during the past 30 years, growing at an average annual rate of 8.5%. The influence of aquaponics can be significantly increased by increasing the commercial potential of hydroponics. An important step in improving the water quality for fish and other aquatic life is the use of biofiltration devices in aquaculture. Since these levels and the availability of nitrogen are directly related to microbial activity, it is now required to monitor and control factors like pH, water temperature, and DO.

Examining the current state and global developments of aquaponics systems with a focus on the IoT, smart systems, and pertinent sensing metrics. The goal of this research is to bridge the gap between biological and electrical engineering expertise in order to advance aquaponics as a sustainable food source.

AI IN AQUAPONICS SYSTEM

The artificial intelligence in the aquaponics system is shown in Figure 6. In this chapter, a wireless sensor is created and installed in an aquaponics grow bed to

Figure 6. Artificial Intelligence in Aquaponics system

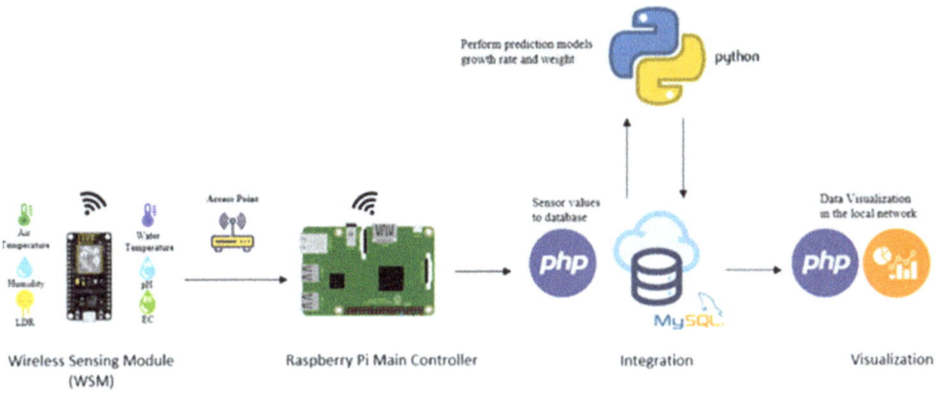

collect data on six distinct water quality and air quality factors. To encourage future implementations of intelligent algorithms and prediction tools, a framework to store and interact with online performance measurements was created. The creation and effective implementation of this work will encourage the development of reliable models to track and forecast the behaviour of aquaponics systems with less human involvement and pave the way for the adoption of smart technologies for optimal parameter autoregulation and precision farming.

Module Development and Experimentation

A robotic prototype that is capable of detecting six different environmental factors, including pH, water temperature, humidity, air temperature, and light intensity. The primary controller (Raspberry Pi), which uses an Arduino microcontroller, receives the data from the sensor wirelessly and stores it in a database there. The primary controller can calculate the growth rate and forecast the fresh weight of the crops using images captured by cameras mounted on the bed. As part of the project's continuous effort, the parts for the hydroponic module's development are being uploaded to the Google Earth database. Additionally, all of the images already gathered from the aquaponic environment are being uploaded to the server via the database. Liquid PH Value Detection using an Arduino UNO USB Microcontroller. The real time artificial intelligence in the real aquaponics setup is shown in Figure 7 (Pi et al., 2017; Rajkumar et al., 2017).

Figure 7. Artificial Intelligence in Real Aquaponics setup

IoT Experimental Setup

The WSM is designed to operate as an algorithm's food supply and to extract data in real-time. It has the ability to connect numerous modules that are installed at various locations and communicate data to a primary controller. The values from the sensors are gathered by an Arduino controller in the WSM every five minutes, and it transmits those values as unique values to the primary controller via the Wi-Fi antenna. The estimation of fresh weight and the assessment of growth rate are two concurrent actions that must be completed by this primary controller. The primary controller prepares and enters sensed values using PHP into a local database as soon as they are received.

The database can be accessed via an IP address and is made to display a unique ID and timestamp. An image from the database that was shown on the server is shown in Figure 7. The surroundings and the aquaculture tank's sensors will be utilised in the future. The main feature of the offered strategy is its ability to link online measurements (regular monitoring systems) to real-time performance indicators. The sensor values (which reflect the state of the systems as they currently exist) in this case are linked to the plant's performance in terms of growth rate and fresh weight, opening the door for further inquiry into the linkages. Here, by altering the existing control systems with the aid of real-time data, the feedback loop duration can be reduced and the process output can be enhanced. Future work may involve calibrating sensors to operate with less error caused by signal disturbance (Elangovan et al., 2018; Gagandeep et al., 2018).

Two cameras, artificial lighting, a humidifier, and sensors are all included in this hydroponic grow bed. Using MDF (medium density fibre) board, the device's frame was put together. The following sensors are employed: brightness, electroconductivity, water temperature, air temperature, and humidity. The main controller is a Raspberry Pi, while the sensor component is an Arduino Nano. The 4-channel relay module is wired to the humidifier, camera ring, and growth lights. To prevent issues, a 2-channel relay module is attached to the pH and electro-conductivity sensors' sensors. Additionally, an ESP8266 Wi-Fi sensor is put in the Arduino, enabling wireless data transmission to the primary controller (Raspberry Pi).

Twinning Interface

By including a visualisation tool in the digital twin process, the concept of a "digital black box" is eliminated, providing them complete management and oversight of the system at both the physical and virtual levels. The system's virtual graphical display is referred to as the "twinning interface" concept. It was created with Visual Studio, and in order to access the database and get the data it needs for the presentation, it is directly connected to it. Real-time sensed values can be retrieved every second thanks to serial communication, which is used to connect to the IoT sensing module. The present size of the crops on the physical side in real time is reproduced in the virtual environment through interactive and resizable plant reproduction models. The user is shown a digital representation of a plant's current growth status. Each plant's weight and rate of growth are shown on the display. An area interval corresponds to each level.

Feedback Loop

In this case study, we demonstrate how a virtual environment can communicate with its physical counterpart using the IoT Core connection in Visual Studio. The scripts are ready for the main controller in the physical system to switch on and off the system humidifiers, ring lights for the cameras, and growth lights as needed. Every choice made by the digital twin is recorded and emailed to the system's operator(J. et al., 2018; Mishra et al., 2018). The feedback controller for the aquaponic system is illustrated in Figure 8.

Figure 8. Feedback system for aquaponic system

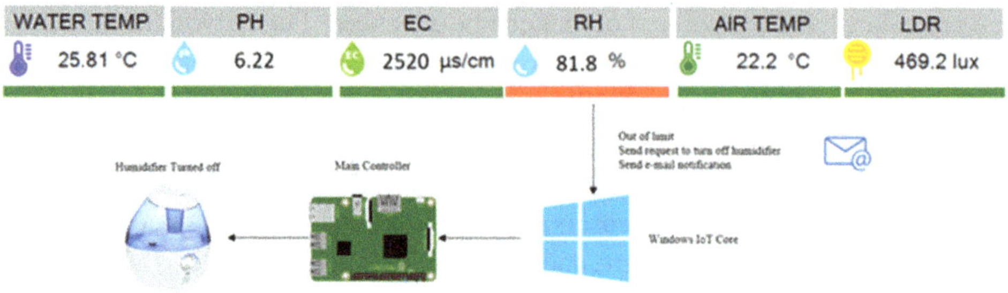

Feedback Control

The graph below demonstrates how, in each case, in the top view or side view, the pixel area retrieved from the computations is growing. With this approach, the user may turn on and off system actuators, such as lights and humidifiers, at whim and is always aware of the condition of the sensors and, as a result, the plants. The system can anticipate the weight of the plants and determine their growth rate by analysing live data practically instantly with KDD algorithms. According to a study, plants grow at an average pace of 28 percent in the morning and afternoon, but only at an average rate of 16 percent in the evening and at night. The water's pH measurements also seem to be a little bit higher in the morning. With the aim of optimal plant growth, various resource consumption and control strategies might be required for various times of the day. The relative humidity behaviour is shown to be higher at night, with values between 15 and 20 percent higher than in the morning(Parmar et al., 2019).

The aquaponics technology is converted to a general digital twin structure; however, the hydroponics beds are the only ones covered. Monitoring the number of fish (density in the tanks), evaluating their health, and taking into account the inherent linked parameters will undoubtedly expand the range of options for the management and improvement of the procedure. As a sustainable way for producing food indoors, aquaponics is gaining popularity. The integration of physical and virtual places has been made possible by the widespread use of digital twin technologies. The framework and application of the digital twin technology for an aquaponic system's hydroponic component are described in this study. The study provides a thorough foundation for digital twinning that will be enhanced to determine how well aquaponic systems function. To facilitate the commercialization of the aquaponics technology, it intends to offer precise performance metrics. The growth rate and

Figure 9. Various layers of the IoT modules in Aquaponics culture

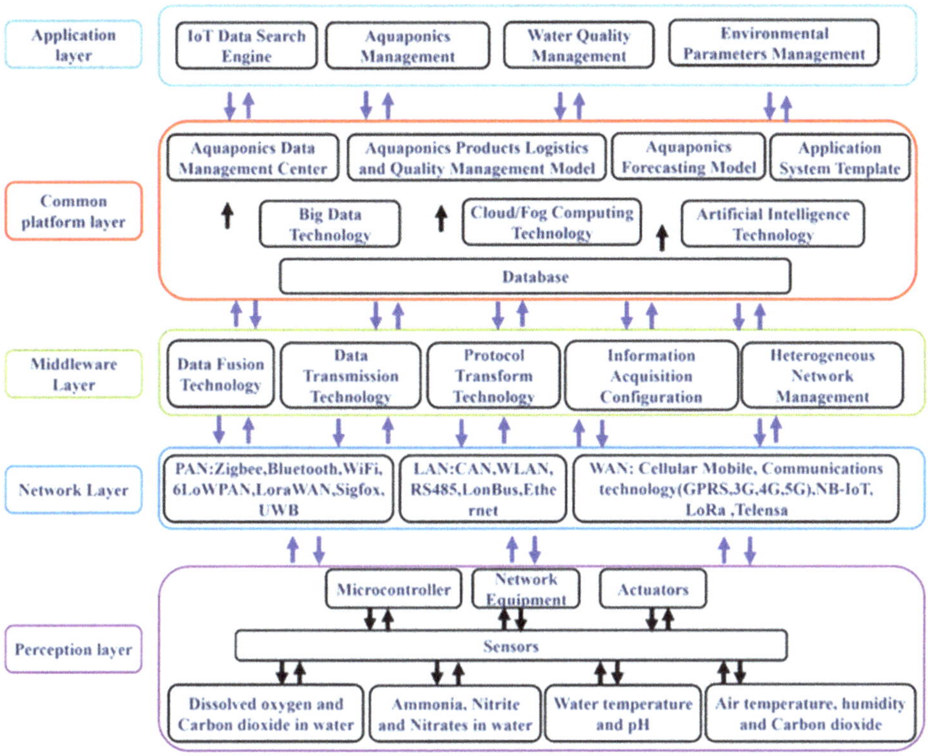

fresh weight of the fish and plants are also calculated using data-driven algorithms in the article (Ezzahoui et al., 2022).

A Digital Twin Framework and IoT for grow beds in Aquaponics Systems

The complete IoT-integrated aquaponic system and its layer-by-layer modules are illustrated in Figure 9. This study's goal is to put out a digital twinning framework that aquaponic systems can use generically. The systematic approach and experimentation in this situation are qualitative methodologies being used for this goal. This chapter's analysis is based on a thorough, systematic examination of the available literature on digital twins.

Aquaponics Digital Twin Framework and IoT

The creation of a digital twin framework aims to answer the requirement for a deeper comprehension of these connections. A complex farming technique in and of itself, aquaponics relies on symbiotic connections between its various components. Here, it is possible to observe how the elements of the general DT framework relate to the aquaponics ecosystem. Building things, machines, or procedures in a virtual world is frequently done with 3D modelling software. Every physical component as well as the virtual processes connected to it are all part of the physical and virtual surroundings. The physical environment, for instance, also consists of biological filters, pumps, feeders, aerators, humidifiers, sensors, etc. The parameters are an essential part of any digital twinning process and are particularly pertinent when virtualizing and simulating an aquaponic system. Aquaponics has a number of characteristics that must be met, including pH, air and water temperatures, dissolved oxygen, electroconductivity, total dissolved solids, salinity, flow, relative humidity, and light intensity (Kumar et al., 2023; Sampath et al., 2023; Taha et al., 2022).

The physical-to-virtual connection connects the real and virtual worlds in both directions. On the one hand, general tools are used to create procedures, transport and store obtained data, and manage internal operations. On the other hand, a number of IoT technologies, such as wireless modules, are now accessible to create reliable connections. In this example, to send and receive information and to be remotely controlled, the communication system will only use the Allen Bradley PLC's communication protocol from the ABB Company. If microcontrollers are used to regulate the physical environment, some of it can be avoided. The development of the connection tools is largely open source, and they may be used with open-source controllers like the Raspberry Pi and Arduino. Many times, aquaponics metrology is built using a variety of sensors for the parameters, cameras, mathematical models, and intelligent software that is installed in the grow beds and fish tanks. Changes occur in aquaponic systems because they involve sluggish dynamical processes.

CONCLUSION

- Aquaponics is a complex farming technique that relies on symbiotic connections between its various components. Every physical component as well as the virtual processes connected to it are all a part of the physical and virtual surroundings. The digital twinning process connects the real and virtual worlds in both directions.
- Systems can use pH test strips, freestanding sensors with an LCD panel, or analogue sensors that can transfer data wirelessly or not to a controller (PLC,

microcontroller, etc.). Automated sensing approaches must be assessed in order to develop a dependable, long-lasting, and financially viable system. The ideal range for fish containing TAN is 0–2 mg/L, and the ideal value for bacteria, specifically AOB and NOB, is 3 mg/l. The acidity or alkalinity of a solution is determined by a solution's pH level. Plant nutrient availability and nitrification rate are influenced by water pH. The ideal temperature for nitrification is between 17 and 34 °C.

- Water that is too hot can limit how much calcium can be absorbed by plants. The water level in tanks can be checked manually using a sight glass or floating objects, although there are several automatic measuring systems on the market. Aquaponics systems must carefully monitor the dissolved oxygen level because it can change significantly in a short amount of time. Temperature, pressure, and salinity have an impact on DO measurements, and some adjustments might be required. Due to interactions between oxygen molecules and the dye when DO is present, the returning wavelengths are constrained or changed.
- The Waterproof Monitoring System (WSM) is designed to extract data in real-time from sensors. Sensors are attached to a hydroponic grow bed and connected via Wi-Fi to a primary controller. The sensor values are linked to the plant performance in terms of growth rate and fresh weight. In this case study, we demonstrate how a virtual environment can communicate with its physical counterpart using the IoT Core connection in Visual Studio. The system's virtual graphical display is referred to as the "twinning interface" concept.
- Every choice made by the digital twin is recorded and emailed to the system's operator. The relative humidity behaviour is shown to be higher at night, with values between 15 and 20 percent higher than in the morning. Monitoring the number of fish (density in the tanks), evaluating their health, and taking into account the inherent linked parameters will undoubtedly expand the range of options for the procedure.

REFERENCES

Calone, R., & Orsini, F. (2022). Aquaponics: A Promising Tool for Environmentally Friendly Farming. *Frontiers for Young Minds*, *10*, 707801. doi:10.3389/frym.2022.707801

Das, S., & Sahu, G. (2020). Soil-Less Farming- An Innovative Way Towards Sustainability. *Agricultural & Food: E-Newsletter*, *2*(10), 82–85.

Elangovan, R., Santhanakrishnan, D. N., Rozario, R., & Banu, A. (2018). Tomen:A Plant monitoring and smart gardening system using IoT. *International Journal of Pure and Applied Mathematics*, *119*(March), 703–710. https://www.researchgate.net/publication/323811801_TomenA_Plant_monitoring_and_smart_gardening_system_using_IoT

Ezzahoui, I., Abdelhouahid, R. A., Taji, K., Marzak, A., & Ghanimi, F. (2022). The Aquaponic Ecosystem Using IoT and IA Solutions. *International Journal of Web-Based Learning and Teaching Technologies*, *17*(5), 1–15. doi:10.4018/IJWLTT.20220901.oa2

Gagandeep, A. D., & Saini, H. S. (2018). Design and implementation of an automatic irrigation feedback control system based on monitoring of soil moisture. *Proceedings of the International Conference on Inventive Computing and Informatics, ICICI 2017*, (pp. 540–544). IEEE. doi:10.1109/ICICI.2017.8365190

Immanuel, J., Patil, R., Sudheer, L. S., Parvathi, C. S., & Bhaskar, P. (2018). Web based Weather Monitoring System using Wireless Sensor Node. *International Journal of Computer Applications, 181*(11), 47–52. doi:10.5120/ijca2018917747

Kumar, P., Sampath, B., Kumar, S., Babu, B. H., & Ahalya, N. (2023). Hydroponics, Aeroponics, and Aquaponics Technologies in Modern Agricultural Cultivation. In Trends, Paradigms, and Advances in Mechatronics Engineering (pp. 223–241). IGI Global.

Mishra, D., Khan, A., Tiwari, R., & Upadhay, S. (2018). Automated Irrigation System-IoT Based Approach. *Proceedings - 2018 3rd International Conference On Internet of Things: Smart Innovation and Usages*. IEEE. 10.1109/IoT-SIU.2018.8519886

Parmar, B., Chokhalia, J., & Desarda, S. (2019). Terrace Garden Monitoring System Using Wireless Sensor Networks. *51 | Research. Journal of Engineering and Technology Management*, *02*(01), 2582–2610. www.rjetm.in/

Pi, R., Paul, S., Antony, A., & Aswathy, B. (2017). Android Based Home Automation Using Raspberry Pi. *International Journal of Advance Engineering and Research Development*, *4*(06), 143–147. doi:10.21090/ijaerd.58329

Rajkumar, M. N., Abinaya, S., & Kumar, V. V. (2017). Intelligent irrigation system - An IOT based approach. *IEEE International Conference on Innovations in Green Energy and Healthcare Technologies,* (pp. 1–5). IEEE. 10.1109/IGEHT.2017.8094057

Sampath, B., Sasikumar, C., & Myilsamy, S. (2023). Application of TOPSIS Optimization Technique in the Micro-Machining Process. In Trends, Paradigms, and Advances in Mechatronics Engineering (pp. 162–187). IGI Global.

Sathyanarayana, S. R., Gangadhar, W. V., Badrinath, M. G., Ravindra, R. M., & Shriramrao, A. U. (2022). Hydroponics: An Intensified Agriculture Practice to Improve Food Production. In Reviews in Agricultural Science (Vol. 10, pp. 101–114). doi:10.7831/ras.10.0_101

Taha, M. F., ElMasry, G., Gouda, M., Zhou, L., Liang, N., Abdalla, A., Rousseau, D., & Qiu, Z. (2022). Recent Advances of Smart Systems and Internet of Things (IoT) for Aquaponics Automation: A Comprehensive Overview. *Chemosensors (Basel, Switzerland)*, *10*(8), 303. doi:10.3390/chemosensors10080303

Yanes, A. R. (2020). *Towards Aquaponics 4.0 : A framework for automated, IoT and smart implementations in indoor farming*. SORACOM.

Chapter 7
Energy Efficiency in the Indian Scenario:
A Critical Review

Vartika Singh
https://orcid.org/0000-0003-4400-8664
Amity University, Noida, India

ABSTRACT

Energy is one of the most basic parts of the framework, and very essential for the national growth and economic development, of any country. The sustainable new sources of energy reproduction from traditional mediums like coal, lignite, petroleum, hydro, and nuclear capacity to non-traditional medium like wind, solar, agriculture, and home waste. Energy interest in the nation was prolonged very fast and is relied upon to rise further in the years to come. To fulfill the expanding need for energy, solar- or sun-based energy will play a real job in helping to this interest. The current scenario indicates that 60 GW of conventional power/energy capacity has been add over the last three years, which is believed to be the highest addition so far. As of April 2021, India has a thermal installed ability of 234 GW, and 53% of the part is covered by coal, diesel, and natural gas. The private sector produces 47.4% of thermal energy, although states and centres generate 27.1% and 25.5% individually. As of now, solar energy in India is exceptionally good.

INTRODUCTION

The advancement of sustained growth in this sector is vital to support the Indian economy. A developing country like India depends on the energy sector. If we compare

DOI: 10.4018/978-1-6684-4118-3.ch007

the historical data of the world, the energy sector India's energy sector is maximum expanded in the whole world. Energy is a crucial component in determining economic growth and development. Economic development and energy consumption, on the other hand, have a significant two-way link. On the one hand, an economy's global competitiveness is dependent on the availability of cost-effective energy sources, while demand for energy puts pressure on our natural resources, environment, and economy. Over the years, however, energy supply has failed to meet the ever-increasing demands imposed by a growing population, rapid urbanization, and progress F. Ahmad, M.S. Alam (2018).

Now a days in developing country like India, government pay special attention in the field of the energy sector. There has been an apparent increase in solar energy concept in the Indian scenario over a couple of years. Solar energy-based applications have helped a great many individuals in Indian towns by meeting their agriculture activities, cooking, household work, lighting, and other energy needs in a climate amicable way. Public Establishment of Solar Energy had been surveyed the whole country and found the result that solar-based ability of around 748 GW accepting 3% of the wasteland region to be covered by Sun powered PV modules. NSM is an important drive of the public authority of India with dynamic investment from States to advance natural practical expansion while tending to India's energy safety challenges. While, arising as a fundamental source of answers to meet the country's energy needs. Energy plays a major factor in economic development and hereby serves as a major step in the development. Energy efficiency refers to the use of less energy to achieve the same amount of energy output. As a result, it is one way to cut down on human-caused greenhouse gas emissions. As a result, energy efficiency is critical to establishing a future energy system that is safe, dependable, economical, and sustainable. In the energy resource, every country wants richness, and the most efficient and cost-effective approach to address energy security, environmental, and economic concerns.

In India, major energy worries remain due to fast rising energy costs, resource depletion, and environmental consequences. These concerns call for a nationwide energy conservation movement. Despite having 18% of the world's people, India wants to fulfill the demand of 6% of the world's major energy and has a per capita energy intake that is one-third of the worldwide regular. As a result, it is expected that as the country improves, with increased income and improved living standard, there would be a greater demand for energy, Sathaye J., and A. Gupta (2010). To meet the supply and the demand chain the country is promoting renewable sources of energy on the demand side where by on the resource side by promoting energy competence and preservation.

The Indian government has embraced a two-pronged method to meet its residents' energy needs while limiting CO2 emissions to ensure that worldwide emissions do not result in irreparable loss to the earth system. In different circumstances, determinations are made to well use of energy in the request side complete with several advanced policy actions with the Energy Preservation, Robichaud, L., Anantatmula, V., (2010). Under the different conditions, government are make several series steps for efficiently use energy in the generation side through larger usage of renewable energy in the fusion of energy, mostly through solar, wind, and coal-based power plants at the same time ever-changing towards supercritical skills.

Due to population expansion and economic development, India's energy consumption is increased at the wildest rates in the world in last few years. Due to population expansion and developmental and economic activities such as agriculture, industrialization, and urbanization, McKinsey & Company (2010). India's energy intake has enlarged at the fastest rates in the world. As a result of the dwindling reserves of fossil fuels, there is growing worry about energy consumption and supply, necessitating energy conservation by boosting total energy efficiency, with the following primary benefits, Phadke A., Sathaye J. and Padmanabhan, S. (2005):

Availability for future generations: Because non-renewable energy sources on Earth are limited in nature and take millions or billions of years to replenish, it is critical to conserve current supplies or use renewable energy sources to ensure that our natural resources are available to future generations.

To protect the environment: Consumption of nonrenewable resources has a direct impact on the environment, therefore consuming less energy protects our ecosystem from:

Pollution: The excessive use of fossil fuels contributes to air and water pollution for example: Burning of fossil fuels in car and other motor vehicle lead to release of carbon dioxide.

Greenhouse effect: Though greenhouse effect is the driving factor of living organisms on earth but the increase in amount of carbon dioxide has lead to excessive warming which is called global warming and led to many disastrous events, Centre for Environment Education CEE (2010).

Acid rain: When coal is burnt, Sulphur dioxide is also released into the atmosphere. Acid rain is formed when Sulphur dioxide interacts with water and oxygen in the clouds to create precipitation. Acid rain has the potential to harm fish and plants, as well as damage limestone structures and sculptures.

Other possible consequences include a threat to human health, environmental consequences such as increasing sea levels, which can devastate coastal regions, and dramatic changes in vegetation growth patterns, which might lead to the extinction of some plant and animal species.

ENERGY EFFICIENCY SCENARIO IN INDIA

India's major energy intake has been growing from over 450 million tons of oil equal in 2000 to almost 770 million in 2012. In 2030, it is anticipated to rise to around 1250 million toe (as per the International Energy Agency) to 1500 million toe (according to the Integrated Energy Policy Report). This rise is due to a variety of causes, the most prominent of which are rising economic growth and wages, which result in increased request for energy facilities like light, cooling, cooking, transportation, industries, and office computerization, among others, US Department of Energy (2008). This expansion reflects India's present low energy supply in 2011 and the normal yearly energy supply of India was just 0.6 toe per capita, compared to the world usual of 1.88 toe per capita. It's also worth noting that no nation has ever been capable to reach a Human Growth Index of 0.9 or higher without having at least 4 toe of yearly energy supply per capita. As a result, there is a significant unmet request for different energy facilities that must be met in order for people to acceptable wages and a fair standard of living.

During the **fiscal year 2016-17**, India's energy saving initiatives saved 10 GW of electricity, Prayas Energy Group (2005) during peak hours. These savings have primarily come from two of the National Mission on Enhanced Energy Efficiency's many programs: the **'Ujala scheme**,' which aims to save energy by swapping glowing bulbs with LED, and the **'Perform, Achieve, Trade (PAT) scheme**,' which encourages specific industries to implement energy-saving actions.

According to the Power Ministry's data, energy savings under the Ujala and PAT schemes were 5,905 MW and 5,635 MW, respectively.

After the United States and China, India is the third biggest creator of greenhouse gases and one of the countries maximum vulnerable towards climate change. As part of its obligations to the United Nations Framework Agreement on Climate Change, which was approved by 195 nations in Paris, in 2015, India intends to decrease its carbon footprint by 33-35 percent by 2030 compared to 2005 levels.

Around 50 percent of India's INDC (Intended Nationally Determined Contribution) emission intensity reduction objectives from the baseline value would come from energy intensity.

REGULATORY FRAMEWORK

In the 52 Year of the Republic of India, on October 1, 2001, the Indian Government passed successfully Energy Conservation Act – 2001, it is work like a bridge to fulfill the gap between demand and supply, significant energy saving potential and benefits, and minimizing environmental impacts by conserving and utilizing it properly. Except for the state of Jammu & Kashmir, this Act applies to the whole country of India. It establishes a legal and administrative framework for energy preservation. The Bureau of Energy Efficiency (BEE) created on March 1, 2002, under the terms of the Act, by combining the old Energy Management Centre of the Ministry of Power, Government of India. The BEE will be in charge of implementing policy programs and coordinating energy saving initiatives across states and India.

The Energy Conservation Act 2001

The Energy Conservation Act of 2001 focuses on:

Reducing energy usage through conservation methods and efficiency.
Reducing the need to build new capability saves resources and reduces greenhouse gas emissions.
Ensure ecologically friendly and long-term growth
Encourage market shifts toward energy- saver products.
In 2002, the Bureau of Energy Efficiency (BEE) was established as the main organization at the federal level, with State Designated Agencies (SDAs) at state level to carry out the Act's provisions.
National Mission for Enhanced Energy Efficiency (NMEEE)

National Mission for Enhanced Energy Efficiency (NMEEE) is known as National Action Plan on Climate Change's eight national missions. The Ministry of Power and BEE have developed a framework for scaling up efforts to build and sustain an energy efficiency market. About the next five years, the Mission is anticipated to save roughly 23 million tons of oil-equivalent fuel in natural gas, coal, and fossil fuel goods, as well as an avoided capacity expansion of over 19,000MW. Carbon dioxide emissions are expected to be reduced by 98.55 million tons per year (BEE, 2010). The following is the implementation structure for the NMEEE's four initiatives:

- **Perform Achieve and Trade (PAT)** it is a system which is created on market and certifies energy savings that may be exchanged in order to expand energy productivity in "Designated Consumers" (big energy-intensive enterprises and buildings).
- **Market Transformation for Energy Efficiency (MTEE)** is a program that accelerates the transition to energy-saving equipment in specific industries by using innovative techniques that make the goods more reasonable.
- **The Energy Efficiency Financing Platform (EEFP)** aims to kickstart the development of mechanisms that will aid in the financing of requirement

based management programs across all sectors by catching upcoming energy savings.

- **Energy Efficient Economic Development (FEEED)** foundation will create two fiscal devices to encourage energy productivity: The Venture Capital Fund for Energy Efficiency (VCFEE) VCFEE BEE (2015) and Partial Risk Guarantee Fund (PRGF).

LEAD ORGANIZATIONS

The organizations working towards the Energy Conservation in India are:

- **The Ministry of Power** controls the Energy Conservation Act, 2001, as well as any modifications to this Act that it is very necessary from time to time to meet the Government's policy objectives. It began operating independently on July 2, 1992, after previously being known as the Ministry of Energy Sources.
- **Bureau of Energy Efficiency (BEE):** BEE (2004) was founded as a central statutory entity below the terms of the Act, by combining the old Energy Management Centre of the Ministry of Power, Government of India. BEE's mission is to:

  Promote energy efficiency by enforcing and monitoring the Energy Conservation Act of 2001.

  Develop and suggest minimum energy performance requirements for equipment and appliances, as well as labelling design.

  Specify the method for certification and accreditation

  Create and promote specialized energy-saving construction codes.

  Recommend labelling any energy user or group of users as "designated consumers."

  Establish Energy Service Companies (ESCOs).

  Take the required steps to raise alertness and spread information about energy efficiency and conservation.
- **The Ministry of New and Renewable Energy (MNRE)** is the Government of India's main ministry for all novel and renewable energy subjects, MNRE (2007, 2008). The Ministry's overall goal is to create and deploy new and renewable energy to augment the country's energy needs.
- **Productivity Council of the United States (NPC)** The National Productivity Council (NPC) is a national body that promotes productivity culture in industry, agriculture, service, infrastructure, and other economic sectors. NPC's definition of productivity includes quality, environmental protection,

and integrated social and economic development, as well as more effective use of energy resources, conservation, management, and auditing.

- **The Energy Resources Institute (TERI)** develops the Green Rating for Integrated Habitat Assessment (GRIHA) for buildings and is working on creative solutions for a sustainable future TERI (2003-04). It also inspires the usage of renewable energy sources.
- **The Petroleum Conservation Research Association (PCRA)** is a government organization in India that promotes energy efficiency and conservation in all aspects of life. PCRA has conducted mass media campaigns in the past using television, radio, and print media. According to a third-party impact assessment survey, overall awareness has increased as a result of PCRA's major campaigns, resulting in the conservation of fossil fuels and the reduction of pollution, Canning, P., S. Rehkamp, A. Waters and H. Etemadnia (2017).
- **The Confederation of Indian Industrial (CII)** strives to develop and maintain an environment in India that is favorable to industry growth, collaborating with both industry and government through advisory and consultative procedures. Through a variety of specialized services and worldwide links, CII collaborates with government on policy problems, improving energy productivity, affordability, and increasing commercial possibilities for industry.
- **The PHD Chamber of Commerce and Industry** is a proactive and vibrant grassroots organization with significant national and international ties. The Chamber helps to promote energy saving by acting as a catalyst. The Chamber maintains an Energy Management Cell (EMC) that provides professional advice and support on energy audits in industrial units as well as energy conservation recommendations. 'Energy conserved is energy created,' is the EMC's selected motto.
- **The Federation of Indian Chambers of Commerce and Industry (FICCI)** works on Resource Conservation and Management (RCM) as one of its initiatives to promote and provide integrated services to industries in order to improve their competitiveness and productivity, particularly through process optimization and improvements, sustainable resource use (raw materials, energy, water, etc.) and effective management. It also undertakes energy audits.

ENERGY EFFICIENCY: AN ENERGY CONSERVATION TOOL

Energy efficiency is a strong and affordable instrument for creating a long-term energy future. It provides added services for the same or less energy input, for example, utilizing a compact fluorescent light (CFL) bulb instead of an incandescent bulb is regarded more energy efficient.

Comparison of Different Types of Bulb

BEE has launched India's first and largest program activity, the "Bachat Lamp Yogna," to promote the adoption of energy efficient CFLs at the household level. BEE have been established Standards and Labeling (S&L) for the equipment/appliances mentioned in the accompanying text, keeping this in mind.

UJALA (LED replacement program) is an example of a flagship initiative that has been effectively implemented using market mechanisms.

STANDARDS AND LABELLING (S&L) FOR EQUIPMENTS

BEE established the Standards and Labeling (S&L) initiative to encourage energy conservation and efficiency by displaying an energy label directly on appliances on May 2006. The Requirements and Labeling program aims to establish least energy performance principles for new energy equipment, uses and machines. It shows:

Energy efficiency levels we represent the levels with different color stars ranging from One Star (minimum energy and money saver) to Five Stars (maximum energy and money saver).

Product information:

- Yearly electricity consumption unit
- Name of appliance • Brand name
- Model name, number, and year of manufacture
- Type
- Gross volume

Bee Star Labeled goods save power expenses and make electricity more accessible to more people.

There are now 12 goods covered by this scheme, four of which are required to be labelled. Frost-free freezers, air conditioners, tubular fluorescent lamps, and transformers are just a few examples. Direct-cool freezers, agriculture pump sets, induction motors, ceiling fans, LPG stoves, electric geysers, color television sets, and washing machines are among the other goods included. Several additional items, as if as personal computers, laptops, and automobiles, are in the process of being implemented (BEE, 2010).

ENERGY EFFICIENCY IN AGRO SYSTEM

In country like India, energy efficiency is an important property of agricultural systems. In the processes along with the value chain, are heavily reliant on the use of gas and petroleum. In the production of agricultural, we directly or indirectly use energy in -intensive manufactured fertilizers, pesticides, tractors and tillage for operating various irrigation systems. It is clear as the ratio of energy out to energy in the system outputs in the food we eat.

If we change the old agriculture methods with the new techniques, we can save energy by suitable technology changes, and improved operations and management system. Diverse types of agriculture systems vary widely in energy efficiency. The energy that the system consumes in order to produce the food has important implications.

ENERGY EFFICIENCY IN BUILDINGS

Finding out where the most energy is consumed in the building is the first stage in an energy efficiency strategy. An energy audit can help you discover the areas with the highest energy expenditures and the most cost-effective ways to reduce them.

We can attain energy efficiency by following ways:

- Use of single glazed recessed windows to enhance indoor thermal comfort.
- Use of water bodies as an evaporative cooling approach
- Plenty of natural light accessible in various places

Buildings might save 10-15% on their energy costs by installing energy efficiency measures, which not only make the atmosphere more comfortable but can also pay off financially in the long run. The recommended energy-saving measures may be used throughout the structure, from the roof, walls, and insulation to the appliances and lighting inside.

Buildings may save 10-15% on energy expenses by implementing energy efficiency measures, which not only make the environment more comfortable but can also save money in the long term. The proposed energy-saving techniques may

Figure 1. Energy Efficiency in an Agricultural System

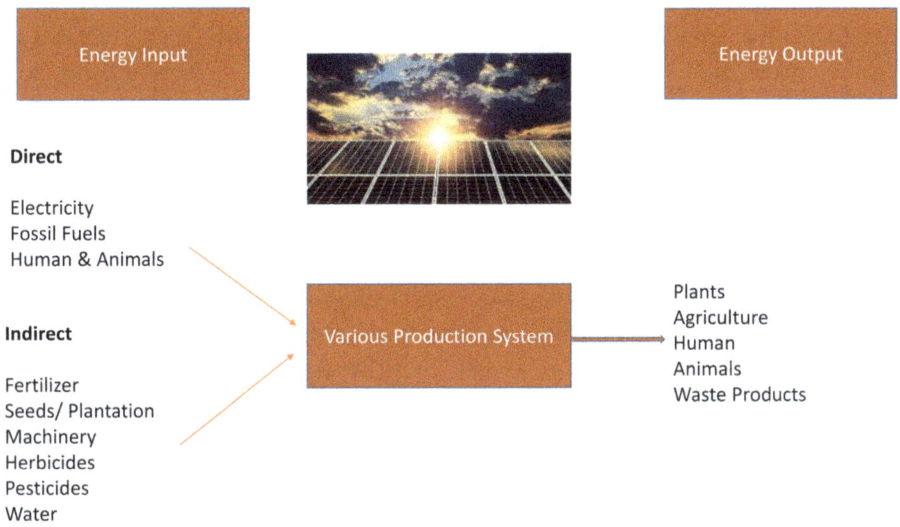

be applied to the entire building, including the roof, walls, and insulation, as well as the appliances and lights on the interior.

High Performance Buildings

Buildings with integrated bioclimatic solar passive architectural design methods and energy efficiency measures, as suggested by ECBC, are classified as high performance buildings (Energy Conservation Building Code). The following Energy Conservation Measures (ECMs) are implemented:

- Passive energy-saving techniques to reduce energy usage and waste.
- Affects the building's lifespan cost.
- Applied to four key areas: building design, building envelope, light, and HVAC (heating, ventilation and air conditioning).
- They were chosen based on the amount of money they could save, the cost of implementation, and the time it would take for them to pay for themselves.
- Cost-effective in terms of electricity and carbon dioxide emissions.

The Energy and Resources Institute (TERI), in collaboration with the Bureau of Energy Efficiency (BEE) of the Indian Power Ministry and White Box Technologies of the United States, has launched a project called "High Performance Commercial Buildings in India" to make profitable buildings in India more energy efficient.

ENERGY CONSERVATION BUILDING CODE (ECBC)

Under the Government of India, the Bureau of Energy Efficiency released the Energy Conservation Building Codes (ECBC) in 2007, which establishes least energy performance requirements for design and structure of major profitmaking constructions.

At the design stage, the Energy Conservation Building Code evaluates for energy efficiency. This is a fantastic idea that will allow high-performance buildings to be designed. It takes into account five climate zones in India (dry and hot, humid and warm, temperate, cold and composite) and establishes minimum energy performance criteria for big commercial buildings or complexes with linked loads of 500 kW. All structures with a acclimatized floor space of 1,000 m2 or more are subject to the code. The code is also suggested for all other structures (HREDA, 2008).

The provisions ECBC apply to:

A number of evaluation programmes have been established across the world to analyse the environmental and energy consequences of buildings. The Environmental Assessment Method of Building Research, the first environmental certification method, was well-known in the United Kingdom in 1990. (BREEAM). The Leadership in Energy and Environmental Design (LEED®) Green Building Rating Methodology, are based on the BREEAM system, was established in 1998. Green Globes was created in 2005 by the Green Building Initiative (GBI), which modified the Canadian version of BREEAM and distributed it in the United States.

Following are the green building rating systems developed in India:

LEADERSHIP IN ENERGY AND ENVIRONMENTAL DESIGN (LEED) is a global green structure grading method created by the US Green Building Council and has been used over twenty-four different countries, including India, for the construction, strategy, and management of green structures.

The Governance in Energy and Environmental Design (LEED-INDIA) Green Building Rating System is well established national and global recognised as standard for the high-performance green building construction, design, and operation.

Building holders, consultants, architects, developers, facility managers, and project managers all benefit from LEED-tools INDIA's for designing, constructing, and operating green buildings. It encourages a whole-building style to sustainability by rewarding achievement in the five major categories listed below:

Sustainable site planning, energy efficiency, water conservation, selection of material, and quality of indoor environmental are all factors to consider.

For each structure type and step of a building's lifespan, the LEED-INDIA grading system provides a path for evaluating and reporting performance. When compared to a typical building, a LEED-certified structure uses 30-50 percent less energy. LEED India for New Construction (LEED India NC) and LEED India for Core and Shell are two specific LEED-INDIA initiatives (LEED India CS)

Green Rating for Integrated Habitat Assessment (GRIHA), the national rating mechanism was designed by TERI (2015) and settled in collaboration by the Ministry of New and Renewable Energy of the Government of India. It's a green building design valuation method' that works for all types of constructions which is suitable for all climate zones through the nation. The grading system's main goal is to aid in the design of green buildings and, as a result, to aid in the evaluation of their 'greenness.' The rating system adheres to best practises as well as national and international regulations that are relevant to attaining the goal of green design.

GRIHA has a total of 100 points spread among 34 criteria, which may be gained by following assurance of the criterion to the point is assigned. The amount of points gained determines the degree of certification (one star to five stars). Certification requires a minimum score of 50. One , two, three and four stars will be awarded to buildings earning 50 to 60, 61 to 70, 71 to 80, and 81 to 90 points, respectively. The maximum rating, five stars, is given to a structure that receives 91 to 100 points.

GRIHA Rating System

The GRIHA grading system offers the following advantages on a macro scale:

Reduced energy usage while maintaining human comfort levels.
Integration of solar energy devices.
Waste recycling and reuse are important.
Reduced pollution in the air and water.
Reduced deforestation and loss of biodiversity in natural areas.
Reduced soil loss due to erosion.
User productivity has increased.
Image and marketability improvements

Solar Passive Building Concept

One aspect of green building design is the Solar Passive Building Concept. The utilization of the sun's energy for heating and cooling areas is stated to as passive solar design. The structure, or a portion of it, uses inherent energy qualities in ingredients and air generated by exposure to the sun in this technique. Passive systems are very simple and small amount of moving components, need little maintenance, and do not require mechanical systems.

Elementary passive solar building design elements

- Extend the building dimension beside the east and west axis
- Turning the structure to face the equator
- Appropriately sized windows that face the noon sun in the wintertime and are cool in the summer, as well as reducing windows on other sides, particularly western windows
- The south face of the structure should be used for interior rooms that require the greatest light and heating and cooling. Spaces that aren't used as much should be situated in the north.
- Installing overhangs or shade components that are the right size for the latitude.
- Using the right quantity and kind of insulation, such as radiant barriers and bulk insulation, to prevent excessive heat gain or loss during the winter.
- During the winter, consuming thermal mass to store extra solar energy (which is then re-radiated during the night).
- Passive system functioning is aided by an open floor arrangement.

Bio-Energy Applications

1. Bioenergy as an alternative source of energy

The carbon dioxide (CO2) delivered from biomass during creation of bioenergy is from carbon that flows the climate in a circle through the course of photosynthesis and disintegration. In this manner, creation of bioenergy doesn't contribute additional CO2 to the climate like non-renewable energy sources.

2. Bioenergy in Renewable energy sector

Bioenergy is an environmentally friendly power that can create numerous extra advantages, the degree of which relies upon a mix of elements including the kinds of feedstocks utilized, how they are delivered and shipped and the effectiveness of the advances conveyed to change them over to bioenergy.

3. Bioenergy in Heat and Electricity sector

Dissimilar to most other environmentally friendly power sources, bioenergy can create both heat and power in a combined heat power (CHP) plant. This would then be able to be utilized for a scope of heating and cooling applications in industry, or for little networks.

4. Bioenergy in Agriculture food-preparing sector

Utilizing biomass can assist with building strength in agriculture, wood and food-processing businesses. Bioenergy provides a use for their waste streams, can assist them with their energy costs and conceivably add another income stream on if they can sell bioenergy-oriented heat and help them with green power provision.

5. Bioenergy for rural and regional dependability and security

Rural and regional energy dependability and security can be upgraded by giving a homegrown energy source that can run persistently, or at busy times as needed by the power market, with more prominent adaptability to increase creation at short notification than huge coal-terminated plants.

6. Bioenergy as an alternative for forest burning

Bioenergy creation can use as an alternative for burning of forest. Mechanical thinning and biomass expulsion for bioenergy can be utilized as a technique to reduce risky fuel levels, particularly in regions where the expense and dangers related with endorsed burning are high.

Water quality advantages have additionally been recorded where fuel decrease consuming is supplanted with biomass harvesting. Thus, biomass harvesting and removal for bioenergy and other small wood applications is a method broadly utilized in woodlands and forests.

Way Forward

India's market would-be good for energy efficient goods is estimated to be Rs. 1.6 lakh crore, according to the World Bank. According to their estimates, the top three demand side management (DSM) markets subsidizing to this potential are domestic end-use appliances, agriculture/irrigation pumps, and municipal infrastructure. With a renewed focus on DSM, it is possible to save 178 billion units of electrical energy each year, resulting in a reduction of 150 million tons of CO2 emissions per year.

Electricity consumption will grow as urbanization accelerates, bringing the urban population to 600 million or more by 2030. The BEE has established the Energy Conservation Building Code (ECBC) are constantly working to promote the concept of energy efficiency in the building sector. ECBC is now in a voluntary phase but will be made mandatory following capacity building activities in rural area, remote area and implementation experience. Building Envelope, Lighting, Air Conditioning System, Heating Ventilation, Solar Hot Water Heating, and Electrical Systems all have minimum performance requirements set by the ECBC.

The industrial sector offers substantial energy reduction potential. To target this sector, BEE (2008) introduced the Perform, Achieve, and Trade (PAT) system under the National Mission on Enhanced Energy Efficiency (NMEEE), which completed a cumulative energy savings of 8.67 million toes against a target of 6.686 million toes in the first cycle of PAT, which ran from 2012-13 to 2014-15. The PAT-1 cycle encompassed 478 designated consumers (DCs) from eight different industries. Between 2016-17 and 2018-19, a goal of 8.869 million toe has been set for the PAT-2 cycle, which will encompass 621 DCs across 11 sectors, with the inclusion of three new areas (refineries, railways, and electricity distribution companies).

BEE has developed various schemes and programs to promote energy efficiency in urban area, industries, and new machines, as well as capacity development programs for all stakeholders involved in energy efficiency, all under the umbrella of the EC Act 2001 and the NAPCC.

BEE has launched a number of initiatives aimed at the following areas:

- Lighting in the Home
- PAT (Perform, Achieve, and Trade) for Big Industries
- Commercial Buildings • Appliance Standards & Labeling
- Municipalities' Demand-Side Management
- Management Program for Agriculture Demand
- SME Program
- DISCOM capacity building
- SDA capacity building
- Energy Conservation Fund for the State (SECF)

In 2006, the BEE launched its Standards & Labelling initiative for equipment and appliances to give consumers more information about the energy-saving and hence cost-saving possibilities of marketed products. Some appliances are required to participate in the program, while others are optional. Appliances have a star rating system that may be compared. The higher the quantity of stars, the more energy efficient the equipment. According to a study by Dr. Kirit S. Parikh and Dr. Jyoti K. Parikh, valued savings in domestic power intake from just four appliances (air conditioners, refrigerators, televisions, and ceiling fans) range from 52 kWh to 145 kWh in 2030, with CO2 drops ranging from 42 Mt to 116 Mt.

Niti Aayog, India's policy think tank, will soon rank states based on their energy efficiency, a step that would help India meet its worldwide goal to reduce energy usage. The key to a successful implementation of all of these energy productivity policies and programs would be effective coordination among all of the diverse stakeholders involved in the development, implementation, and enforcement of these policies and program, which all take place at different levels, Punjab ENVIS

Figure 2. Benefits of Energy Efficiency and Renewable Energy

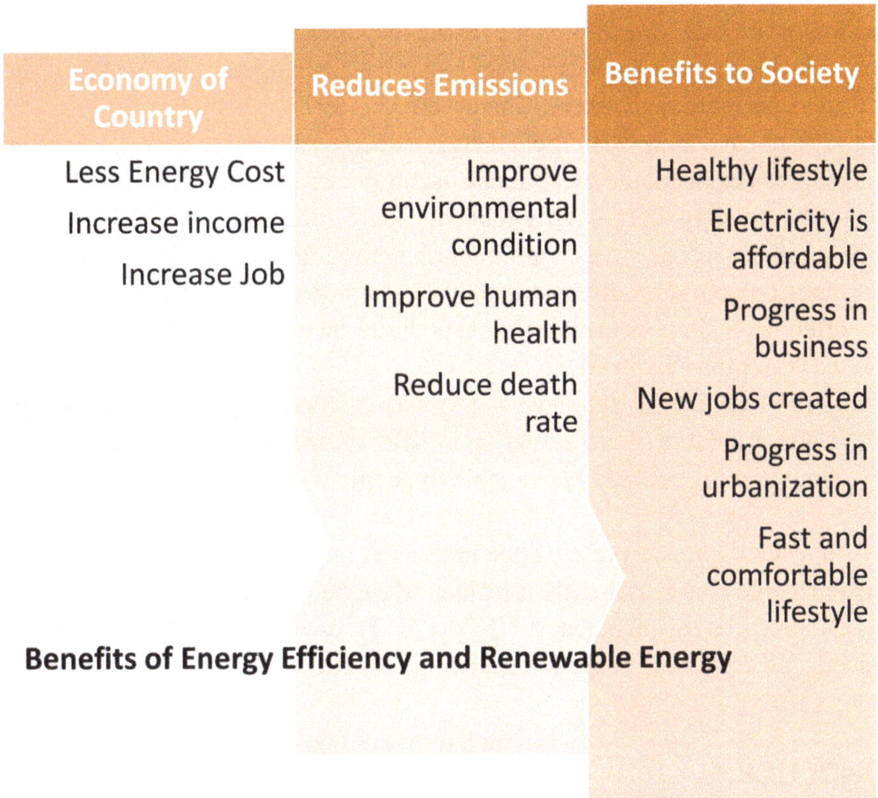

Centre (2008), Gangwar, K.K. and Singh, R. (2010). This would encompass various ministries of the Indian government, government of state, citizens, industries, and so on, all of whom would be required to share possession.

Despite numerous efforts/initiatives, the rate of change for energy efficiency requires a strong push to overcome hurdles that continue to stymie India's market transformation for energy efficiency. The following are some of these impediments:

Price distortions, such as subsidized energy (including electricity) prices, particularly for agricultural. People have less incentives to limit and/or reduce their consumption when prices are low. This increases the fiscal strain on the country's budget. As a result, the recent deregulation of gasoline is a step in the right direction.

In the absence of knowledge was a stumbling block in the beginning, the government's efficient tack helps of mass media, in which print and electronic media, play important role to spread energy efficiency message to the common man.

Economic developments, on the other hand, have benefited urban and semi-urban areas far more than others, Chaturvedi, P.(1997).

Rural regions have energy saving challenges in the agricultural sector due to inefficient methods exacerbated by an inconsistent power supply. Non-technical losses, such as energy theft, have remained a source of worry.

While small-scale conservation programs in households require little or no financial input, comparable initiatives in industrial, construction, and some other sectors often necessitate considerable expenditure owing to the scope and nature of the operations involved.

At both the federal and state levels, a variety of funding programs are available. While it is too early to assess their efficacy, there is worry that information about them is not readily available to people who may benefit from it.

As a result, there is a pressing need for better understanding of the link between climate change concerns and their influence on the national economy and society, with a focus on energy efficiency, UNDP (2011). It will take a lot of consumer-pull to increasing demand for energy-efficient goods and services to have a big impact.

CONCLUSION

The Mission's goal is to stablished India as strong innovator in field of solar powered energy by creating the sustainable method for solar based new creation. The Mission objectives hosting 100 GW network associated sunlight-based force plants continuously 2022. India's Intended Nationally Determined Contributions (INDCs) focus to accomplish around 40% combined electric force introduced limit from non-petroleum product-based energy assets and to lessen the emanation power of its Gross domestic product 33 to 35 % from 2005 to 2030.

To accomplish the given target, Legislature of India have dispatched different plans to empower age of solar-based force in the nation like VGF Schemes, CPSU Scheme, Solar Park Scheme, Bundling Scheme, Defence Scheme, Canal bank Scheme, Grid Connected Solar Rooftop Scheme etc.

REFERENCES

Ahmad, F. & Alam, M. S. (2018), Economic and ecological aspects for microgrids deployment in India. *Sustainable cities and society*, *37*, 407-419

Bureau of Energy Efficiency (BEE). (2004). ESCO: Cashless Alternative for Energy Conservation. *IEEMA Journal.*

Bureau of Energy Efficiency (BEE). (2008a). *Awards Booklet: Municipalities.* BEE. www.emtindia.net/eca2008/AwardBooklet2008/f_Municipalities.pdf

Bureau of Energy Efficiency - BEE. (2010). *BEELINE, Quarterly Newsletter*, 1. Bureau of Energy Efficiency.

Bureau of Energy Efficiency (BEE). (2008). *Roadmap for Energy Efficiency in Buildings.* Australia- India Technical & Policy Exchange Power.

Canning, P., Rehkamp, S., Waters, A., & Etemadnia, H. (2017), *The Role of Fossil Fuels in the U.S. Food System and the American Diet, Economic Research, Economic Research Report no. 224.* United States Department of Agriculture.

CEA. (2016, December 30). *Draft National Electricity Plan.* CEA. http://www.cea.nic.in/reports/committee/nep/nep_dec.pdf

Central Energy Regulatory Commission (CERC). (2010). *The Electricity (Supply) Act, 1948.* Government of India. gov.in/ElectSupplyAct1948.pdf

Centre for Environment Education – CEE (2010). *Low Carbon Life Styles - A trainers Guide to low carbon life styles.* CEE.

Chaturvedi, P. (1997). *Energy Management: Policy, Planning Utilization.* Concept Publishing Company Ltd.

Gangwar, K. K., & Singh, R. (2010). *Guidebook on Energy Education for Students, Published by Punjab ENVIS Centre, Punjab State Council For Science & technology.* Chandigarh and Punjab Energy Development Agency.

Garv. (2017, April 03). *Rural households electrification.* Garv.gov: http://garv.gov.in/dashboard/ue

HREDA. (2008). *Energy Conservation Hand Book for Buildings and home.* Haryana Renewable Energy Development Agency.

Jerath, N. Saxena, S.K., gangwar, K.K. & Singh, R. (Eds.). (2008). Renewable Energy. Punjab ENVIS Centre Newsletter on Status of Environment and related Issues, 6(2-3), 32. Punjab State Council for Science & Technology.

McKinsey & Company. (2010). India's urban awakening: building inclusive cities, sustaining economic growth. McKinsey.

MNRE. (2007). In: Akshay Urja. Ministry of New & Renewable Energy Government of India, 3(1), 48.

MNRE. (2007-08). In *Annual Report, Published by Ministry of New & Renewable Energy* (p. 112). Government of India.

MOEF. (2010). *India: Greenhouse Gas Emissions 2007*. Ministry of Environment and Forests, Government of India.

Phadke A., Sathaye J. and Padmanabhan, S. (2005) Economic benefits of Reducing Maharashtra's electricity shortage through end-use efficiency improvement. *LBNL Report 57053*.

Prayas Energy Group. (2005). Demand-side Management (DSM) in the Electricity Sector: Urgent Need for Regulatory Action and Utility-Driven Programs. Prayas Group.

Robichaud, L., & Anantatmula, V. (2010). —Greening Project Management Practices for Sustainable Construction||. *Journal of Management Engineering*, *27*(1), 48–57. doi:10.1061/(ASCE)ME.1943-5479.0000030

Sathaye, J. & Bouille, D. (2001). Barriers, Opportunities, and Market Potential of Technologies and Practices. In Metz, O. Davidson, and R. Swart Cambridge (eds.) Climate Change Mitigation. BUniversity Press for the Intergovernmental Panel on Climate Change.

Sathaye J. & Gupta, A. (2010). Reducing Electricity Deficit through Energy Efficiency in India: An Evaluation of Economic and Emissions Benefits. *LBNL Report*.

Sivaharan, T, Blair, G., & Coulson, G (2005). GREEN: A Configurable and Re-configurable Publish Subscribe Middleware for Pervasive Computing. *Lecture Notes in Computer Science*. Springer

TERI. (2015). *Energy Security Outlook: Defining a secure and sustainable energy future for India*. TERI Press.

TERI (*2003-04*). Beacon, Published by The Energy and Resource Institute.

UNDP (2011). *Improving energy efficiency in the Indian Railways System*. UNDP India.

US Department of Energy. (2008). Energy Savers. National Renewable Energy Laboratory, a DOE National Laboratory, Pp.32.

VCFEE BEE. (2015). Venture Capital Fund for Energy Efficiency (VCFEE). BEE India. https://beeindia.gov.in/sites/default/files/VCFEE.PDF

Chapter 8
Sustainable Developments of Hybrid Floating Solar Power Plants:
Photovoltaic System

Vinay Kumar Domakonda
Vignan's Foundation for Science, Technology, and Research, India

Shaik Farooq
Vignan's Foundation for Science, Technology, and Research, India

Sekhar Chinthamreddy
Narasaraopeta Engineering College, India

R. Puviarasi
Saveetha School of Engineering, Saveetha Institute of Medical and Technical Sciences (SIMATS), India

M. Sudhakar
Sri Sai Ram Engineering College, India

S. Boopathi
https://orcid.org/0000-0002-2065-6539
Muthayammal Engineering College, India

ABSTRACT

In this chapter, the various sustainable development activities of hybrid floating solar photovoltaic (FSPV) technology with various emerging technologies are discussed. The basic principles, benefits, and important components of emerging FSPV techniques are illustrated. The developments of various environments, recent developments, classifications, and performance improvements by integrating various existing renewable energy resources have been elaborated. FSPV hybridization with hydro-power plants, hydro-storage, offshore, wave, and air energy storage are also demonstrated using ground-mounted, floating, and tracking integrated technologies.

DOI: 10.4018/978-1-6684-4118-3.ch008

INTRODUCTION

Energy demand has increased steadily and reflentlessly as a result of population growth. Between 2020 and 2040, the world's energy consumption is anticipated to rise by 30%. Although large-scale solar photovoltaic systems need a lot of area to function, their land use is comparable to or even less than that of conventional energy sources. Plants for floating photovoltaics (PV) occupy less land that could be used for other, more lucrative uses. While coal power operations have an environmental impact, those of a PV power plant don't. Farmers who didn't want to replace their crops with PV panels mounted the first floating PV plants on irrigation tanks. One of the most extensively used solar PV technologies in use today is solar energy. With no moving parts, solar cells function close to room temperature and enable generation of any size. Theoretically, a 10-square-kilometer and a 10-square-meter array are equally efficient per unit area (Sreenath et al., 2021).

A lot of off-grid systems also use batteries and charge controllers to store energy during the day and supply on-demand power at night. Low operating costs (no fuel), no wearable moving parts, and high temperature operation (no corrosion or safety issues at high temperatures). Low starting costs, high initial (installed) costs, and erratic hourly or daily output are advantages. There are drawbacks, such as a lack of cost-effective energy storage and a shaky level of long-term predictability. Farmers who lack the acreage to dedicate to a PV system may be able to attain energy independence by installing photovoltaic panels floating on irrigation tanks. The rate of water evaporation in these agricultural reservoirs is another benefit that a floating installation on an irrigation tank would offer, especially in warm and well-insulated environments. Over 366 GW of installed electrical production capacity is currently available in the nation. When it comes to the deployment of RE-based installations, India has excelled, increasing over 3.55 times in the previous 5–6 years (Sahu et al., 2016). The Jawaharlal Nehru National Solar Mission's launch is primarily to blame for this rise in solar energy.

SOLAR POWER PANEL IN FSPV

Solar photovoltaic systems generate electricity by absorbing light photons, releasing them, and separating the electrons from their atoms. Direct current (DC) power is produced through the conversion of light energy by smaller components known as solar photovoltaic cells. The photovoltaic solar power panel is shown in Figure 1. Before being used in a national or local grid, DC electricity generated by solar panels must be converted into safer, more reliable alternating current (AC) electricity. Sunlight is converted into power using polar PV cells. PV cells are made of carefully processed semiconductor material that combines the characteristics of an insulator and a metal. The PV cells' positively and negatively charged electric fields regulate the direction of electron flow. Silicon is a semiconductor material that is used in 90% of solar panels currently on the market. The crystal lattice structure of the silicon atoms contributes to the marketability of the material. Thin-film photovoltaics and organic photovoltaics are two more semiconductors used in solar cells. Photovoltaic cells make up the majority of the solar panel, which typically has 40 of them. A solar array can be created by connecting several solar panels together. For an array of solar panels to provide enough electricity to run a typical home, between 10 and 20 panels are needed. The two factors that most significantly affect a photovoltaic cell's performance are temperature and sun irradiation. Figure 1 below displays a graph outlining the properties of a PV cell functioning normally. The relationship between the current and voltage under the current conditions of temperature and solar irradiation can be determined using PV cell I-V characteristic curves (Solomin et al., 2021). How much of the energy captured by the PV cell will be transformed into electrical energy depends on a solar panel's efficiency. Various types of solar panel structures in FSPV are used to improve the efficiency of FSPV system (Figure 2).

FLOATING SOLAR PHOTOVOLTAIC TECHNOLOGY (FSPV) – BENEFITS

Higher gains in energy production: Depending on the floating platform's design, the distance between the water's surface and the PV array, and other factors, modifications to sea-based solar photovoltaic arrays can boost their energy output by up to 10%. It had previously been reported that the improvement over land-based PV systems may reach 10%. Land neutral: When compared to solar PV plants that are located on the ground, FSPV plants require little to no land because they are positioned on water surfaces (Acharya, 2020). For developing nations like India, where land is rare and land acquisition causes obstacles for the deployment of solar PV installations, this is especially important.

Easy installation and deployment: Compared to land-mounted solar PV plants, a conventional FSPV plant is easier and simpler to install. A floating platform made up of individual modular floats is used to float solar PV arrays on the surface of the water. Once finished, boats are used to tow the entire platform to the precise spot on the water body.

Reduction in water evaporation: water resources evaporating owing to evaporation is a well-known phenomenon, with estimates of as high as 40% worldwide. Its effects are particularly pronounced in dry and airy places. Due to their placement on the water's surface, FSPV plants offer shade, which lowers

Figure 1. Schematic diagram of solar power panel

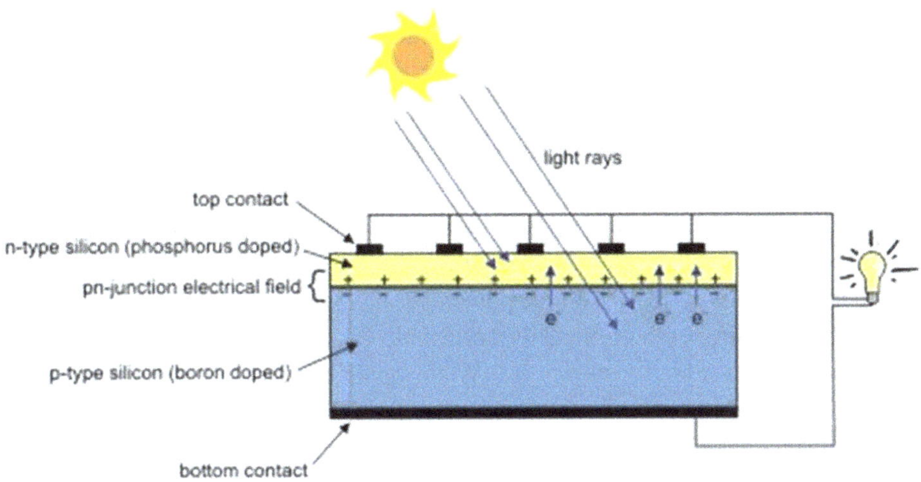

the quantity of solar radiation that reaches the water's surface and lessens wind interaction with it. The combined advantage of these could lead to a decrease in the losses from water evaporation.

Complementary operation with hydroelectric power plants: According to the Indian Institute of Hydroelectric Power (IIHPP), hydroelectric floating platforms can be used in India to produce power during periods of peak demand and serve as energy storage, employing solar power during the day and hydropower at night.

Figure 2. Various types of solar panel structures in FSPV

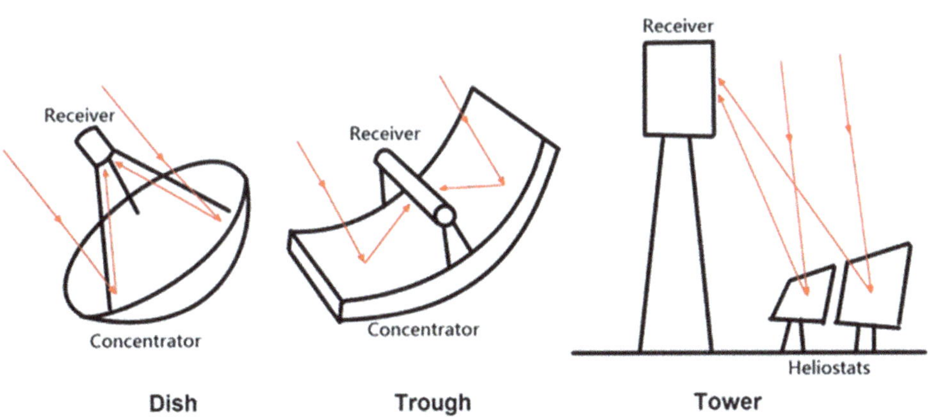

Possibility of sharing existing electrical infrastructure: Many inland waterbodies, particularly reservoirs used for irrigation, water supply, hydroelectric plant reservoirs, and so on, already have grid connections. Utilizing the infrastructure that is already in place, the deployment of FSPV plants in such circumstances may reduce investment costs.

FSPV as a new source of revenue: Fossil-powered solar photovoltaic plants can be installed on water features to produce energy, which their owners can use to offset their own use or sell to the grid. In China's Poyang Lake, for instance, Jinko18 erected a 120-MW FSPV-based plant on a fishing farm that covers 2000 acres of water and generates 137.70 GWh of power.

Ease of cleaning: Water is easily accessible for cleaning in the case of FSPV plants installed on inland waterbodies. Water quality must be examined and must not exceed the permitted level as specified by PV manufacturers.

Reduction in algae growth: The amount of sunlight that reaches the water's surface and the temperature both affect how quickly algae develop. Less sunlight reaching the water's surface due to FSPV plants could result in less algae development. However, it is still unclear how placing such plants may affect the nearby aquatic ecosystem.

Less soiling loss: FSPV plants are exposed to less dust than land-mounted solar PV facilities because the wind over water includes less dust than the wind over land.

FLOATING SOLAR PHOTOVOLTAIC TECHNOLOGY COMPONENTS

The basic and crucial part of such a project are solar PV modules, which can be made of either monocrystalline or thin-film solar cells or photovoltaic cells (Kampker et al., 2017). performance of solar panels, particularly in conditions of high moisture and high humidity, including power tolerance, efficiency, and temperature coefficient Solar panel dependability: conformance to dependability standards such as IEC 61215 Certifications for solar panel quality such as ISO 9001 (wind loading). Main components of FSPV system is shown in Figure 3.

Inverters: For a solar PV project, a developer may decide to use several string inverters or central inverters. Inverters can be positioned on a distinct floating platform or on the ground, depending on their size and proximity to the beach. Similar to a traditional solar plant, a series of combiner boxes transport the DC power produced by solar PV modules to the inverter.

Hydrelio from Ciel and Terre: The Ciel & Terre, a French technological business, has been providing cutting-edge floating solar solutions since 2011. It owns the majority of the market share in terms of the overall number of installations across 25 countries, accounting for more than 300 MW of installed capacity. The connecting pins are made of fiberglass and polypropylene, whereas Hydrelio floats are made of HDPE and HDPE.

Pure-floats design: The technique is utilized by many manufacturers, some of which have already included it into their designs. It uses a specially designed float that can safely hold PV panels in place. Because it is built in a modular way, a huge platform can be made by joining the entire system using pins or bolts.

Figure 3. Main components of floating solar Photovoltaic Technology

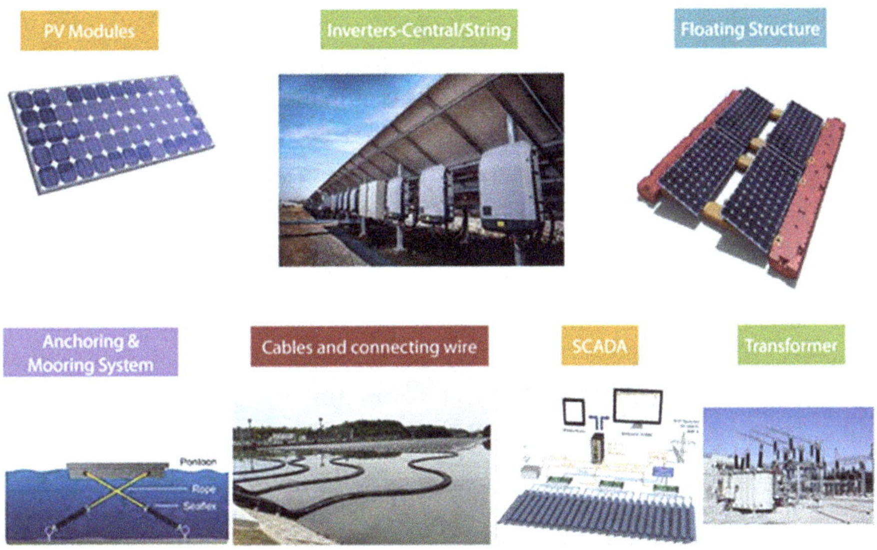

Sumitomo Mitsui Construction: Similar in concept to Ciel & Terre, but formed more frequently for simpler shipping and denser packing. It is bound with binding bands and filled with polystyrene foam to reduce the possibility that it may sink even if damaged.

Floating Platform: Most FSPV power plants worldwide use HDPE, which is the most widely used material. For the floating platform, additional materials like ferro-cement, medium density polyethylene (MDPE), and FRP are also used. The following is a description of many floating platform designs.

Pontoon and metal structures design: One common design uses pontoons to provide buoyancy and a metal structure mimicking a land-based system to do away with the need for specialized floats. This type of design has the advantage that floats are easier to make and may thus be made locally. However, such designs make it challenging to reach some locations for use and maintenance.

Yellow Tropus: A design and engineering company called The Yellow Tropus, which has its headquarters in India, specializes in building FSPV power plants. The company offers three different types of technologies for land-neutral hydropower, wind power, and other forms of renewable energy.

Anchoring and Mooring System: FSPV plants may experience problems if there are any variations in water levels brought on by the monsoon, wind speed, or an increase or decrease in the volume of water because they are located close to water bodies. Mooring systems include structures like quays, wharfs, jetties, piers, anchor buoys, and mooring buoys. The mooring system for a floating platform is commonly attached using nylon polyester or nylon nautical ropes. Bank anchoring, bottom anchoring, and piles are the three mooring techniques. Bank anchoring is the most cost-effective option, but whether it is suitable depends on the conditions along the shore. Divers are required for bottom anchoring, and careful design consideration is necessary. After being drilled into the sea, the floating platform is secured to the piles.

Stingray Technology: It aims to lower evaporation losses by enclosing a bigger area of water surfaces. The use of technology is best suited for large waterbodies like thermal power plants, steel mills, and chemical industries. One unit of this consists of a walkway float, two solar PV panel floats, and two solar photovoltaic panels.

Cabling: Effective cable management and routing in FSPV operations require planning. The movement of a floating platform on a water surface causes cable length to change, unlike solar PV installations on land. If this is overlooked, inadequate cable length could result in cables breaking and snapping under tension. It is advised to use wiring trunks to shield the cables from direct sunlight because they need to be UV-resistant. The appropriate cable ties or clamps need to be used to prevent DC wires or conduits from coming into contact with water. The main infrastructure on land can be connected to using either separate, specialized floats or subsea cables for AC lines.

EMERGING FLOATING SOLAR PHOTOVOLTAIC (FSPV) TECHNOLOGY

Tracking: Additionally, floating platforms can be designed to rotate on one or both axes to track the sun. A simple technique for tracking in a single axis is vertical-axis azimuth tracking. Some designs omit the core pile altogether and instead use a fixed outer ring to encircle the floating platform (Pizarro et al., 2019).

Concentrated FSPV: The solar PV cell/panel, reflecting mirror/concentrating lens assembly, and floating platform are used in this technology, which is similar to ground mounted concentrated solar PV (CSPV). Similar to this, a pilot project in India uses a liquid solar array, a plastic concentrator.

Membranes and mats type: This kind of platform often uses rubber mats to entirely cover the water's surface. Particularly suitable for small artificial waterbodies is this variety (size less than 0.2 km^2). The Norwegian company Ocean Sun has also proposed a floating platform made of rafts that is 70 meters long.

Submerged FSPV: The was put into use in 2010 by MIRARCO Mining Innovation and has a 0.57 kW capacity. Submersible PV modules are placed just beneath the surface of the water. The technology, which is still in its early stages, presents various concerns concerning the reliability of PV modules over the long term as well as electrical safety in underwater environments.

HYBRID FLOATING SOLAR POWER PLANTS

Hybridization: A "hybrid energy" system aims to collect energy simultaneously and at different places from both conventional and renewable power sources. Devices using hybrid floating solar and other variable renewable energy sources have better energy densities than those using conventional renewable energy sources. The requirement for additional fossil fuel-based energy sources can be decreased when water resources are used in tandem as a hybrid to make up for each other (Kohnert, 2007).

By 2050, the demand for power will have doubled worldwide. One of the challenges is the sustainable use of land resources. As an alternative, floating PV (FPV) plants are being created more frequently. The hybrid FPV technologies with hydro energy and solar input were among the most promising ones that might be employed for effective power generation.

Demands for housing, agricultural land, and energy have all significantly increased. Most places on earth have access to at least one commercially feasible renewable energy source. The use of a significant number of land resources is necessary for the deployment of solar energy on a large scale. The performance of solar panels in converting sunlight to energy increases as the temperature decreases because they have a negative temperature coefficient. To make use of this phenomenon, solar arrays are erected over bodies of water in a novel application called floating solar PV (FPV). Large water surfaces are used by floating photovoltaics to generate energy, and water is used to control the plant by serving as a cooling medium. Eventually, it might be less expensive than purchasing expensive land for the installation of PV. The majority of floating photovoltaic systems, with power output ranging from 4 kW to 20 MW, have been erected over water reservoirs in Japan. Between the lake and the land, there is a measured temperature difference of 1-3 degree centigrade.

Solar panel installation on water is roughly 15% more expensive than installation on land. The lowering of floating building costs and the acceleration of growth in developing nations point to a rapid uptake of FPV technology. The natural water-cooling action can partially increase the modules' effectiveness and lengthen their lifetime, offsetting some of the higher expenses associated with floating solar. In order to address the problem of the food-water-energy nexus, the use of a floating solar system in aquaculture is considered a viable solution (Naveed, 2021). The mining industry has looked into a unique technique for using FPV modules to produce electricity and solar energy to heat water in water ponds. The limited onshore usable land and the offshore region around the island nations are the driving forces behind the construction of solar panels on the ocean's surface. An inventive idea that allowed the FPV's azimuth angle to be changed while maintaining a fixed tilt increased power output by 27.68% compared to a typical floating PV system.

CONSTRAINTS AND CHALLENGES OF HYBRID FSPV

Due to the strict module specifications, installing a floating solar panel costs a lot of money, but as technology advances, costs should go down. Hydro dam water level fluctuations in hot, tropical areas provide a serious technical concern. HFPV's need for a complicated mooring mechanism poses a serious obstacle. Because of its vulnerability to strong winds, it requires a significant number of mooring points to be kept in place. In order to prevent panel destruction during flash floods, rigorous mooring and anchoring are required. HFPV systems suffer from unfavourable effects from wind, waves, currents, and snow, as well as instability. The constant bobbing motion beneath the solar cells makes it feasible for microcracks to form in the solar cells. The amount of water surface that may be covered and the amount of space needed for leisure and recreational activities have an impact on the quality of the water. Hybrid FPV powered by hydropower can only operate in certain regions. The majority of hydro plants are peaking plants, heavily dependent on the monsoon seasons (Norani et al., 2021). Some reservoirs may be too dry or unusable at certain times of the year to support hybrid floating PV. Combined implementation: The industry, including operations, scheduling, and optimization, has not yet fully understood the advantages and potential of the hybrid system's joint implementation.

Limited expertise and knowledge: Data-driven decision-making will be hampered by the current knowledge gap regarding hybrid FPV systems. There is little experience with floating solar and hydropower systems. This is a unique innovation; hence, more specific installation knowledge is needed. To confirm the breadth of these benefits, larger installations across various geographies are required over time. In many cases, these benefits will outweigh any increase in capital costs. According to the Royal Institution of Chartered Engineers, FPV should not be used for hydroelectric projects in the UK because of the social and environmental consequences. To ensure adherence to societal values, further application of FPV on dam surfaces would need to undergo significant discussion and evaluation. The main flaw in the HFPV system is the absence of installation rules and regulations. Another significant problem is the aggressive and inconsistent policies regarding the installation of HDTV set-top boxes. According to the US Department of Energy's Office of Energy, Environment, and Natural Resources (EER) in the White House, a hybrid installation of solar and hydropower facilities needs the appropriate structure and a contractual agreement between multiple parties in order to be successful. According to a study that was published in the Journal of Solar Energy Engineering and Design, a hybrid solar-floating photovoltaic (FPV) system will be a more effective alternative to the conventional floating system for power generation (JSE).

DEVELOPMENTS OF HYBRID FSPV IN VARIOUS ENVIRONMENT

Figure 4. Various FSPV Environments

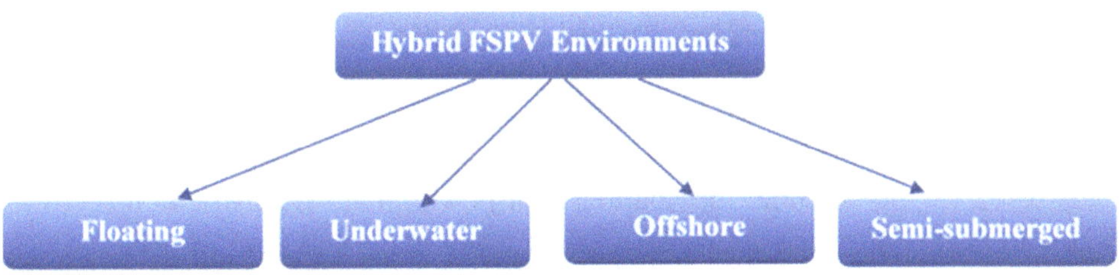

The classification of different hybridization environments for FSPV is illustrated in Figure 4.

FSPV: In contrast to placing solar panels on terrestrial systems, floating photovoltaics entails putting them over natural or artificial bodies of water. Based on their supporting architecture, FPV systems can be divided into three main categories: fixed tilt arrays, tracking, and flexible tilt arrays. Additionally, they can be divided into three categories based on the implementation scale: small scale (a few kW), medium scale (kW to MW), and big scale (MW to GW). Although floating-track photovoltaic (FPV) systems have a greater net capital cost than fixed-track PV systems, they produce more electricity.

Underwater/Submerged/Semi-Submerged: The use of PV modules in both deep and shallow water is suggested as a way to clean panels, cut down on reflection losses, and boost efficiency by getting rid of thermal drift. On a small-scale solution using thin-film panels, semi-submerged solutions with both flexible and rigid structures were devised and tested.

Offshore/Marine Environment: The concept behind offshore PV power generation is to collect solar energy in oceans and seas using an FPV system. Due to the rough waves produced by the strong winds, FPV designs in offshore situations differ from those in ordinary lakes.

RECENT DEVELOPMENTS, AND PERFORMANCE OF HYBRID FSPV

Since the development of centralised power distribution systems, hydropower has played a significant role in the world's energy grid. To provide towns and cities with drinking water, flood control to prevent disasters, and locks and dams to pass rivers and streams, mankind has constructed tens of thousands of dams and water reservoirs. The main energy source for flood control, water management, entertainment, and grid stability are hydropower. Since the turbine cannot use the water vapour to generate energy, the evaporation of water from reservoirs in hydropower facilities represents a significant loss. Hydropower significantly harms the environment and eliminates natural habitats. Hydropower facilities use the force of falling water to produce energy. The installed hydropower capacity was 1308 GW in 2019, and the generation peaked at 4306 terawatt-hours. 328 GW of these are run-of-the-river plants, and 138.7 GW of that is hydroelectric pumped storage. According to World Bank research, the potential for floating solar power plants on artificial reservoirs is at least 400 GW. Large hydropower plants with dams can have their electricity production capacity matched by floating solar panels that cover between 1 and 30% of the reservoir. Instead of having a direct connection to the earth's surface, like through cables and turbines, the majority of modern commercial projects on hydroelectric reservoirs rely on co-location (Chavan et al., 2017).

HFPV might act as a more powerful motivator for future installation than conventional floating solar PV. Additionally, a number of developers are making efforts to take advantage of the popularity of floating solar and hydropower facilities globally. Development teams may have a significant impact on the future of offshore energy production if they can effectively create cost-effective hybrid solutions.

The employment of an acceptable scientific methodology to assess the effectiveness of the HFPV systems is a topic of intense dispute within the scientific community. The sustainability index (SI) is a metric created to compare the state of a region's and a nation's electrical industry. NPV, LCOE, CF, and EY can be used to analyse project performance. Investments in renewable energy projects have been valued in terms of their internal rate of return (IRR) and net present value (NPV). Energy delivered by FPV systems in terms of the peak power of the module is known as energy yield (EY). Alternative technologies can be compared using the LCOE at different scales and operating levels. A specific generation and storage mix's cost-effectiveness can be determined by looking at its LCOE. The potential to reduce greenhouse gas (GHG) emissions is used to assess how much CO_2 emissions will be reduced by the suggested technology. In terms of societal advantages, solar energy systems are anticipated to raise living standards and open up employment opportunities for communities.

On manmade water bodies like reservoirs, floating photovoltaic (FPV) PV systems could provide alternatives to scaling up renewable energy and ease pressure from competing land uses. FPV offers an almost zero-risk method for siting new PV plants without obstructing natural vegetation and safeguarding dams from insulating. The biggest advantage of hybrid FPVs is that they don't need any fields or other open spaces. They deal with the issue of the energy-water-land nexus.land is used. The quantity of land needed is decreased by installing PV modules on the existing water infrastructure (dams, ocean, etc.). In comparison to land-based solar systems, water-cooled solar photovoltaics can offer a number of advantages, including increased energy efficiency and less water evaporation from hydropower generation. The performance of the device is additionally enhanced by water cooling beneath the panels. Utilizing current electrical transmission infrastructure, hydropower projects are situated near demand centres (in the case of water supply reservoirs). Due to the road access to the hydropower project, building and transportation costs are probably lower. By joining a shared substation, a hybrid system would save money on transmission. The distance between two water-cooled photovoltaic (PV) panels, which is reduced when they are positioned on an inclined surface, increases solar energy production. The natural cooling effect of the water, which lowers temperature increases in the solar modules, is the cause of the enhanced energy production. The ratio of the total energy produced to the maximum energy that could be produced is known as a hydroelectric power plant's capacity factor. Hybrid hydropower-connected F-PV plants in existing dams could compensate for the loss of electrical output caused by changes in water levels. Water savings and adequate and economical management of power and water usage are both made possible by the hybrid operation of the dams with FPV. Due to the pre-existing grid infrastructure, HydroFPV-Hybrid has a reduced capital expenditure (CAPEX), however, it is dependent on the site conditions. Co-locating with hydropower has advantages for the electricity system and lower capital costs because interconnection expenses are lower. By using the same connection to the grid as conventional photovoltaics, hybrid F-PV and hydropower projects could lower the cost of grid integration. For such a plant, modernising the current infrastructure could enhance its performance, dependability, and safety. Hydropower facilities are adaptable and may run on demand with variable output. Due to the fact that a portion of the power is generated from erratic, non-dispatchable solar energy, this is useful in nations with a weaker grid. By operating in "load-following" mode, hydropower can tame fluctuating solar performance. Water quality may be enhanced, water loss can be decreased, and hydropower generation can be increased using solar energy from photovoltaic (PV) panels. As a substitute energy source for societies in need of renewable resources like drinking water, it can help rejuvenate polluted or ineffective water bodies. By lowering thermal excursion and raising dam stability, solar panels are used to increase the stability of hydroelectric dams. Direct solar radiation, which might undermine the stability of the dam if it is exposed to severe winds or heavy rain, is protected from the dam surface by PV panels.

Water is a sign of gravity energy storage, especially when it is present in conjunction with hydroelectric basins. Pumps in hydropower stations with pumped storage can be run on PV energy. In one scenario, hybrid system operators could use pumped storage hydroelectricity to store excess solar energy. This type of combined hybrid FPV, often known as "hydropower," is rising in popularity throughout the world, especially in developing nations. There may be further advantages to combining hydropower and floating PV. Solar energy has the greatest potential in the dry seasons and can help hydroelectric plants with black starts. This is true both during the wet seasons and the diurnal period (during the day and at night).

SUSTAINABLE DEVELOPMENTS OF HYBRIDIZATION AND INTEGRATION OF HYBRID FSPV

A floating object, mooring system, PV module, DC/AC wires, and connections make up an FPV system. There are seven elements that may show whether or not a system is created properly: modular design, dependability, durability, protection, appropriate support structure size, simple installation, and cost reduction. If the system is an FPV, the PV is mounted on a floating body (floater and structure). To keep the entire system afloat, it offers the necessary buoyant force. High-density polyethylene is frequently used to make floating objects (HDPE or MDPE). A floating platform can be stationed in any water depth with the use of mooring mechanisms. A mooring line connects an anchor on the ocean's bottom to a floating building. Additionally, the system is capable of adapting to changes in water levels. A buoyant platform is intended to hold an array of floating photovoltaic (FPV) devices. PV modules and other power converting devices turn electrical energy into electrical energy (Ahamed

Figure 5. Hybridisation and Integration of Hybrid FSPV

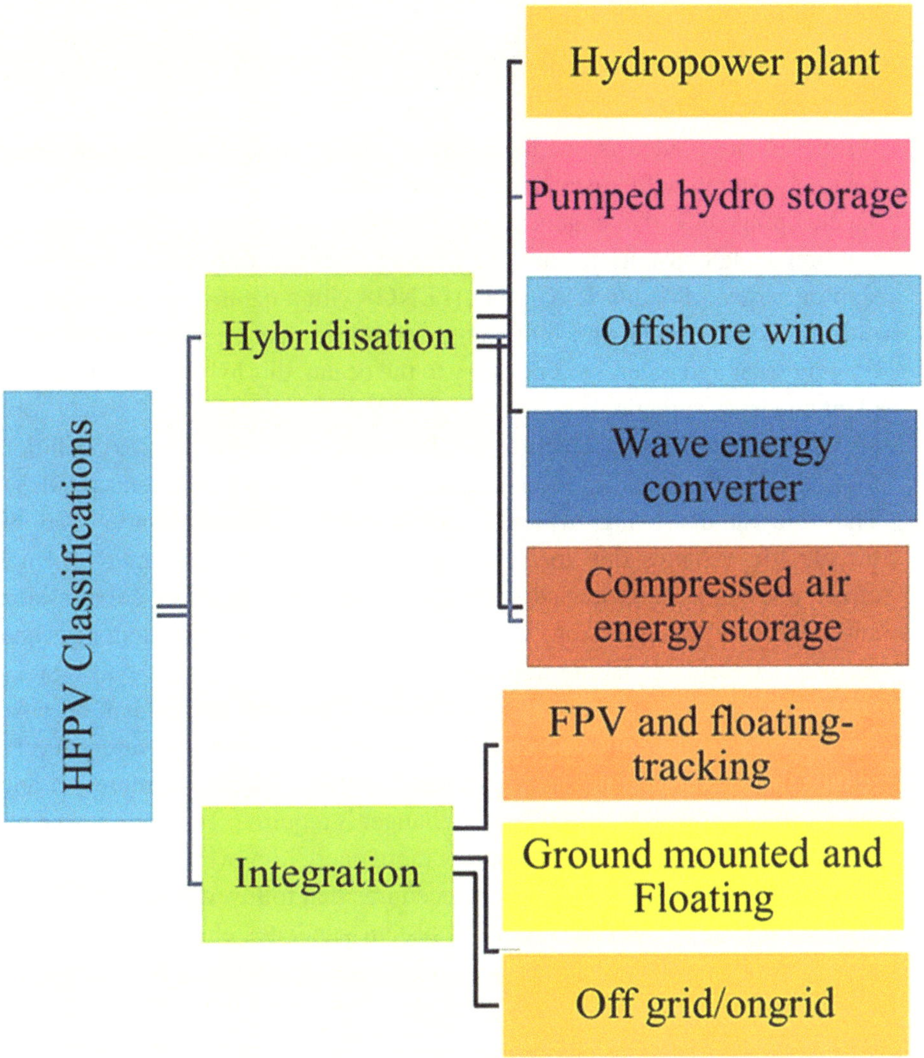

et al., 2020; Belyakov, 2019; Etter, 2017). Underground cables are often used to deliver the PV system's electricity to the substation. Underwater wires need to be leak- or shack-proof. Similar to a conventional ground-based system, the inverters are located on the shore side of the system. a second floating system that is used to transport the same amount of water used daily by the hydroelectric turbines. The various hybridizations and integrations of hybrid FSPV are illustrated in Figure 5.

- **Floating Solar Integration with conventional plant:** The viability of an FPV system working with a diesel generator (DG) and battery bank to satisfy an island's off-grid needs is investigated. An analysis was done on the viability of deploying a floating photovoltaic array (PV array) to produce power as well as the integration of floating PV with a traditional gas-fired system. A novel idea is put forth to replace the fossil fuels used in home heating systems with sustainable power generated by floating solar farms. To encourage effective cogeneration systems, a cost-optimal district heating solution using conventional and PV heating electrification is presented. On a ship or barge, a floating LNG plant (FLNG) with a natural gas-powered gas turbine, LNG storage, and offloading capabilities can be constructed. With floating solar and natural gas delivery to the ocean, this hybrid floating power plant concept is practical.
- **FPV with Compressed Air Storage:** The most practical technology solution for use in electric vehicles is compressed air energy storage systems (CAES). They are rapidly being explored for use in fuel-efficient power stations for electric vehicles due to their high efficiency, minimal environmental effect, and high energy density. The choice between adiabatic compression and isothermal compression, which both call for effective heat storage and recovery, is crucial. Isothermal compression makes it possible to store a significant quantity of energy. The process remains isothermal because the majority of the collected energy is transferred to the heat reservoir. To reduce the irreversible process, care must be taken with the compressor and pneumatic motor, and a robust heat exchanger is required. The same is true for the expansion process, which requires more energy on overcast or nighttime days and takes longer overall. The procedures that follow describe FPV with CAES's operating system. The FPV system generates electricity using PV modules, which it then transfers via an inverter to the grid or a water pump. The pressurised water is used to raise the air pressure in the pipes after the water pump compresses the water up to 20 MPa in a suitable storage tank.
- **Floating PV and Wind Technology:** Hybrid solar and wind farms aboard boats are still mostly an untested idea. The combined wind farm and FPV in this offshore renewable energy field would increase the amount of power produced per square foot of sea space. In contrast to wind energy, the production of solar energy is anticipated to be constant throughout the research zone. Therefore, compared to offshore wind farms, deep-water PV installations are not anticipated to boost solar energy output.
- **Floating PV with Hydro and Wind Technology:** Hydropower and floating solar panels are used in HFPV to produce electricity. For nations with a lot of hydropower plants with dams, floating solar farms (HFPV) are an appropriate

choice. In order to fulfil peak demands, a battery energy storage system is coupled with the intermittent floating solar resource. If the floating solar farm is situated close to hydropower plants, HFPV may also be able to utilise current transmission cables. HFPV is a versatile and auxiliary technology. The generation of such assets will increase, and the generation curve will be smoothed through colocation with hydropower units. The erratic energy production from hydro and FPV can be offset by the addition of a wind turbine. A key local resource is hydropower. Asia has the biggest concentration of superpowers on the globe.

- **Floating Solar PV with Pumped Hydro Energy Storage (FPV + PHES):** When solar radiation is strong, the FPV panels' power output is directly sent to the grid, while the reservoir stores water that can be used later when solar radiation is low. Depending on where the reservoir is located, higher coverage ratios could be taken into consideration to improve efficiency and lower water usage. Two reservoirs of differing heights, with the lowest reservoir attached to a water source, are needed for a PHES system. Hydroelectric dams might still produce electricity in this way, albeit at a lower rate. In this scenario, the reservoir's water storage will serve as a reliable energy storage system. Using pumped-storage hydropower, hybrid system operators might also store any extra solar energy produced.
- **Hybrid Floating Solar with Wind and Ocean Energy System:** Heavy Duty-No Foundation, Maximum Robustness, Scalability, and Little Maintenance are the distinctive characteristics of the design. It is possible to maximise the utilisation of various maritime resources while lowering deployment, maintenance, and decommissioning costs by integrating devices into an integrated framework or strategically placing diverse technologies in a farm plan.The idea of a fully oceanic hybrid platform that can transform wind, solar, and wave energy was created by Sinn Power. As a component of an off-grid energy solution, floating photovoltaic arrays have been created. The system is made to be modular and simple to connect to and expand. It has the ability to support the global installation of offshore wind farms and offer clean energy to islands all over the world.
- **FPV with Micro-Hydrokinetic Turbines:** It is possible to employ low-lying solar panels that float in aqueducts, canals, tidal flows, and rivers. Even the arrival of darkness has no impact on energy production. Although the technology was initially intended for installations on canal tops, it can be modified to work in a floating solar-energy environment.
- **FPV and Wind Energy System and Battery Storage:** After being converted by an inverter, FPV and wind turbines can be powered by either direct current (DC) from the PV arrays or wind turbines. To maintain a stable and regulated

output, the system controller continuously modifies the load and the output from both PV and/or wind systems. A transformer transmits and feeds the distribution system with electricity that quickly satisfies demand. If there is insufficient electricity supply to meet demand, energy is imported via a connection to the main grid. Imports of electricity serve as a supplement to total renewable energy.

- **Floating Solar Island with Hydrogen Energy Storage:** These floating platforms must be built to withstand challenging weather, such as strong currents and powerful winds. The system integrates a number of technologies, including battery storage, wind farms, and floating solar. Additionally, it will produce green hydrogen using an electrolyzer, which can be used as an additional storage method. This might provide the maritime sector with a viable substitute for burning fossil fuels in the future's fuel-efficient ships.
- **Floating Solar and Waterlily (Solar Tree) system:** The system consists of a floating platform linked to a network of cables, buoyancy sensors secured to the ocean floor by an underwater concrete anchor, and cables. For a hydroelectric power plant to produce electricity from the sun's rays, a solar tree-based water lily design has been suggested. The submersible propellers that ensure the water lilies rotate are driven by a tracking algorithm for the movement of the sun.
- **Hybrid Floatovoltaic Aquaculture System:** Integrated offshore floating platforms increase the output of aquatic species while also opening up new economic prospects. Utilizing water for the production of both food and energy is the goal of aquavoltaics. The off-grid aquaculture potential can be met using the electricity produced by the FPV system. The scientific community does not fully comprehend concerns about biofouling, FPV interaction with aquatic ecosystems, or political and regulatory challenges.
- **Floating Solar Methanol Island:** For the creation of H_2 and the extraction of CO_2 from seawater, the idea of a "solar methanol island" based on offshore floating PV has been put forth. A floating platform with many different sustainable energy producing choices makes up an energy island. Synthetic MeOH fuel is produced by the catalytic reaction and can be used to produce energy. Engineering analyses and optimizations specify the operating parameters for large-scale solar methanol islands. A central location with electrolysis and desalination cells for processing H_2 receives power from the islands. Batteries for storing electrical energy temporarily are among the other island components.
- **Floating Photo-Thermal (FPT) and Seawater Desalination:** Desalination can be done with the use of photothermal devices, which are sun concentrators and receivers. The process starts with the creation of low-temperature water

vapour, which is followed by the condensation of saltwater vapour. The use of FPV and FPT systems to power the seawater desalination process may be a desirable choice. Technology costs, design limitations, implementation requirements, and design optimization will all affect how this field develops in the future.

CONCLUSION

- A floating platform made up of individual modular floats is used to float solar PV arrays on the surface of the water. Depending on the design, modifications to sea-based solar photovoltaic (PV) arrays can boost their energy output by up to 10%.
- Floating platforms can be designed to rotate on one or both axes to track the sun. Some designs omit the core pile altogether and instead use a fixed outer ring. Submerged FSPV involves submersible PV modules placed just beneath the surface of the water. An analysis was done on the viability of deploying a floating photovoltaic array (PV array) to produce power. Compressed air energy storage systems are being explored for use in electric vehicles. Hybrid solar and wind farms aboard boats are still mostly an untested idea. Hydropower and floating solar panels are used in HFPV to produce electricity. In order to fulfil peak demands, a battery energy storage system is coupled with the intermittent floating solar resource.
- The erratic energy production from hydro and FPV can be offset by the addition of a wind turbine. Hydroelectric dams might still produce electricity in this way, albeit at a lower rate. In this scenario, the reservoir's water storage will serve as a reliable energy storage system. Using pumped-storage hydropower, hybrid system operators might also store any extra solar energy produced.
- Low-lying solar panels can float in aqueducts, canals, tidal flows, and rivers. The Solar Tree system consists of a floating platform linked to a network of cables. It will produce green hydrogen using an electrolyzer, which can be used for fuel-efficient ships. Aquavoltaics involves using water for the production of both food and energy. Off-grid aquaculture potential can be met using the electricity produced by the FPV system. A solar tree-based water lily design has been suggested for a hydroelectric power plant.

REFERENCES

Acharya, M. (2020). Floating Solar Photovoltaic (FSPV): A Third Pillar to Solar PV Sector? *Lecture Notes in Civil Engineering*, *15*(3), 68. doi:10.1007/978-981-13-8743-2_21%0Ahttps://

Ahamed, R., McKee, K., & Howard, I. (2020). Advancements of wave energy converters based on power take off (PTO) systems: A review. *Ocean Engineering*, *204*, 107248. doi:10.1016/j.oceaneng.2020.107248

Belyakov, N. (2019). Sustainable power generation: Current status, future challenges, and perspectives. *Sustainable Power Generation: Current Status, Future Challenges, and Perspectives*, 1–620. doi:10.1016/C2018-0-01215-3

Chavan, S. A., & Anwar, P. M. A., & R, P. S. B. (2017). Review of Effect of Process Parameters on Mechanical Properties of 3D Printed Components. *Internation Journal of Advance Research in Science and Engineerng*, *6*(5), 461–467.

Etter, P. C. (2017). Underwater Acoustic Modeling and Simulation, (Fourth Edition). In Underwater Acoustic Modeling and Simulation, Fourth Edition. CRC press. doi:10.1201/b13906

Kampker, A., Kreisköther, K., & Reinders, C. (2017). Material and Parameter Analysis of the PolyJet Process for Mold Making Using Design of Experiments. *International Journal of Materials and Metallurgical Engineering*, *11*(3), 242–249.

Kohnert, D. (2007). On the Articulation of Witchcraft and Modes of Production among the Nupe, Northern Nigeria. *Hexenglauben Im Modernen Afrika: Hexen, Hexenverfolgung Und Magische Vorstellungswelten (Witchcraft in Modern Africa: Witches, Witch-Hunts and Magical Imaginaries)*, *48*(2), 39–61. doi:10.1227/01.NEU.0000349921.14519.2A

Naveed, N. (2021). Investigate the effects of process parameters on material properties and microstructural changes of 3D-printed specimens using fused deposition modelling (FDM). *Materials Technology*, *36*(5), 317–330. doi:10.1080/10667857.2020.1758475

Norani, M. N. M., & Abdollah, M. F., Abdullah, M. I. H. C., Amiruddin, H., Ramli, F. R., & Tamaldin, N. (2021). 3D printing parameters of acrylonitrile butadiene styrene polymer for friction and wear analysis using response surface methodology. *Proceedings of the Institution of Mechanical Engineers, Part J: Journal of Engineering Tribology*, *235*(2), 468–477. doi:10.1177/1350650120925601

Pizarro, F., Salazar, R., Rajo-Iglesias, E., Rodriguez, M., Fingerhuth, S., & Hermosilla, G. (2019). Parametric Study of 3D Additive Printing Parameters Using Conductive Filaments on Microwave Topologies. *IEEE Access: Practical Innovations, Open Solutions*, *7*, 106814–106823. doi:10.1109/ACCESS.2019.2932912

Sahu, A., Yadav, N., & Sudhakar, K. (2016). Floating photovoltaic power plant: A review. *Renewable & Sustainable Energy Reviews*, *66*, 815–824. doi:10.1016/j.rser.2016.08.051

Solomin, E., Sirotkin, E., Cuce, E., Selvanathan, S. P., & Kumarasamy, S. (2021). Hybrid floating solar plant designs: A review. *Energies*, *14*(10), 2751. doi:10.3390/en14102751

Sreenath, S., Sudhakar, K., & AF, Y. (2021). 7E analysis of a conceptual utility-scale land-based solar photovoltaic power plant. *Energy, 219*, 119610. doi:10.1016/j.energy.2020.119610

Chapter 9
Comprehensive Characterisation of Biomass and Weather for Deployment of the Renewable Energy Systems

Yusto Mugisha Yustas
https://orcid.org/0000-0002-8291-4062
Sokoine University of Agriculture, Tanzania

Valerian Cosmas Silayo
Sokoine University of Agriculture, Tanzania

ABSTRACT

The inefficient use of traditional biomass energy sources for cooking and scarce electricity are challenges in the least developed countries. This chapter presents the comprehensive characterisation approach for renewable energy (RE) related resources and sources, which support the measures to address the challenges. The approach's effectiveness is demonstrated in rural semi-arid areas. The demonstration involved sampling and laboratory analysis of cow dung (CD), data collection and analysis of solar irradiance, wind speed (WS), and ambient temperature (AT). In addition, weather data were collected using the novel onsite weather station. Results show that the area has quality CD (C/N ratio of 19), annual average solar insolation and WS of 4.5 kWh/m2/day and 4.7 m/s, respectively. Also, AT varies by 4oC within a day. The results show that the development of RE systems is feasible in the areas. Therefore, the approach is recommended to researchers and RE systems project developers.

DOI: 10.4018/978-1-6684-4118-3.ch009

INTRODUCTION

Establishing renewable energy systems is a process, which requires a robust procedure, framework or scientific approach. The approach involves gathering information that will assist in designing, installing, and operating the systems. Information gathering is essential as it provides detailed evidence of specific features or trend conditions at a specific location where the system has to be installed. The performance of renewable energy systems depends on the design, which is based on the accurate prior knowledge of availability trends of the resources, and sources at a specific site where the system is intended to be established. The quantity, quality and availability of the resources and sources at a particular location must be determined using a well organised and concise scientific approach. The resources and sources may include solar, wind, and biomass (e.g. cow dung). The collection of information of these resources and sources in the remote rural areas, particularly in the least developed countries, is a challenge that needs to be addressed. The energy from the resources and sources can be harnessed by a system exploiting a single resource or combining resources and sources. Despite the type of system to be installed in the particular areas, the approach has to be manageable and robust in terms of expertise with less risk of producing erroneous information. When used as they are to develop the energy system, the erroneous information will result in the poor performance of that energy system. This chapter proposes a comprehensive scientific approach for characterising biomass (e.g. cow dung) and weather-related resources and sources for energy generation. Also, it aims at demonstrating application of the approach with test results for easier adaption and adoption by other researchers and renewable energy projects developers.

BACKGROUND INFORMATION

Poor access to beneficial energies and their low utilisation hinder the sustainable development of any community (Oyedepo, 2014; Mohammed et al., 2013a). Such hindrances may be related to energy resources availability and energy infrastructure to harness and distribute it to a community. Also, they may be based on lack of awareness of available energy technology to support access and utilisation of energy to a community. The hindrances are handled differently between developed and developing countries (Vandaele and Porter, 2015). This study considered handling these deterrents in developing countries.

Most rural communities lack reliable access to energy for their domestic uses (Mohammed et al., 2013a; Mohammed et al., 2013b). Besides, in households of these communities, a large portion of energy consumption is taken by cooking activities.

In contrast, the other portion is taken by lighting and powering of electrical and electronic gadgets. Moreover, the primary energy source in these communities is traditional biomass, including fuelwood, charcoal and agricultural wastes. Therefore, the communities utilize the sources in addition to fossil fuel, solar energy (mainly PVs), and rarely grid connection (Suberu et al., 2013; REN21, 2017). Also, the use of traditional biomass is augmented by the low financial capability of the community in question.

Heavy dependency on traditional biomass as the source of energy and rapid population growth have increased pressure on the dwindling forest stocks in the rural semi-arid areas, especially in most agro-pastoral communities. As a result, deforestation and land degradation are noted in these areas, such as central Tanzania (Felix & Gheewala, 2011; Wawa, 2012).

Several actors have put in place specific and holistic strategic measures regarding the above-explained hindrances and challenges, including researchers and policymakers (Bukari et al., 2020; Scheid et al., 2018). These measures aim to improve availability, accessibility and adoption of energy for all deprived areas in Tanzania and elsewhere with similar conditions (Bekele and Tadesse, 2012; Wawa, 2012). There are preparatory stages for developing robust and innovative technological systems. The systems are meant for addressing the above-stated challenges. Scientific publications related to this matter were intensively reviewed. Some researchers characterised weather factors and biomass to offer vital information in developing the innovative technological system for energy and other development issues (Zhou and Yang, 2016; Bekele and Tadesse, 2012; Haluschak, 2006; Peters et al., 2003). The Characterization approach is a process that provides essential information to gain practical knowledge about materials or conditions (Ananthapadmanaban, 2018). It can also be adopted to correlate the energy system' performance with trends of the behavior observed for the environment around that system. However, with this fact, a comprehensive characterization of renewable energy-related resources and sources research approach has not been addressed.

Weather data may be collected from remotely sensed information systems, meteorological stations and weather stations installed at the micro-scale level, e.g. a village. In most cases, the remotely sensed data is, in coarse spatial resolution, 1° x 1° global grid. The coarse spatial resolution does not precisely capture the weather factors' data, such as wind at the microclimate level (Suri et al., 2008; Johnston and Matlock, 2011). Meteorological stations are scattered in most developing countries but do not cover many rural areas. However, weather stations installed at the micro-scale level collect valuable and precise data for designing renewable energy systems. In most cases, they offer high temporal and spatial resolutions data. Nevertheless, the former two data collection means can be considered if precise data are not available.

The characterization of weather factors and biomass is essential because the results obtained are essential in designing and modeling integrated renewable energy systems. These systems may involve solar, wind, and biogas technology. Specific primary data needed for the designing and modeling are not readily available in the rural central region of Tanzania at the micro-scale level. The primary data included solar radiation, wind speed, and ambient temperature at the maximum of one-hour interval. The one-hour interval enhances visualisation of the variation of weather parameters, which influence the performance of renewable energy systems. Such system may include biogas plants, solar PV systems and wind turbines. Information on qualities of cow dung suitable for biogas technology in rural semi-arid areas is not readily available.

Therefore, this study aimed at formulating and applying a comprehensive characterization approach, which helps assess the quantity and quality of the locally available renewable energy resources and sources. The renewable energy sources are solar radiation and wind, while the resource is cow dung. Results of the assessment of these resources and sources were intended to be applied in developing and modeling the integrated renewable energy system. The system was to be installed at an individual household of rural semi-arid central Tanzania, aiming at producing biogas for cooking under ideal constant temperature and generation of electricity.

THE COMPREHENSIVE CHARACTERIZATION APPROACH FOR COW DUNG AND WEATHER-RELATED RENEWABLE ENERGY RESOURCES AND SOURCES

The characterization approaches or frameworks of various types are available to facilitate executing several issues or tasks in the different fields. For instance, there are characterization approaches for guiding the deployment of software technologies (Carzaniga et al., 1998). Also, there is the approach for studying forensic issues of physical devices, which is essential in verifying the trust and authenticity of data and the device that created it (Khanna et al., 2006). Another approach is used for developing a computerised predictive system that estimates the aquatic toxicity of chemicals (Blackburn and Stuard, 2014). Additionally, there is a framework for investigating a physical/chemical composition and assessing the potential of municipal solid wastes (Azam et al., 2020). The wastes fall in the group of biomass material, so they are renewable energy sources. Another approach supports the system-level voltage and frequency scaling in the microchip design (Papadimitriou et al., 2021). Finally, there is the approach for establishing the energy consumers' profile to help decision-making in the electrical sectors (Ramos et al., 2021). However, the availability of a comprehensive characterization approach for simultaneously investigating the biomass (cow dung) and weather-related renewable energy resources and sources (solar, wind, and ambient temperature) is limited. So far, there is no such approach in the literature that has been demonstrated to generate the findings useful for development of renewable energy and model performance of the system.

The approach being considered supports the theme of this book, which is stated as the Human Agro-Energy Optimization for Business and Industry. The support lies in the 'human agro-energy optimisation issue'. Agriculture involves the production of useful products/produces and generates agricultural wastes (biomass wastes, e.g. cow dung). The wastes can be converted into useful materials (bio-fertiliser) and energy (biogas). The conversion of the wastes into energy may require understanding the weather's behaviour trends at the specified site. Understanding weather trends is imperative for the development of systems to convert wastes into biogas that will be installed on or above the ground. Biogas generation is affected by many factors, which include the ambient temperature. Therefore, the system for biogas generation installed above the ground, can be controlled, and its performance improved by integrating it with other systems exploiting solar and wind energy resources and sources.

The integration of these systems requires the use of optimisation techniques. The techniques usually need the input variables, including the information on the characterization of the biomass wastes and weather-related renewable energy resources and sources. The characterization has to be done comprehensively with minimum risk of obtaining erroneous information. The integrated renewable energy system, developed based on the reliable, comprehensive characterization approach, has the least risk of performing poorly. Also, the system touches on improving the living standards of human beings. It may produce bio-fertiliser for crop production, biogas for cooking energy, and electricity for powering electric and electronic gadgets. Therefore, guaranteeing high performing renewable energy systems will increase the adoption of the renewable energy systems, and for that matter, boost the renewable energy business and industry.

This section of the chapter presents the novel generic comprehensive characterization approach and demonstrates its application. The demonstration of the application of the approach included the determination of information of the properties of cow dung, which is one type of biomass materials. It also involved searching for information about the trends of weather-related renewable energy resources and sources in the semi-arid areas at the micro-scale level. Solar, wind and ambient temperature conform to sources and cow dung is a resource.

The demonstration produced comprehensively detailed findings. The findings are useful for developing and modeling the intended integrated renewable energy system (IIRES) in rural semi-arid areas. The approach framework is presented in Figure 1.

Figure 1. The comprehensive characterization framework

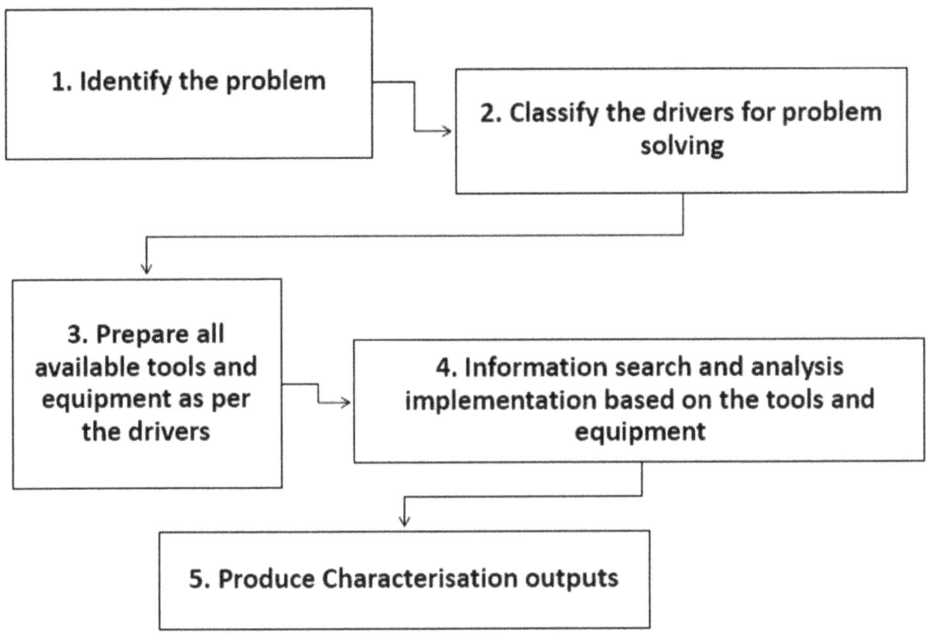

The figure presents the five generic steps of the approach. The first step is to identify the problem, which may include the energy need for the community, lack of knowledge of the environmental behaviour trends where the energy systems can be installed, and understanding the energy systems' performance. The second step involves the classification of the drivers useful for solving the problem. Such drivers may include the types of available energy resources and sources, the types of energy conversion technologies to be applied, the scope of the area to host the technology, and the energy systems available in the targeted area. The third step is to prepare all the available tools and equipment as per the drivers to establish the needed information for characterization. The equipment may include, data collection devices, weather stations, lab equipment, and data processing tools. The fourth step is the implementation of the information searching and analysis of the data obtained based on the tools and equipment available. In this step, the layout of tools or equipment for data collection, onsite sampling tools and lab equipment are prepared; then, they are activated for data collection, sampling, sample analysis, data analysis, and processing the results. Also, the interpretation of the results follows to prepare the characterization findings report. The fifth step, which is the last step in the sequence of the approach's steps, is the output presentation. This step involves the

Figure 2. Location of the study site

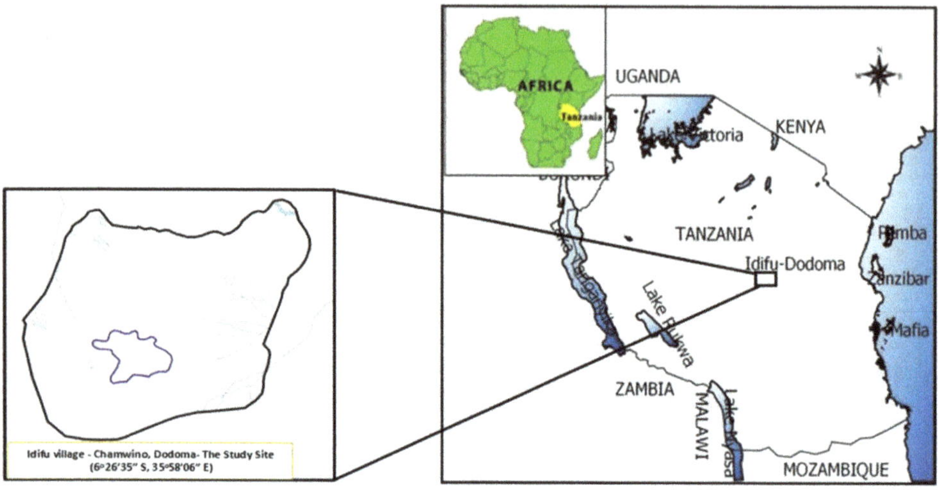

presentation of the characterization findings reports in soft or hard copies based on the above four steps. The reports may include the figures and tabulated data results.

The information in the reports can be referred to for developing the energy system and or developing the system performance model and environmental behaviour model. In accomplishing this task, the solutions of the identified problem or problems in step one are obtained, as shown in Figure 1. The solutions may include the developed renewable energy system to generate energy, the established model for understanding the environmental behaviour, and the developed model for evaluating the system's performance.

The study site (Figure 2), and the comprehensive research characterisation of the collected biomass (cow dung) and weather parameters approach at the study site are explained in this section. Furthermore, the approach demonstration is detailed in the sub-sections titled Cow Dung Characterisation and Weather Parameters characterisation of this section. Finally, the findings, based on the approach mentioned above are presented and discussed in the sub-section titled Results and Discussion of this section.

Demonstration of the Comprehensive Characterization Approach Application

The challenges were to generate sustainable, reliable, affordable and clean energy in the rural semi-arid areas. The energy was needed primarily for cooking and powering electric and electronic gadgets in rural households. The research was conducted at

Idifu village located in the Chamwino district in Tanzania's Dodoma region (Figure 2). The site is geographically located around the point coordinate values of 6°26'35" S, 35°58'06" E. The district represents the semi-arid climate where considerable insufficient energy is evident for households' purposes. It has one rainfall season with an average of 350-500 mm per annum. In addition, the prevalence of high food insecurity characterises it.

Idifu village and the neighbouring areas face several constraints, including inadequate access to sustainable and reliable energy. The inhabitants of this village are agro-pastoralists, who grow pearl millets and keep cattle under the half-stable system. Cattle are grazed in the village and the neighbouring communities' fields and kept in the kraal during the night. The food system is mainly cereal-based, with pearl millet as the preferred staple (Mutabazi, 2013).

The village was a beneficiary of the interventions of the Trans-SEC (Trans-SEC – *Innovating Strategies to Safeguard Food Security Using Technology and Knowledge Transfer: A People-Centred Approach*) project of introducing upgrading strategies for the improvement of community livelihood, especially on food security-related issues as reported by Mwinuka *et al.* (2015). One of the strategies and a basis for this study was the beneficial conversion and utilisation of agricultural and animal wastes for generating useful energy and bio-fertilizers.

Moreover, the study involved the laboratory analysis of biomass (cow dung). The cow dung sample analysis was done in the Departments of Animal Science, Soil Science and Geology laboratories at Sokoine University of Agriculture (SUA), located in Morogoro, Tanzania, 301 km South-East of the study site.

Cow Dung Characterization

The potential minimum amount of cow dung that could be collected for biogas generation at the study area was quantified. The cow dung amount was estimated using Equation 2.0. Kahimba *et al.* (2015) and Mwinuka *et al.* (2015) found that the study area had 234 households owning cattle. On average, each household owned ten cattle, implying that the total number of cattle in the area was 2340.

$$CD_{min} = P\_CD_{c} \times CD\_ac \times P\_ac \times T\_hc \quad \text{... Eq.2.0}$$

Where CD_{min} is the minimum potential amount of cow dung collected per day ($kg.day^{-1}$); P_CD_{c} is the percentage of cow dung collected based on the type of grazing system of cattle, which was taken as 60% (%); CD_ac is the amount of cow dung excreted by adult cattle in a day ($kg.adult\text{-}cattle^{-1}.day^{-1}$); P_ac is the percentage of adult cattle in the study area (%); and T_hc is the total number of cattle in the study area (-).

Zhu *et al.* (2020) revealed that 40% of cow dung is lost to the grazing area in the East Africa semi-arid area. The phenomenon occurs when the half-stable grazing system of keeping cattle is used, which is also the case in the study area. Also, Zhu *et al.* (2020) reported that adult cattle could produce 25 kg of cow dung and 21 litres of urine per day. Realistic quantification of cow dung amount was based on the assumption that the herd of cattle has 75% of adult cattle. Therefore, the amount of cow dung obtained using Equation 2.0 was the minimum of the total possible amount of cow dung, which may be collected in the study area.

Samples of cow dung for laboratory analysis were collected from 35 households in the study area. The households involved in the purposive survey were selected based on owning at least two cattle. In addition, cattle ownership was depicted from secondary data of baseline survey conducted by the Trans-SEC project between January and February 2014. The survey was reported by Mwinuka *et al.* (2015) in the scope area of the project, including the study area.

The household sample size was computed using the online open-source statistical calculator, the OpenEpi-SSPropor version 3 (Sullivan *et al.*, 2009). The input data to the statistical calculator were population size (234 households), hypothesised percentage frequency of outcome factor in the population (30±5%), confidence limits (5%) and design effect factor value (0.3). The researcher set the population size and confidence limits, while the calculator's remaining input data were default values. The results obtained from the calculator included the 35 households as the sample size at a 90% confidence level. The sample size was selected, as the confidence level was adequate for this study. The sample size included at least one household selected randomly from each of the seventeen sub-villages of Idifu village (the study site) with ownership of at least two cattle to make a fair representation of the administrative units.

Sampling of cow dung was done as per the sampling technique and procedures of solid manure recommended by Peters *et al.* (2003), which stresses the homogeneity of the sample from the stockpile of cow dung. The homogeneity was ensured by taking ten sub-samples from different locations around the pile at least 45 cm below the surface. The sub-samples were mixed to form a composite sample weighing 450 g, which was then kept in a plastic zip-lock bag after squeezing out entrapped air.

After sampling, samples were sealed inside a double-walled plastic bag, kept at a temperature below 4°C in a cold conditioned container. Low temperature inhibited the biodegrading of the samples. Then the samples were transferred to the laboratories at SUA in Morogoro within 19 hours. The 19 hours' period is below the maximum holding time of seven days recommended for samples at 4°C before specific analysis (Peters *et al.*, 2003). The analytical results obtained were compared with published results of samples from feedstock used for biogas technology. The comparison aimed at checking whether the available cow dung at the study site qualified to be the feedstock for an IIRES involving biogas production. The physicochemical analyses of the cow dung samples are described in the following subsections.

Determination of Organic Carbon (OC) by Wet Titration

Organic Carbon determination of cow dung samples was done using the adapted Walkley-Black chromic acid wet oxidation method (Moberg, 2001). Four samples from the same batch were used in this analysis.

Determination of Organic Nitrogen (ON) and Other Cow Dung Properties

Organic Nitrogen (ON) or total nitrogen, ammonium nitrogen (NH_4-N) and pH of cow dung samples were determined by the standard method illustrated by Peters *et al.* (2003). However, moisture content (MC), dry matter (DM) or total solids, volatile Matter (VM) or volatile solids and ash content were found by use of the standard method described by García *et al.* (2012). Carbon Nitrogen ratio (C/N) was computed by dividing the values of OC with ON on a mass basis (Yadav *et al.*, 2013). The four samples from the same batch of cow dung were used to analyse each cow dung property mentioned.

Weather Parameters Characterization

Weather parameters considered in this study are solar irradiance, wind speed, wind direction and ambient air temperature. The parameters' data were obtained from a weather station installed at the study site. The data was collected from 1st December 2016 to 31st January 2018. However, data collected from 1st January 2017 to 31st December 2017 were considered in the analysis, while data out of this range were used for comparison purposes only. The weather station for data collection consisted of the following devices:

1. A pair of Umwelt-Geräte-Technik GmbH (UGT) Pyranometer CMP 3 with measuring range of 0.00-2000.00 W / m^2, the sensitivity of 5 - +20 μv / W / m^{2}v, the spectral range of 0.31 - 2.8 μm, the accuracy of 0.01 W/m^2 and UGT data-logger with the designation of DL 389 (www.ugt-online.de) set to sample solar radiation values after one-hour intervals;
2. Seeeduino Stalker Microcontroller Version 2.3 interfaced with a computer by use of Arduino software Version 1.0.5-r5 and set to sample wind speed (m/s)

Figure 3. Layout of the installed weather station at the study site

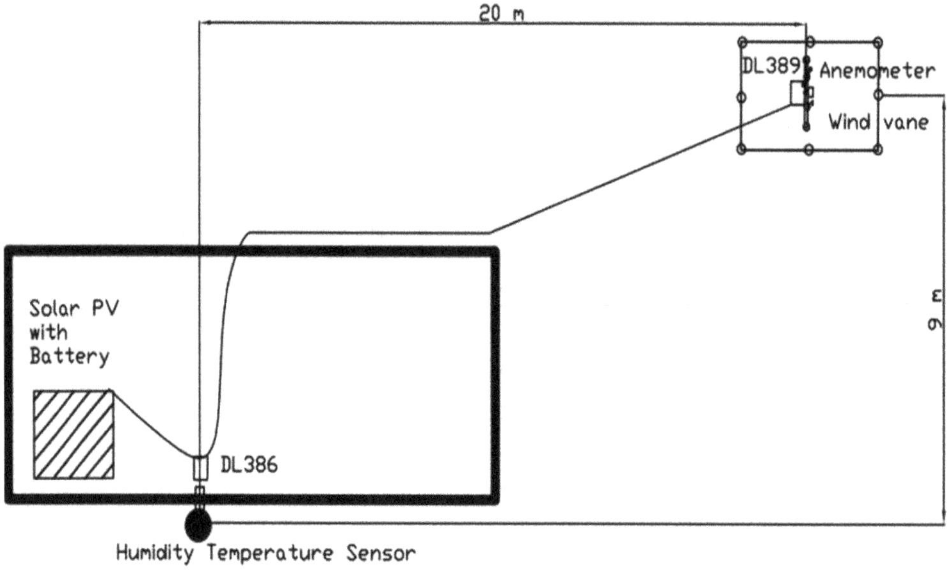

from an anemometer with an accuracy of 0.01 m/s and wind direction (degree) from a wind vane with the accuracy of 0.50 at 30 minutes' intervals (Kiobia, 2015), and

3. UGT Humidity and temperature (thermocouple Type-K) sensor with Relative Humidity (RH) measuring range of up to 100%, humidity accuracy of ± 2% and temperature measuring range of -30.00 - +70.00°C with the accuracy of 0.01oC. It was connected with UGT DL 386 data logger. The sensor was set to sample ambient air temperature values at one-hour interval.

The weather station was powered by a solar PV module (30 Wp) with a backup battery system of 12V- 14Ah. The layout of the weather station is shown in Figure 3.

The detailed procedures for characterizing the above weather parameters are illustrated in the subsequent parts of this sub-section.

Solar Radiation Characterization

The solar radiation measurements were downloaded from the UGT data logger (DL 389), cleaned, and used to compute daily hourly average, maximum and minimum solar irradiance. The solar irradiance trends from remotely sensed information systems do not differ much from those collected on the ground (MNRC, 2015). Therefore, it was relevant to use the secondary data from a different source to get the monthly average insolation prevailing at the study site. The criteria of choice of secondary data source were that data should have been collected for more than ten years and are immediately freely available online or offline. Two sources were selected based on these criteria. They include Solar Electrical Handbook data and Ret-Screen plus data. On plotting data from these sources, they were overlapping with each other. Therefore, the average value of insolation data from these sources was found and plotted. This average data was taken as insolation data for the study site. The insolation data was checked for validity in terms of the trend using data collected at the study site in the specified period. Converting solar irradiance into solar insolation is explained in Wald (2007). The conversion formula is shown in Equation 2.1.

I= ⟦Sr⟧ _a×h_d .... Eq. 2.1

Where ***I*** is monthly average insolation (kWh/m^2/day), $\boldsymbol{Sr_a}$ is the monthly average solar irradiance based on recorded daily data in the specific month at the study site, and $\boldsymbol{h_d}$ is an assumed monthly average daily hours of bright sunshine, i.e. 9 hours (range of 7-11 hours), at the study site for each month.

The Solar Electrical Handbook data is based on over 22 years up to 2017 (temporal coverage) to provide average monthly insolation (temporal resolution). Its spatial coverage is global; the spatial resolution was not given (Boxwell, 2017). RetScreen Plus data is based on the available parameter of insolation on a horizontal surface at temporal coverage ranging from July 1983 to the near present. Also, it has a daily temporal resolution. Furthermore, its spatial coverage is global, while its spatial resolution is 1° x 1° (Johnston and Matlock, 2011). The lowest monthly average insolation was taken as design insolation. It is meant for computing the minimum energy harnessed by the solar PV panel. The solar PV unit was intended for solely generating electricity for powering electric and electronic gadgets in the household. Also, the design insolation was used in predicting the amount of heat energy to be collected by other energy systems involving temperature-sensitive microorganisms. The design heat energy collected or designed electrical energy produced by the IIRES's solar PV panel was computed using Equation 2.1(a) with an assumed solar collector area of 1 m^2.

$Hc = 0.67\ Y_d \times I$.. Eq. 2.1(a)

where is the design heat energy collected or designed electrical energy produced by the IIRES's solar PV panel, Y_d represents days in a year, i.e. 365 days, *I* is the design insolation, and 0.67 is an intuitively assumed factor, which takes care of days in a year when the sun is available to generate solar radiation at the study site; usually, sunshine is guaranteed between April and November (https://en.climate-data.org/).

Also, annual heat energy collected was determined using Equation 2.1(a). Annual average insolation, in this equation, replaced the design insolation during analysis. The annual average insolation was based on the average values of monthly insolation from secondary data sources. Therefore, characterising solar irradiance was important for designing an IIRES for rural semi-arid areas of central Tanzania, which will involve a solar energy component.

Wind Speed and Wind Direction Trends

Seeeduino Stalker microcontroller stored wind speed and direction data from the anemometer and the wind vane. Then wind speed data was downloaded, cleaned and transformed, averagely, from 30 minutes' intervals into one-hour intervals. However, wind direction data was not transformed. After that, the transformed wind speed data was used to determine hourly and monthly average, maximum and minimum wind speeds. The Ret-Screen Plus wind speed data source was used for extracting the monthly average wind speed of Dodoma. Dodoma Region is where the study site is located. This data was used for comparison reasons. The monthly average wind speed data was used to compute the annual wind speed at the study site from both data sources. The Weibull cumulative distribution function (Bidaoui *et al.*, 2019) was used to determine the probability (Equation 2.1b) that wind speed events equal to or above the annual average wind speed at the study site using the experimental data. Finally, all data determined were plotted.

$$F(v) = P(v \geq v_m) = e^{-(\frac{v_m}{C})^k} \quad \text{Eq.} \quad (2.1b)$$

where $\boldsymbol{v}$ is the wind speed, which is equal to or above the average wind speed at 4 m hub height at the study site v_m, $\boldsymbol{e}$ is an exponential function, $\boldsymbol{c}$ is the characteristic wind speed or scale factor at the study site (m/s), and $\boldsymbol{k}$ is the shape parameter (dimensionless). Note: that v_m was extrapolated from the reference height of 2.5 m, where the anemometer in the weather station was positioned. The extrapolation was done using Equations 2.2 and 2.2a.

Parameters $\boldsymbol{k}$ and $\boldsymbol{c}$ were determined using Equations 2.1c and 2.1d, respectively (Kutty *et al.*, 2019).

$$k = \left[\frac{\sigma}{v_m} \right]^{-1.086} \quad \text{Eq. (2.1c)}$$

$$C = v_m \left(0.568 + \frac{0.433}{k} \right)^{-\frac{1}{k}} \quad \text{Eq. (2.1d)}$$

where σ is the standard deviation of wind speed values from the mean wind speed.

Moreover, the applied wind direction data generate the wind roses, which show the predominant wind blowing directions monthly and wind speed bins, i.e. wind speed intervals, at the study site. They are expressed in wind speed (m/s) and frequency percentage (%). The anemometer and wind vane were installed at 2.5 m to capture data under the influence of pearl millet canopy (maximum height around 2.5 m) during the cropping and off-cropping seasons (Kanyeka *et al.*, 2007). It is because the IIRES that was intended to be installed in the rural semi-arid area of central Tanzania was supposed to include a wind turbine with a hub height of around 4 m.

In comparison with 10 m usually used in wind measurements, the hub height of 4 m adopted in this study was low. Also, most wind data for wind turbine design is referenced at 10 m high above the ground. However, 4 m hub height was chosen to take the advantage that most of the time throughout the year, the study site has short grasses, which may not affect the wind regime. In addition, the same hub height was assumed to reduce the cost of the system to be installed. The guiding formula was used to extrapolate data from 2.5 m to 4 m hub height. The extrapolation formula applied was the power-law based formula. It is represented by Equation 2.2 (Safari and Gasore, 2010). The extrapolation of unknown wind speed (v) in m/s at a given level of height (Z) in metres from reference height (Z_0) in metres with known wind speed (v_0) in m/s above ground may be computed by Equation 2.2.

$$v = v_0 \left(\frac{Z}{Z_0} \right)^{\alpha} \quad \text{Eq.2.2}$$

where v_0 is the wind speed measured at anemometer height Z_0, v is the wind speed to be calculated at the height Z, α is the power-law exponent depending on the surface roughness and was obtained using Equation 2.2a (Bidaoui *et al.*, 2019, Elamouri and Amar, 2008).

$$\alpha = \left(0.37 - 0.088 ln v_0\right) / [\left(1 - 0.088 ln\left(\frac{Z_0}{10}\right)\right] \text{.................... Eq.} \qquad (2.2a)$$

Note that the lowest, highest average monthly wind speed and maximum (gale) wind speed are essential parameters derived from the collected wind speed data. Therefore, they are useful for conceptual designing the wind turbine system to generate energy.

Ambient Temperature Trends

The procedure described in the sub-section titled solar radiation characterisation was used for downloading data from the UGT data logger (DL 386). The data was cleaned and used to determine both hourly and monthly averages, maximum and minimum ambient air temperature. However, secondary data was not considered for this case. The ambient air temperature data is vital as temperature influences the biogas microbes' performance to generate biogas in the anaerobic digestion systems (Schnurer and Jarvis, 2010). As the IIRES was supposed to incorporate a unit for anaerobic digestion of cow dung, characterizing this weather parameter data at the study site was crucial.

Results and Discussion

Based on the methodological approach explained in the subsection titled Demonstration of the Comprehensive Characterisation Approach Application, results were obtained as presented in this subsection. The results include cow dung sample properties and weather parameters.

Availability of cow dung and its properties

Quantification of the amount of cow dung using Equation 2.0 showed potential availability of a minimum of 26,330 kg of cow dung to be collected per day in the study area. Assuming about 20% biogas is generated from one kilogramme of cow dung and the biogas' density is 1.15 kg/m^3 the available cow dung per day at the study area can generate about 4,600 m^3 in a day. Also, the cooking stove can be kept on for more than 85 days using the amount of biogas collected in one day. The value of 85 days is derived by conversion of the number of hours into days

Table 1. Laboratory analysis results of the sampled cow dung

Property	Amount found	Standard deviation	Number of samples
Moisture content (MC)	77.11%	0.52%	4
Dry matter (DM)	228.93 g/kg of feed	5.21 g/kg	4
Volatile matter (VM)	744.25 g/kg of DM	53.20 g/kg	4
Organic nitrogen (N)	2.24% by TKN in DM	0.09%	4
Ammonium Nitrogen (NH4-N)	21.06% in DM	0.64%	4
Organic carbon (C)	43.37% in DM	2.87%	4
pH	6.90	0.04	4
Carbon Nitrogen ratio (C/N ratio)	19.30	0.56	4
Ash Content	21.20%	5.35%	4

by dividing the hours by 24. The number of hours is obtained by multiplying the assumed 0.45 m^3/h biogas flow rate of the cooking stove and the potential biogas amount obtained per day.

The characterisation results of cow dung sample properties revealed that volatile matter (VM) and carbon-nitrogen ratio (C/N) was 744± 53 g/kg of dry matter and 19±0.6, respectively. These properties fall within the ranges of 60% to 89% on a dry matter basis and 15 to 30 for VM and C/N, respectively, which are suitable for feedstock (cow dung) utilization in biogas generation as argued by Weiland (2010) and Raposo *et al.* (2012). Moreover, other important cow dung parameters Characterised, and their results are presented in Table 1.

The cow dung characteristic results indicate that quality feedstock is available at the study site. The findings justify the establishment of IIRES involving biogas technology in the region.

Weather Characteristics

The weather parameters assessed include solar irradiance, wind speed, wind direction and ambient air temperature. First, the solar irradiance analysis results are presented in Figure 4. The first daily result item consists of hourly average, hourly maximum and hourly minimum solar irradiance, as illustrated for the year 2017. Figure 4 shows that the maximum solar radiation values in a day reached 1280 W/m^2 at 13:00 hours; also for five hours, i.e. 10:30-15:30 hours, the maximum solar irradiance was above 1000 W/m^2.

Besides, the hourly average solar irradiance was above 200 W/m^2 for nine hours a day. On the other hand, the hourly minimum solar irradiance was 25 W/m^2 for

Figure 4. Hourly average, maximum and minimum solar irradiance for the year 2017

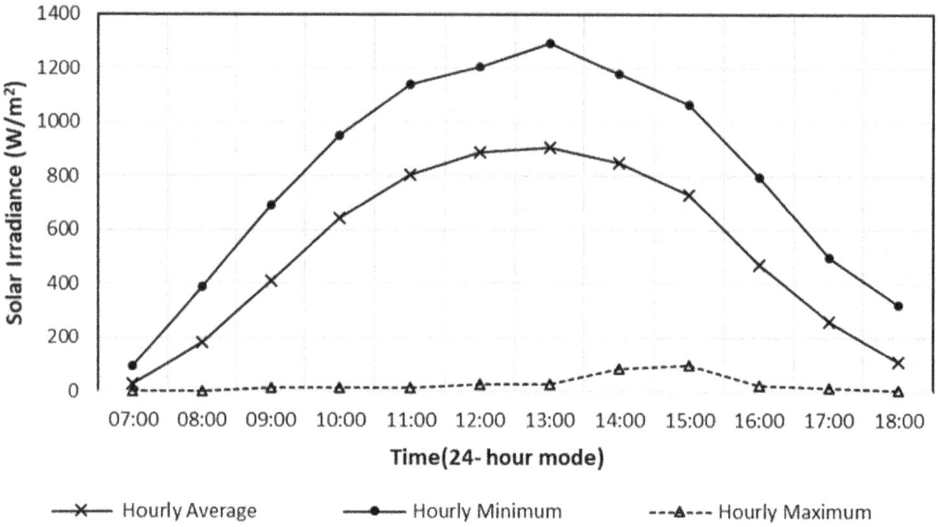

three hours a day. However, comparative data of the Dodoma region, from other researchers, on hourly solar irradiance, aggregated daily, were not available.

The second result item is the monthly average solar irradiation or monthly average insolation based on primary and secondary data, as shown in Figure 5. This figure reveals that the monthly average solar insolation was at the minimum in May. For all three data sources of experimental test, Ret-Screen and Solar Electrical Handbook solar insolation data values were 3.76 kWh/m^2/day, 5.38 kWh/m^2/day and 5.19 kWh/m^2/day, respectively. The average solar insolation was 5.29 kWh/m^2/day. It was based on secondary data sources and taken as the design solar insolation at the study site. The value is found in the curve of Average Data (Dodoma) [ADD] in Figure 5.

The annual average solar insolation based on ADD and experimental data was 5.9 kWh/m^2/day and 4.8 kWh/m^2/day, respectively. The difference between these values is almost 23%, which is enormous. This enormous difference was caused by the change of monthly average insolation computation procedures, which produced these values, as explained in the subsection titled solar radiation characterisation of this Chapter. Therefore, the 5.9 kWh/m^2/day value was taken as the annual average solar insolation of the study site. The revealed information indicates that solar radiation is abundant and can be harnessed with these values.

Equation 2.1(a) gives design solar insolation and annual average solar insolation values, which were obtained at the study site. They were used to produce the design heat and average annual heat to be collected at the study site. The design heat and averagely heat collected at the study site were 1.3 MWh and 1.4 MWh, respectively.

Figure 5. Monthly average solar radiations at the Study site (Jan-Dec 2017)

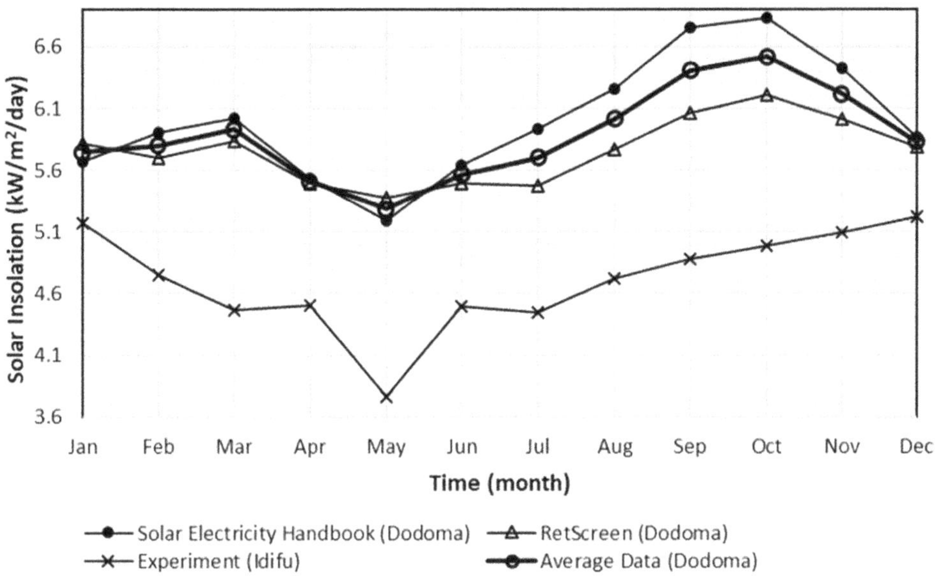

This procedure was adopted to find the designated electrical energy produced by the assumed IIRES's solar PV panel with 1 m^2 and efficiency of 13%. The value obtained was 168.2 kWh.

In the case of wind speed, the hourly and monthly average, maximum, and minimum daily wind speeds are presented in Figures 6 and 7.

Figure 6 shows that the average wind speed was above 4 m/s for 16 hours from 7:30 hours. The maximum wind speed of 19.2 m/s was recorded at 21:00 hours. Also, zero wind speed events were minimal for the window of 9:30 to 16:30 hours, except at 13:00 hours. Furthermore, in most cases, the lowest average wind speed of 2.83 m/s was recorded in the morning at 6:00 hours.

Figure 8 shows that the experimental monthly average wind speed was above 4 m/s from August through December. However, in the same figure, the Ret-Screen Plus wind speed data indicate that wind speed was above 4 m/s from June through October. The gale (wind speed above 16 m/s in this case) occurred in March, July, October and November by referring to maximum wind speed data in Figure 7. In some of the days of December through May and July, and August, wind speed events of 0 m/s were occurring. Generally, the study site has a predominant wind speed that falls in a range of 4-6 m/s. The lowest average monthly wind speed was recorded in March with a 2.3 m/s value. It is close to the Ret-Screen Plus data, which indicated that the lowest monthly wind speed was 2.5 m/s in March. Therefore, it shows that

Figure 6. Hourly average, maximum, and minimum wind speed (Jan.-Dec. 2017)

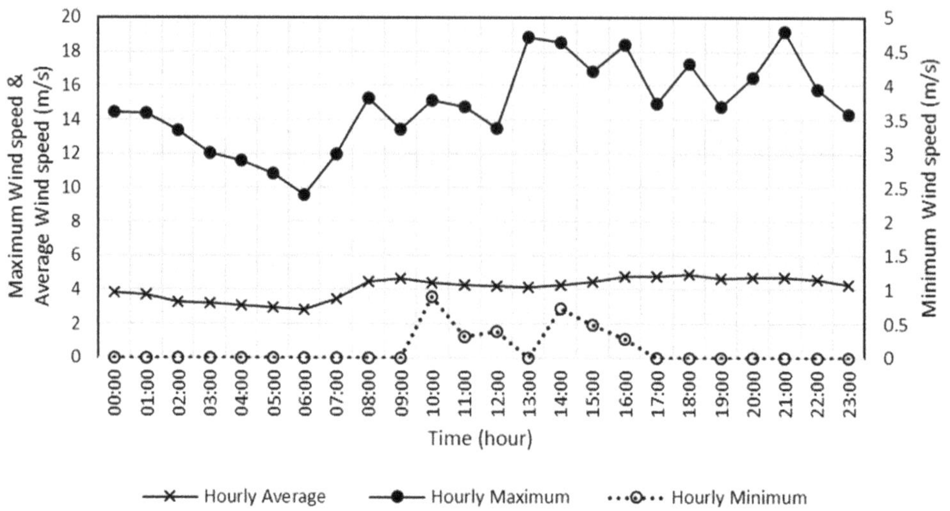

Figure 7. Experimental testing and Ret-Screen Plus monthly wind speed data in 2017

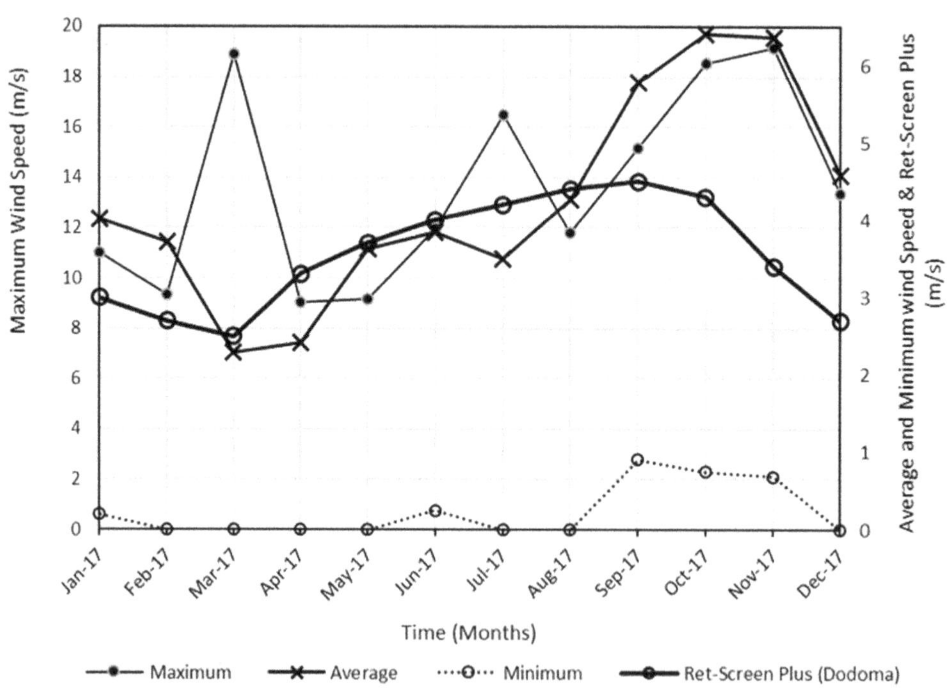

Figure 8. Weibull cumulative distribution analysis results of wind speed at 2.5 m hub height in 2017

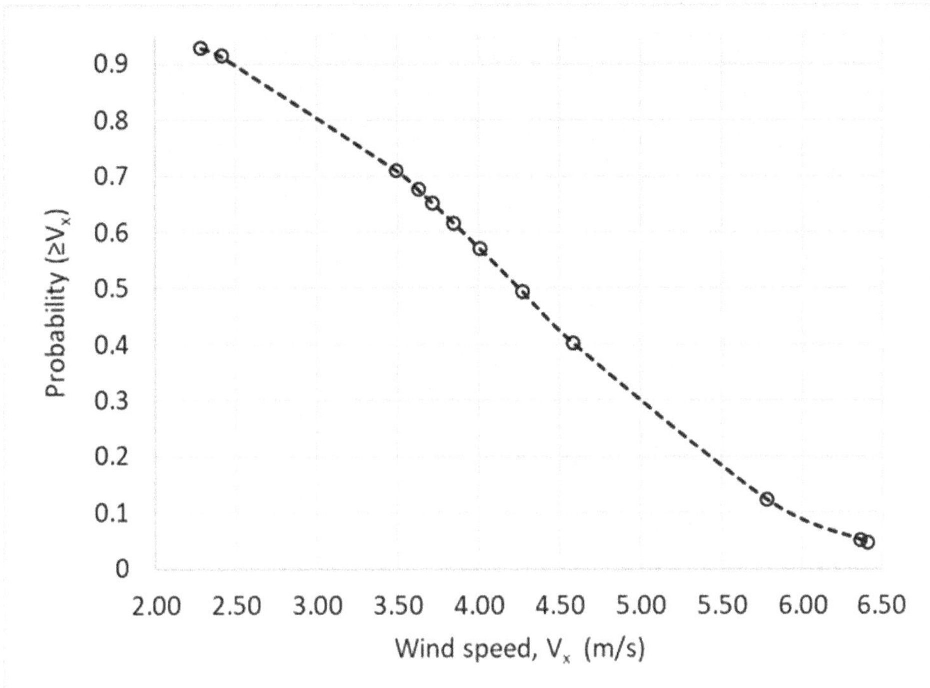

the experimental data are reliable as they fall into the same data trend produced by a proven institution.

The experimental monthly wind speed data were used to extrapolate the monthly wind speed at 4 m hub height. The Weibull cumulative distribution analysis of data at hub heights of 2.5 m and 4 m was done. The results are shown in Figures 8 and 9, respectively. A cut-in wind speed range of 3.00 - 4.00 m/s is for small wind turbines based on Arnett *et al.* (2013) research work article. Assuming the IIRES will incorporate a small wind turbine with a cut-in wind speed above 4 m/s, say 4.5 m/s, then from Figures 8 and 9, the probability of having wind speed events equal to or above 4.5 m/s at 2.5 m and 4 hub-height in a year will be the same, which is 0.441. This probability value indicates that the system operating above that wind speed may need to be integrated with other systems for generating energy at the probability value of above 0.5. The 4.5 m/s may be taken as the design wind speed for further engineering design works.

All wind speed data collected in a year at the study site were used to generate the wind roses. The wind roses were derived from the extrapolated wind data at 4

Figure 9. Weibull cumulative distribution analysis results of wind speed at 4.0 m hub height in 2017

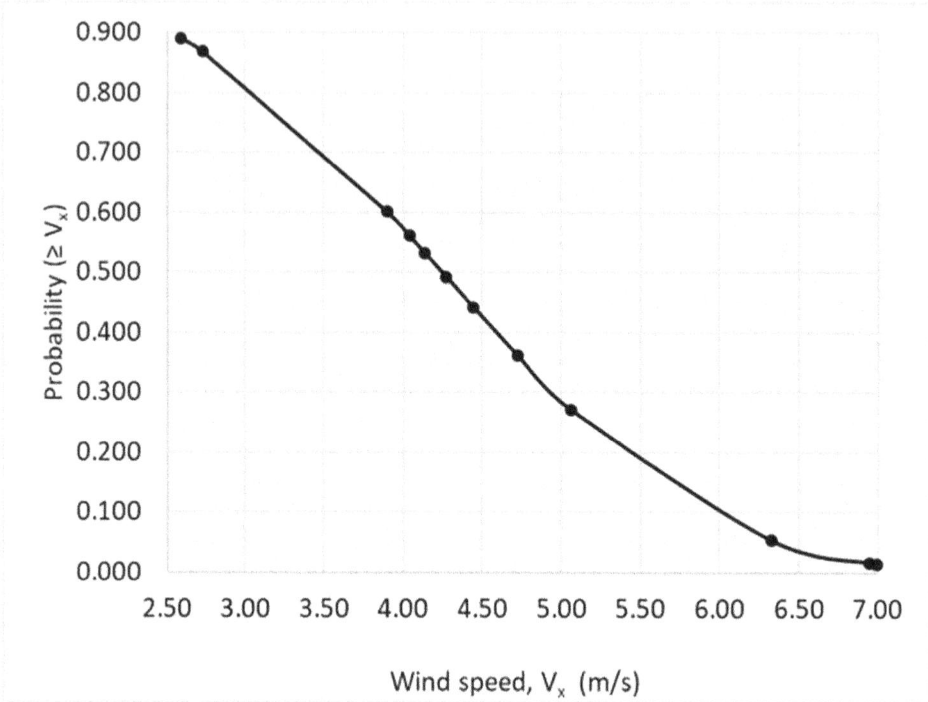

m hub height, referenced from 2.5 m hub height. The wind roses present the wind direction data at the study site.

The average wind speed in a given month and wind speed against the wind direction are presented in Figures 10 and 11, respectively. The frequencies of wind blowing directions in percentage at a given month and wind bin against wind direction are presented in Figures 12 and 13, respectively.

Figure 7 shows that wind speed is above 4 m/s and high in June through October. During these months, Figure 10 shows that wind was blowing from the ranges of the direction of North (N) to North East (NE) and South (S) to South West (SW), clockwise. Also, it blew from the West (W) direction. The wind speed in these directions at 4 m hub height was blowing at an above annual average wind speed of 4.69 m/s.

Figure 11 shows that the annual average wind speed (4.7 m/s) falls in the wind speed bin of 4.44-6.56 m/s at 4 m hub height. This wind speed bin occurs in the following ranges of directions: clockwise: North (N) to North East (NE) and South

Figure 10. Average wind speeds in a month per wind direction at 4 m hub height in 2017

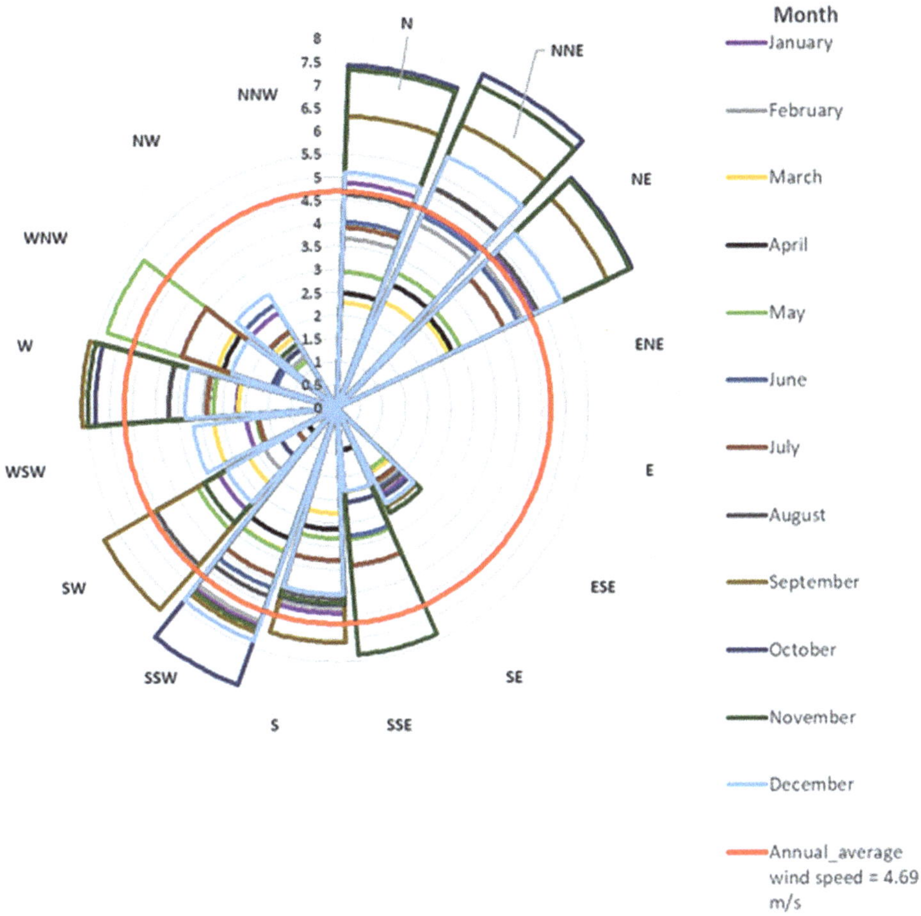

East (SE) to South West (SW). Also, the wind blew from the West (W) and West North West (NW) direction from the same figure.

The frequency percentage of direction from which wind blows monthly at the study site is presented in Figure 12. The figure shows that almost all months' wind predominantly blew from North East (NE) direction followed by the North (N) direction. Other directions in which annually wind blowing frequency is significantly noted are South(S), South-South West (SSW) and West South West (WSW). Figure 13 shows a similar trend as Figure 12 regarding wind directions when considering the whole wind speed bins. However, the wind speed bin of 4.44-6.56 m/s is predominantly falling in the North East (NE) direction, and the wind is blowing from there. This wind speed bin includes the annual average wind speed at 4 m

Figure 11. Wind speed bin's average wind speed per wind direction at 4 m height in 2017

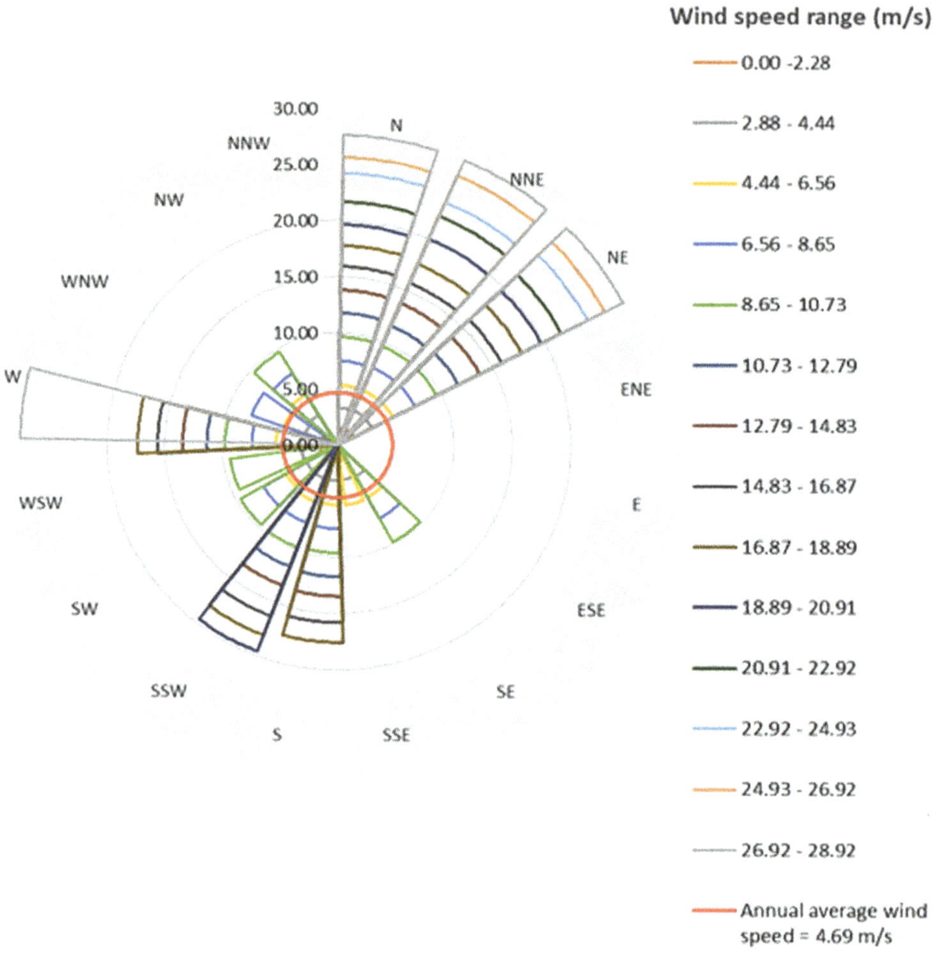

hub height at the study site, as was noted in the previous paragraphs. Therefore, it is essential to consider the wind blowing direction data if the installation of IIRES may need the information on the wind blowing direction.

Another weather parameter characterised was ambient air temperature. Characterisation resulted in hourly and monthly maximum, average, and minimum ambient temperature trends shown in Figures 14 and 15. Figure 14 shows that the maximum ambient air temperature was 38°C recorded at 16:00 hours. The minimum ambient air temperature was 12 °C recorded at 6:00 hours. For the whole data collection period, the average ambient air temperature was 25°C, with a standard deviation of 4°C within a day. The ambient air temperature was averagely above

Figure 12. Monthly wind blowing events' percentage frequency per wind direction in 2017

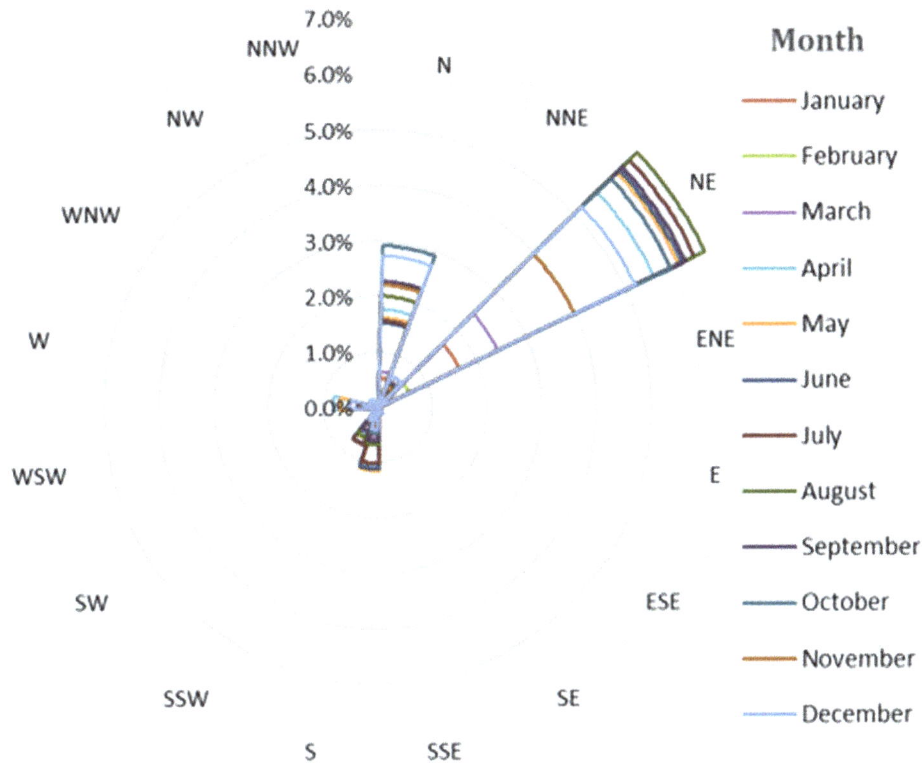

25°C for ten hours from 10:00 hours through to 20:00 hours. For the same ten hours, the difference between maximum and minimum ambient air temperature values was averagely 13°C, with a standard deviation of 3°C. The temperature variation at the study site within a day was expressed in the standard deviation (4°C) by considering the average ambient air temperature values for 24 hours.

Additionally, from Figure 15, it is depicted that the monthly average ambient air temperature was recorded to be below 25°C from April through September. However, the lowest and highest ambient air temperature values were recorded in July (12°C) and October (38°C), respectively. Furthermore, the months of October,

Figure 13. Wind blowing events' percentage frequency in a wind speed bin per wind direction in 2017

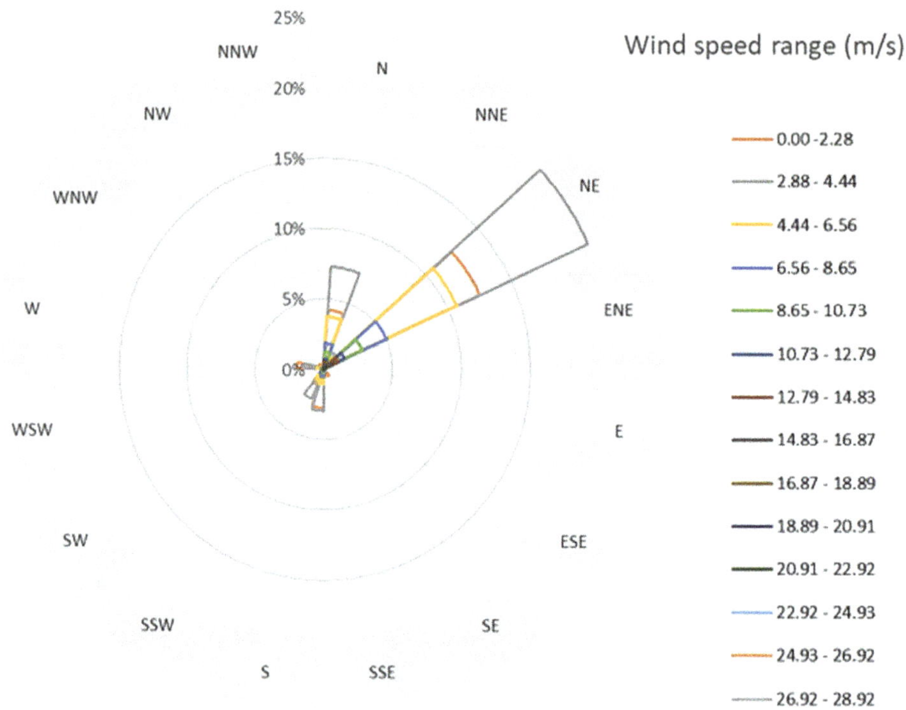

November and December had high average, maximum and minimum ambient air temperature records.

This trend is similar to the one obtained from RETScreen Plus software for the Dodoma region (Ganoe *et al.*, 2014). However, the RETScreen Plus data shows that ambient air temperature was the lowest of all months in June. Therefore, the information obtained is useful in planning and developing a renewable energy system that can endure temperature variations and be installed on the ground outside the respective household.

The IIRES involving biogas technology performs well at a constant temperature ranging from 25°C to 55°C, depending on the state of operating temperature. Therefore, designing the IIRES requires collecting and storing heat at high temperatures during the day to raise the temperature in the morning. This phenomenon would keep the temperature of the biogas digester of the IIRES significantly constant. Also, a strategy should be set for operating the IIRES during the low-temperature period, which is around July.

Figure 14. Hourly average, maximum, and minimum ambient air temperature in 2017

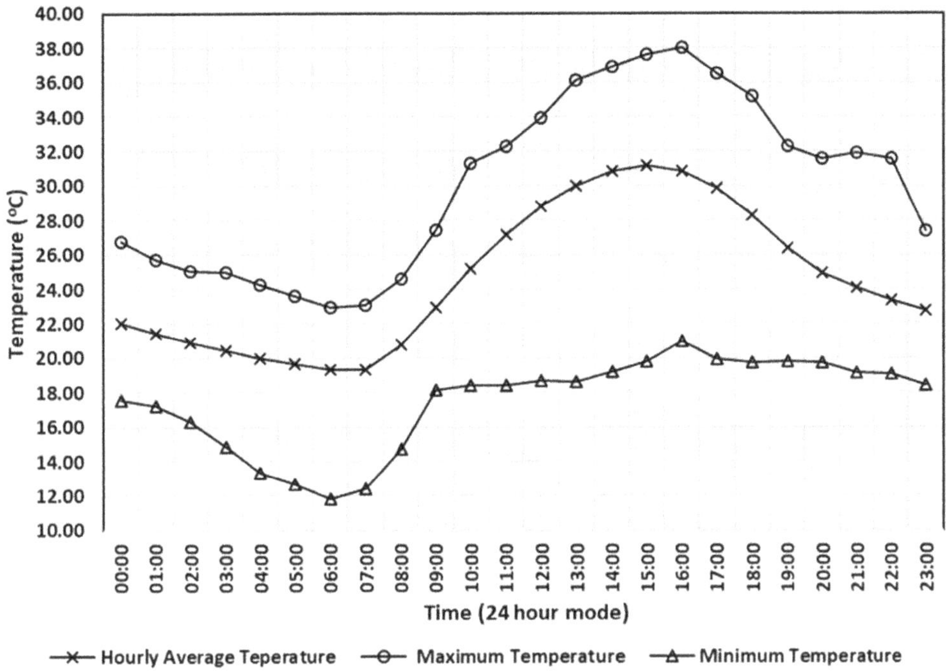

SOLUTIONS AND RECOMMENDATIONS

A comprehensive characterisation of the locally available renewable energy resource (cow dung) and sources (weather factors) research approach has been formulated, and its application demonstrated. The demonstration has produced the findings revealing that locally available cow dung per day at the study site is abundant (production of 26,330 kg/day). The cow dung amount can generate biogas amounting to 4,600 m^3/day. Also, the cow dung has properties (744±53g/kg and the C/N ratio of 19) suitable for biogas generation. Furthermore, characterisation results indicate that there was abundant solar radiation (annual average solar insolation of 5.9 kWh/m^2/day and solar insolation above 4.5 kWh/m^2/day for seven months) at the study site. Any existing solar energy technology can harness this amount of solar energy.

Moreover, the study site has a predominant wind speed, which falls within the range of 4-6 m/s in a year, while the annual wind speed is 4.7 m/s at 4 m hub height (Figures 10 to 13). In addition, wind speed is lowest in March (2 m/s) and highest in October (6 m/s). These data show that the wind speed available at the study site is adequate to transform wind energy into usable energy. Also, wind direction at the

Figure 15. Monthly average, maximum and minimum ambient air temperature in 2017

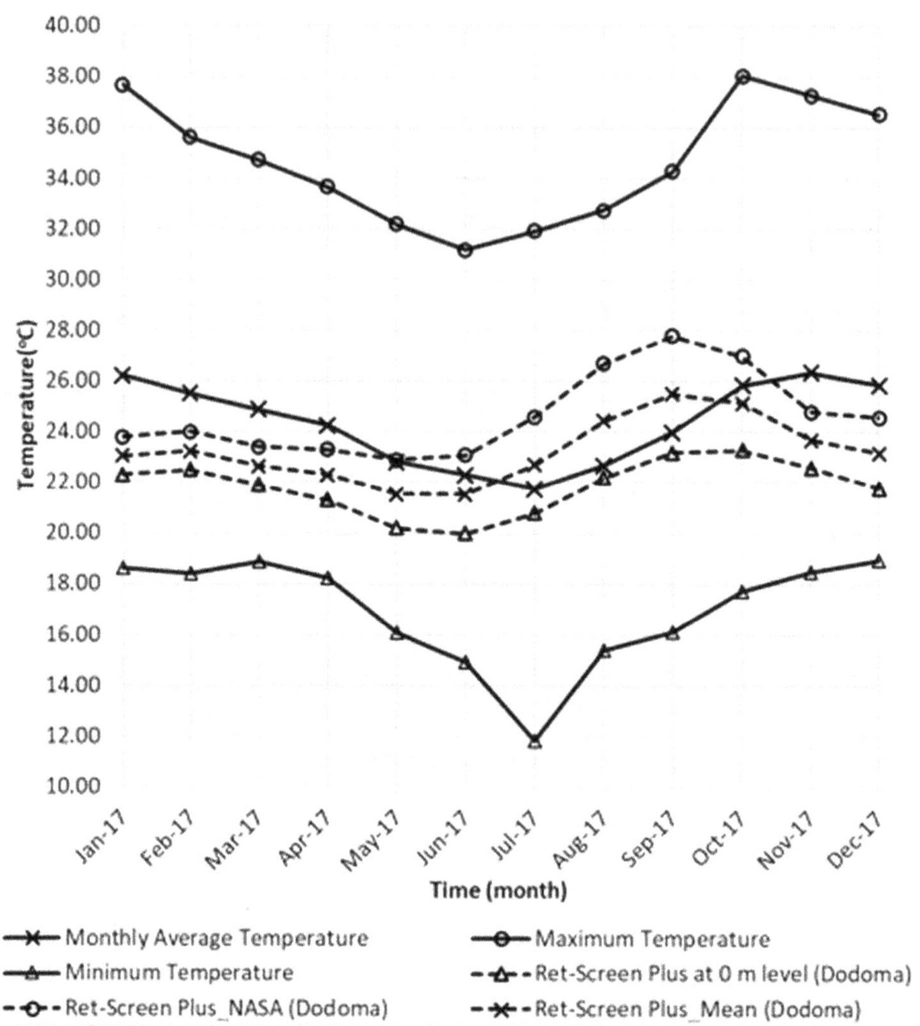

study site is predominantly blowing from North East (NE) for almost all months in a year.

Generally, ambient temperature variation is high (±4°C within a day) at the study site. This variation challenges biogas generation systems installed on or above ground without heat insulation features. The underground installed digesters offer reduced temperature variation; however, the challenges associated with their default troubleshooting are discouraging.

All the above-described findings are useful in developing and modelling energy systems. These systems can involve solar and wind energies and biogas generation units at the study site and nearby areas under similar semi-arid conditions. Therefore, it is recommended that the comprehensive characterisation research approach formulated and demonstrated be adopted by researchers and project developers engaged in renewable energy systems assessment and development in similar rural semi-arid areas conditions or elsewhere appropriate. Also, it is recommended to use heat insulation features to protect the biogas system installed on or above the ground from ambient temperature variation in the rural semi-arid areas.

FUTURE RESEARCH DIRECTIONS

Solar, wind, and biomass (cow dung) are abundantly available at the study site. They can be converted into biogas and electricity. With this note, investigating optimal integrated renewable energy systems to exploit these resources to produce clean energy is an exciting research topic. The investigation of the performance of the developed system on generating biogas or both electricity and biogas using the results from the approach demonstration is an essential research topic. Although rainfall was not characterised, it may affect the grazing of cattle, which produces cow dung and thus feedstock to the digester. Therefore, further research on the relationship between rainfall and the availability of cow dung is a recommended point of research. These recommended research topics will immensely support this book's theme, human-agro-energy optimisation for business and industry.

CONCLUSION

The effort to address insufficient, unreliable, unsustainable energy access in rural semi-arid areas is needed. This effort has been strengthened by formulating the novel generic comprehensive characterisation approach of renewable energy resources and sources, including cow dung, solar, wind, and ambient temperature. The application of the approach has been demonstrated to prove its relevance. The results of the demonstration are sound and promising. Therefore, the approach is vital in cementing the need for robust human agro-energy optimisation for business and industry initiatives. The initiatives will guarantee the improvement of the human living standard in energy-deprived areas around the globe.

ACKNOWLEDGMENT

This research was supported by the German Federal Ministry of Education and Research (BMBF) and the Federal Ministry for Economic Cooperation and Development (BMZ) through the Trans-SEC project, with Project ID No.: 031A249 A-G. The above are ushered with gratitude for this vital support. However, the views expressed in this chapter are purely those of the author and may not in any circumstances be regarded as stating an official position of BMBF or BMZ.

REFERENCES

Ananthapadmanaban, D. (2018). Summary of Some Selected Characterisation Methods of Geopolymers. In Mazen Alshaaer and Han-Yong Jeon (Ed), *Geopolymers and Other Geosynthetics IntechOpen,* doi:. https://www.intechopen.com/chapters/64656 doi:10.5772/intechopen.82208

Arnett, E. B., Johnson, G. D., Erickson, W. P., & Hein, C. D. (2013). *A Synthesis of Operational Mitigation Studies to Reduce Bat Fatalities at Wind Energy Facilities in North America.* Bat Conservation International.

Azam, M., Jahromy, S. S., Raza, W., Raza, N., Lee, S. S., Kim, K. H., & Winter, F. (2020). Status, characterisation, and potential utilisation of municipal solid waste as renewable energy source: Lahore case study in Pakistan. *Environment International, 134*, 105291. doi:10.1016/j.envint.2019.105291 PMID:31730999

Bekele, G., & Tadesse, G. (2012). Feasibility study of small Hydro/PV/Wind hybrid system for off-grid rural electrification in Ethiopia. *Applied Energy*, *97*, 5–15. doi:10.1016/j.apenergy.2011.11.059

Bidaoui, H., El Abbassi, I., El Bouardi, A., & Darcherif, A. (2019). Wind speed data analysis using Weibull and Rayleigh distribution functions, case study: Five cities northern Morocco. *Procedia Manufacturing*, *32*, 786–793. doi:10.1016/j.promfg.2019.02.286

Blackburn, K., & Stuard, S. B. (2014). A framework to facilitate consistent characterisation of read across uncertainty. *Regulatory Toxicology and Pharmacology*, *68*(3), 353–362. doi:10.1016/j.yrtph.2014.01.004 PMID:24457134

Boxwell, M. (2017). *Solar electricity handbook.* Greenstream Publishing Limited. [http://solarelectricityhandbook.com/ solar-irradiance.html]

Bukari, D., Tuokuu, F. X. D., Suleman, S., Ackah, I., & Apenu, G. (2020). *Ghana's energy access journey so far: a review of key strategies*. International Journal of Energy Sector Management., doi:10.1108/IJESM-02-2020-0008

Carzaniga, A., Fuggetta, A., Hall, R. S., Heimbigner, D., Van Der Hoek, A., & Wolf, A. L. (1998). A characterisation framework for software deployment technologies. Colorado State Univ Fort Collins, Dept of Computer Science. doi:10.21236/ADA452086

Elamouri, M., & Amar, F. B. (2008). Wind energy potential in Tunisia. *Renewable Energy*, *33*(4), 758–768. doi:10.1016/j.renene.2007.04.005

Felix, M., & Gheewala, S. H. (2011). A Review of Biomass Energy Dependency in Tanzania. *Energy Procedia*, *9*(2), 338–343. doi:10.1016/j.egypro.2011.09.036

Ganoe, R. D., Stackhouse, P. W. Jr, & DeYoung, R. J. (2014). *RET Screen Plus Software Tutorial. Technical Report*. International Clean Energy Decision Support Centre.

García, R., Pizarro, C., Lavín, A. G., & Bueno, J. L. (2012). Characterisation of Spanish biomass wastes for energy use. *Bioresource Technology*, *103*(1), 249–258. doi:10.1016/j.biortech.2011.10.004 PMID:22050839

Haluschak, P. (2006). Laboratory Methods of Soil Analysis. Canada-Manitoba Soil Survey. Manitoba.

Johnston, R. Z., & Matlock, M. D. (2011). Geospatial Climate Data. University of Arkansas. The Sustainability Consortium, Food, Beverage and Agriculture.

Kahimba, F. C., Mbaga, S., Mkoko, B., Swai, E., Kimaro, A. A., Mpanda, M., Liingilie, A., & Germer, J. (2015) Rain-fed crop, livestock, and agroforestry systems in Semi-arid and sub-humid Tanzania: A Baseline report, Trans- SEC Project, ZALF. http://project2.zalf.de/trans-sec/public/media/ upload/ prod uct/pdf/83905803646d86cb3ba323cdc201860b.pdf.

Kanyeka, E., Kamala, R., & Kasuga, R. (2007). *Improved Agricultural Research Technologies Recommended in Tanzania*. Ministry of Agriculture Food Security and Cooperatives.

Khanna, N., Mikkilineni, A. K., Martone, A. F., Ali, G. N., Chiu, G. T. C., Allebach, J. P., & Delp, E. J. (2006). A survey of forensic characterisation methods for physical devices. *Digital Investigation*, *3*, 17–28. doi:10.1016/j.diin.2006.06.014

Kiobia, D. O. (2015). *Design and development of a low-cost acoustic device to detect pest infestation in stored maise*. Virginia Tech. [http://hdl.handle.net/10919/56588]

MNRC, Ministry for Natural Resources Canada. (2015). *Clean Energy Project Analysis: RETScreen Engineering and Cases* (3rd ed.). Natural Resources.

Moberg, J. P. (2001). Soil and Plant Analysis Manual. Revised Edition, the Royal Veterinary and Agricultural University. Chemistry Department, Copenhagen University.

Mohammed, Y. S., Mustafa, M., Bashir, N., & Mokhtar, A. S. (2013b). Renewable energy resources for distributed power generation in Nigeria: A review of the potential. *Renewable & Sustainable Energy Reviews*, *22*, 257–268. doi:10.1016/j.rser.2013.01.020

Mohammed, Y. S., Mustafa, M. W., & Bashir, N. (2013a). Status of renewable energy consumption and developmental challenges in Sub-Sahara Africa. *Renewable & Sustainable Energy Reviews*, *27*, 453–463. doi:10.1016/j.rser.2013.06.044

Mwinuka, L., Schneider, I., Maeda, C., Mutabazi, K. D., Makindara, J., Graef, F., Sieber, S., Swai, E., Mbwana, H., & Swamila, M. (2015). Comparing stakeholder views for mutual acceptable food value chain upgrading strategies in Tanzania. *African Journal of Agricultural Research*, *10*(12), 1376–1385. doi:10.5897/AJAR2014.8901

Oyedepo, S. O. (2014). Towards achieving energy for sustainable development in Nigeria. *Renewable & Sustainable Energy Reviews*, *34*, 255–272. doi:10.1016/j.rser.2014.03.019

Papadimitriou, G., Kaliorakis, M., Chatzidimitriou, A., Gizopoulos, D., Favor, G., Sankaran, K., & Das, S. (2021). A system-level voltage/frequency scaling characterisation framework for multicore CPUs.

Peters, J., Combs, S. M., Hoskins, B., Jarman, J., Kovar, J. L., Watson, M. E., Wolf, A. M., & Wolf, N. (2003). *Recommended Methods of Manure Analysis*. Publication No. 3769. Coop. Ext. Publishing.

Ramos, S., Soares, J., Cembranel, S. S., Tavares, I., Foroozandeh, Z., Vale, Z., & Fernandes, R. (2021). Data mining techniques for electricity customer characterisation. *Procedia Computer Science*, *186*, 475–488. doi:10.1016/j.procs.2021.04.168

Raposo, F., De la Rubia, M. A., Fernández-Cegrí, V., & Borja, R. (2012). Anaerobic digestion of solid organic substrates in batch mode: An overview relating to methane yields and experimental procedures. *Renewable & Sustainable Energy Reviews*, *16*(1), 861–877. doi:10.1016/j.rser.2011.09.008

REN21. (2017). Renewable Energy Policy Network for the 21st Century. Renewables Global Status Report. REN21 Secretariat.

Safari, B., & Gasore, J. (2010). A statistical investigation of wind characteristics and wind energy potential based on the Weibull and Rayleigh models in Rwanda. *Renewable Energy*, *35*(12), 2874–2880. doi:10.1016/j.renene.2010.04.032

Scheid, A., Hafner, J., Hoffmann, H., Kächele, H., Sieber, S., & Rybak, C. (2018). Fuelwood scarcity and its adaptation measures: An assessment of coping strategies applied by small-scale farmers in Dodoma region, Tanzania. *Environmental Research Letters*, *13*(9), 095004–095015. doi:10.1088/1748-9326/aadb27

Schnurer, A., & Jarvis, A. (2010). Microbiological Handbook for Biogas Plants. Swedish Waste Management U2009:03.

Suberu, M. Y., Mustafa, M. W., Bashir, N., Muhamad, N. A., & Mokhtar, A. S. (2013). Power sector renewable energy integration for expanding access to electricity in sub-Saharan Africa. *Renewable & Sustainable Energy Reviews*, *25*, 630–642. doi:10.1016/j.rser.2013.04.033

Sullivan, K. M., Dean, A., & Soe, M. M. (2009). On Academics: OpenEpi: A web-based epidemiologic and statistical calculator for public health. *Public Health*, *124*(3), 471–474. PMID:19445426

Suri, M., Remund, J., Cebecauer, T., Dumortier, D., Wald, L., Huld, T., & Blanc, P. (2008). *First steps in the cross-comparison of solar resource spatial products in Europe.* In EUROSUN 1st International Conference on Solar Heating, Cooling and Buildings, Lisbon, Portugal.

Vandaele, N., & Porter, W. (2015). Renewable energy in developing and developed nations: Outlooks to 2040. *Journal of Undergraduate Research*, *15*(3), 1–7.

Wald, L. (2007). Solar radiation energy (fundamentals and theory). Solar Energy Conversion and Photoenergy Systems. In J. Blanco and S. Malato (eds.) Encyclopedia of Life Support Systems (EOLSS). Eolss Publishers.

Wawa, A. I. (2012). *The Challenges of Promoting and Adopting Biogas Technology in Semi-arid areas of Tanzania: The Case of Kongwa and Bahi Districts in Dodoma Region.* [Doctoral Thesis, The Open University of Tanzania].

Weiland, P. (2010). Biogas production: Current state and perspectives. *Applied Microbiology and Biotechnology*, *85*(4), 849–860. doi:10.100700253-009-2246-7 PMID:19777226

Yadav, A., Gupta, R., & Garg, V. K. (2013). Organic manure production from cow dung and biogas plant slurry by vermicomposting under field conditions. *Recycling of Organic Waste in Agriculture*, *2*(1), 1–21. doi:10.1186/2251-7715-2-21

Zhou, K., & Yang, S. (2016). Understanding household energy consumption behaviour: The contribution of energy big data analytics. *Renewable & Sustainable Energy Reviews, 56*, 810–819. doi:10.1016/j.rser.2015.12.001

Zhu, Y., Merbold, L., Leitner, S., Pelster, D. E., Okoma, S. A., Ngetich, F., Onyango, A. A., & Butterbach-Bahl, K. (2020). The effects of climate on decomposition of cattle, sheep and goat manure in Kenyan tropical pastures. *Plant and Soil, 451*(1), 325–343. doi:10.100711104-020-04528-x

ADDITIONAL READING

Demirbas, A. (2005). Potential applications of renewable energy sources, biomass combustion problems in boiler power systems and combustion related environmental issues. *Progress in Energy and Combustion Science, 31*(2), 171–192. doi:10.1016/j.pecs.2005.02.002

Gielen, D., Boshell, F., Saygin, D., Bazilian, M. D., Wagner, N., & Gorini, R. (2019). The role of renewable energy in the global energy transformation. *Energy Strategy Reviews, 24*, 38–50. doi:10.1016/j.esr.2019.01.006

Jiang, Y., Heaven, S., & Banks, C. J. (2012). Strategies for stable anaerobic digestion of vegetable waste. *Renewable Energy, 44*, 206–214. doi:10.1016/j.renene.2012.01.012

Kuleape, R., Cobbina, S. J., Dampare, S. B., Duwiejuah, A. B., Amoako, E. E., & Asare, W. (2014). Assessment of the energy recovery potentials of solid waste generated in Akosombo, Ghana. *African Journal of Environmental Science and Technology, 8*(5), 297–305. doi:10.5897/AJEST2014.1663

Panwar, N. L., Kaushik, S. C., & Kothari, S. (2011). Role of renewable energy sources in environmental protection: A review. *Renewable & Sustainable Energy Reviews, 15*(3), 1513–1524. doi:10.1016/j.rser.2010.11.037

Sohni, S., Norulaini, N. N., Hashim, R., Khan, S. B., Fadhullah, W., & Omar, A. M. (2018). Physicochemical characterisation of Malaysian crop and agro-industrial biomass residues as renewable energy resources. *Industrial Crops and Products, 111*, 642–650. doi:10.1016/j.indcrop.2017.11.031

Spataru, C. (2021). Business Models and Policies to Support Renewable Energy Community in Islands. In Research Anthology on Clean Energy Management and Solutions (pp. 1526-1543). IGI Global. doi:10.4018/978-1-7998-9152-9.ch065

Twidell, J., & Weir, T. (2015). *Renewable energy resources*. Routledge. doi:10.4324/9781315766416

KEY TERMS AND DEFINITIONS

Characterisation Approach: A systematic procedure of finding the information about something with aim of getting better understanding of that something behaviour in a given condition.

Insolation: A term describing a phenomenon, where a unit area of surface receives solar radiation intensity in a period of time; it may be expressed in the SI Units of $kWh/m^2/day$.

Pearl Millet Canopy: A uppermost leaves and flowers, which form a continuous foliage of the pearl millet plant.

Rural Semi-Arid Areas: In the least developed countries of sub Saharan Africa context, a land of marginal crops production potential, quick maturing cereals, and carrying natural vegetation of Acacia-Themeda association but including dry Brachystegia woodland. It is potentially productive grazing area, limited by bush encroachment, leached soils, and inadequate water.

Solar Irradiance: A instantaneous solar radiation intensity striking normal plane surface of a given unit area; it may be expressed in the SI Units of kW/m^2.

Weibull cumulative distribution function: A cumulative distribution function that provide the percent of time that the wind speed is greater than or equal the stated wind speed value.

Wind Roses: A web like representation figure, which shows the event frequencies of wind direction at a given area.

Chapter 10
Artificial Intelligence Techniques in Water Purification and Utilization

Suthanthira Vanitha
https://orcid.org/0000-0002-9579-7651
Muthayammal Engineering College, India

Radhika K.
https://orcid.org/0000-0003-4849-8680
Muthayammal Engineering College, India

Sampath Boopathi
https://orcid.org/0000-0002-2065-6539
Muthayammal Engineering College, India

ABSTRACT

In this chapter, the various water monitoring systems—internet of-things, wireless network systems, cloud computing, optimization systems, and reporting systems—are illustrated. The development of artificial intelligence in water management has the potential to change and manage water resources. The use of AI and cutting-edge technologies such as data analytics, regression models, and algorithms simplifies the water management process. AI-based systems will be sustainable, cost-effective, and capable of predicting potential damage in order to optimize water management solutions with minimal waste. Machine learning algorithms support the water management process and improve the accuracy of sensor-based systems in agriculture.

DOI: 10.4018/978-1-6684-4118-3.ch010

INTRODUCTION

The United Nations (United Nations Department of Economic and Social Affairs, 2015) predicts that there will be 8.5 billion people on the planet by 2030, with India having the largest population. Countries all around the world will experience water scarcity due to this population density combined with shifting weather extremes and occurrences. Another factor making water scarcer is the ineffective management of water resources. Another industry where there has been an increase in water demand is agriculture. According to the FAO, this sector is responsible for over 70% of all worldwide water withdrawals. By 2050, as the world's population grows, both the need for water and resource conservation measures will be on the rise.

Water bodies such as lakes, rivers, oceans, and groundwater become contaminated when pollutants are discharged into them without being treated to remove harmful compounds. Water pollution has a negative impact on not only aquatic plants and animals, but also on humans and ecosystems. Water pollution is caused by a variety of factors. Household sewage, garbage, and liquid waste are discharged into lakes and rivers. These wastes contain hazardous chemicals and toxins, rendering the water toxic to aquatic animals and plants. Solid waste and litter dumped in bodies of water cause major issues. Glass, plastic, aluminum, and Styrofoam are all examples of litter. Different substances degrade at varying rates in water. They impact aquatic plants and animals. Industrial waste includes pollutants such as asbestos, lead, mercury, and petrochemicals, all of which are extremely harmful to both people and the environment. Oil spilled from ships and tankers while travelling pollutes the sea. The oil spilled does not dissolve in water and instead forms a thick sludge that pollutes the water. Acid rain occurs when acidic particles in the atmosphere caused by air pollution mix with water vapor. Temperature increases caused by global warming kill aquatic plants and animals. The increased bloom of algae in water depletes the oxygen in the water, negatively affecting fish and other aquatic animal populations.

There is an urgent need to put new technology into use that helps with water conservation. AI might be a key component in resolving this problem. The use of AI technologies in water resource management can effectively address managing water resources, building efficient supply systems, maximizing current water resources, and planning infrastructure for water resources and safe water mechanisms. In the subject of water management, artificial intelligence has been used by scientists to cut down on resource waste. Researchers have documented ground-breaking situations that support the use of AI in reducing water waste. Several researchers, utilized the efficient, successful AI as a substitute for conventional hydrological methodologies. when climate change consequences exacerbate over the future decades, as many scientists expect they will, they will lead to an increase in extreme weather events and disasters. Along with climate change issues, the next ten years will see revolutionary steps needed to address the loss of biodiversity, conservation efforts, ocean health, clean air, and water security (Feng et al., 2008).

In general, the use of artificial intelligence in water supply has a number of policy implications to enhance the performance of water utilities and quality service liberation. Applications of AI play an important role in improving environmental quality from all factors. Water is considered a primary source of daily life that requires intelligent management systems. Water resource management involves the formulation of the latest paradigms which are capable of combining water resource protection through innovative intelligent systems. With the support of AI technology, it increases the efficiency of water resources; network performance; distributed system monitoring; plant treatments; and easy access to quality control. AI plays an astounding role in a persistent situation as a result of the strengthening of water and environmental monitoring systems. Hence, computer-based intelligence can be viewed as an approach to making helpful models or strategies to play out a complex task typically done by people.

Applications that use data to make them "intelligent" have disrupted ordinary life. This digital technology revolution can help cutting-edge water utilities increase their performance. Water utilities may use knowledge and data available to make better decisions while boosting service delivery and lowering costs by leveraging the power of artificial intelligence algorithms and big data analytics.

VARIOUS METHODS FOR MONITORING WATER QUALITY

Wireless Network Method

Environmental monitoring, with a focus on water quality, is becoming increasingly important in WSN deployment. The Microcontroller Units monitored the research factors as well as the sensor node structures (MCU). Various authors discuss a variety of articles based on the display of wireless communication standards, distribution, data protection, power structures, autonomy, and usability. The proposed design contributes to the development of a real-time, low-cost water quality monitoring system. Future research could be linked to these systems to measure more quantities, such as chloride and fluoridation conductivity. The data collected were not saved in the IoT cloud storage. Only a warning SMS alert was sent to the Pollution Control Board. Water Network Sensors (WSNs) detect water pollution in rivers, streams, and coastal areas. A data averaging algorithm was developed to reduce the size of communication traffic. According to the experimental results, the average water quality data transmission loss ratio was less than 1%. Based on WSN technology, a Romanian team created a system for monitoring water eminence parameters in some Romanian river waters. The monitoring system makes use of the Cyber water platform, which aims to assist decision-makers when accidental pollution occurs in rivers (Hassan et al., 2020).

Because the system does not support local data analysis, it will be offline whenever a mobile network outage occurs. The system was able to receive and plot pH, temperature, and turbidity measurements on the MATLAB (Matrix Laboratory) GUI. Because of the ZigBee module, the signal coverage area was reduced. The Water Quality Monitoring System (WSN) is an underwater sensor system that monitors the quality of water. The prototype was powered by a solar cell and included ZigBee to transfer data from pH, turbidity, and oxygen level sensors to the base station. To analyze water parameters, a Visual Basic 6.0 Graphical User Interface (GUI) was used. Mini-boats are equipped with sondes, probes, and a Wireless Sensor Network (WSN) and guided by a GPS-aided INS-based navigation algorithm. It collects water quality parameters on a regular basis. A group of portable water quality probes mounted on buoys at fixed positions analyze water quality in the second system. The use of WSNs in water pollution alarm systems is based on fusion rules based on the Maximum posteriori (MAP) algorithm for the Water Contamination Warning Systems (WCWSs) application. False alarms from typical graphical algorithms were not suitable for regular WCWSs to give the same priority, which also missed detection.

Fuzzy System

Mobile-based IP (Internet Protocol) cameras with fuzzy logic were proposed to monitor the quality, nutrition, and water contaminants in an overhead tank. Like an IP camera, the mobile camera can transfer video to a central server. The images captured by the cameras served as the input. Water quality monitoring using fuzzy logic and graphical programming. Sensors, microcontrollers, and LabVIEW (Laboratory Virtual Instrument Engineering Workbench) software comprise the system. The sample testing time period was extremely long. The system requires more manpower to operate. The primary goal of this study was to maintain the standard parameters of neutralized wastewater by maintaining the appropriate pH level and temperature of industrial waste water using a fuzzy controller.

The entire system output could be viewed remotely through the Data Dashboard app on a smartphone or tablet. The extended FAHP framework demonstrated its capability as a robust tool for decision making in complex water problems. To find a suitable location for land filling, An integrated Fuzzy logic and Analytic Network Process (i.e. F-ANP). The extended FAHP framework demonstrated the capability of the extended FAHP as a robust tool for decision making in complex water problems. an integrated Fuzzy logic and Analytic Network Process (i.e., F-ANP) to find a suitable location for land filling. The expanded Fuzzy Analytic Hierarchy Process

(FAHP) contributed significantly. FAHP framework, and sensitivity analysis of the results demonstrated the extended FAHP's capability as a robust decision-making tool in complex water problems(Hassan et al., 2020).

Figure 1. Elements of Internet of things

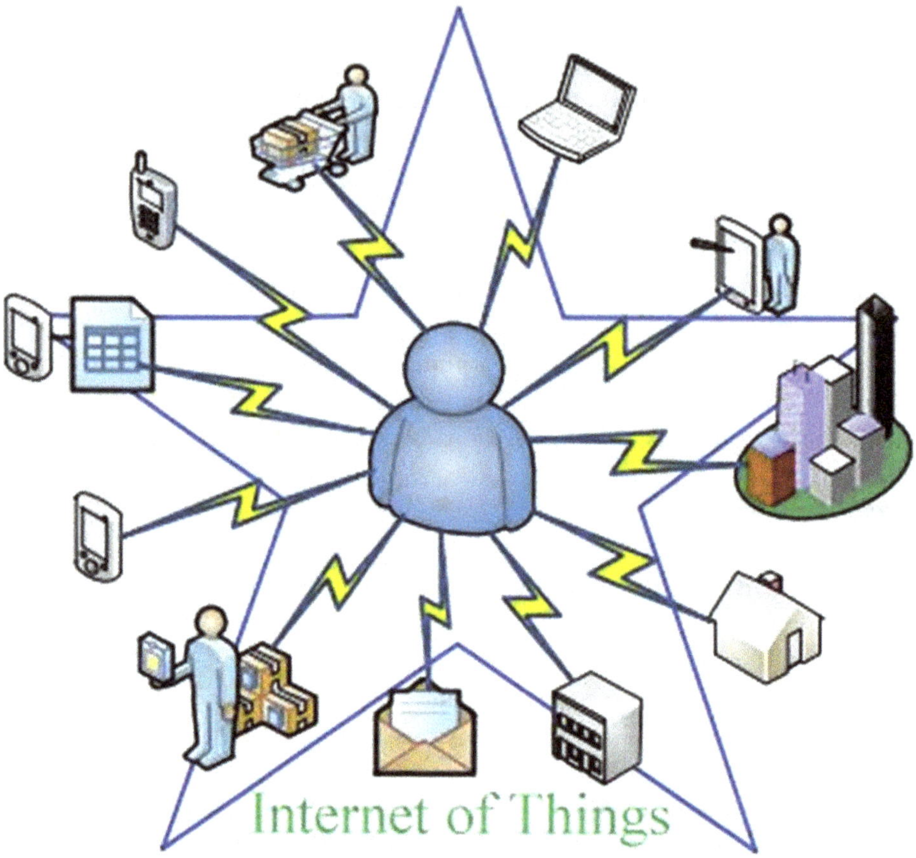

IOT and Cloud Computing Method

The device manufacturer may be able to correct its design without introducing hardware triggers or artificialities with known credible flaws. The Genetic Algorithm was used to determine the best configuration for various types of VMs when resources were limited. The vision for future networking was that all things, services, and media would be connected and integrated, resulting in an Internet of Everything. The elements of IoT system is shown in Figure 1.

IoT systems aim to connect and scale billions of devices in various domains such as transportation, industry, smart home/city, medical services, and energy systems. Electronic healthcare has emerged as a promising application domain with significant potential to improve medical care quality. The rapid integration of IoT, Wireless Body Area Networks, and cloud computing. Finally, a set of design challenges for achieving high quality and secure patient-centric monitoring schemes were identified. The Industrial Internet of Things (IoT) will be able to combine normal physical objects with their own and be monitored by a single system. A cloud platform offers a series of processing challenges for data storage. Population data can be generated quickly, the data block can be large, and it can contain a variety of data types.

Optimization and Internet of Things

The internet of things (IoT) wireless sensors and actuators are critical to the design of advanced Cyber-Physical Systems (CPSs). Tariff delay and scale must meet stringent requirements when delivering tariffs and tight security for tight energy communication lines in such complex networks. A research group has presented a camera-based wireless sensor node with a self-optimizing end-to-end computation and communication design aimed at surveillance applications. Support for multiple feature extraction and classification algorithms, as well as tunable processing depth and power amplifier gain, was demonstrated. The vital secrecy outage performance of wireless communications under eavesdropper cooperation. The techniques of Laplace transform, keyhole contour integral, and Cauchy Integral Theorem were combined to work around the difficult multifold convolution problem involved in such analysis(Hassan et al., 2020).

The Internet of Things as a common platform and service for consumer electronics. This study developed an interactive framework for population creation in interior environments such as the home and small office. The virtual sensors and actuators that have general physical things and their virtual behaviors on top of their networks were the framework building blocks. The Internet of Things (IoT) is a smart technology that connects anything, anywhere, at any time. The IoT's pervasiveness was responsible for draining energy from its resources. This study proposed three layers of energy power design: consciousness and control, information processing, and delivery. The application of Compressed Sensing Theory to data sampling and acquisition in wireless sensor networks and the Internet of Things could provide new insights (IoT). A CS-based framework for the IoT was proposed, in which end nodes measure, transmit, and store sampled data. An efficient cluster-sparse reconstruction algorithm was also proposed for in-network compression.

Reporting System

Water distribution systems present a significant challenge for structural monitoring. Maintaining the integrity of these networks is critical for providing safe drinking water. This study describes the WaterWiSe@SG project in Singapore, with a focus on the use of Water WiSe as a tool for monitoring, detecting, and predicting abnormal events. Zigbee is based on the TI CC2530 System on Chip, which includes critical sensors such as relative humidity and light-intensity sensors. For a five-minute period, this system provided partial real-time monitoring of environmental parameters. It also provided a surveillance diagram and the average value of surveillance on a given day. Cloud computing and hybrid wireless networks must develop a programme to integrate these new technologies into existing systems. In cloud services-based enterprise information systems, The proposed method combines access control measures in a mixed structure that displays Cloud Service Access Requirements and Customer Security Certificates to users.

The energy information systems, which manage energy consumption over the internet, The study's goal was to produce an innovative result and cloud computing software system that will combine technologies to develop and use an energy expert, consumer, energy services practice, and modern web. The ASIC is made up of a Microcontroller Unit (MCU), a Power Management Unit (PMU), reconfigurable sensor interfaces, and communication ports that control a wireless transceiver, as well as an integrated passive Radio-Frequency (RF) receiver with energy harvesting capability. For energy efficiency operations, the MCU and PMU provided the most flexible communication and power control methods. A wireless sensor network-based water environmental monitoring system was proposed. It is divided into three sections: data monitoring nodes, database station, and remote monitoring center. The system included a large-scale water environment with reservoirs, lakes, rivers, marshes, and shallow or deep underground water. A sensor network environment could be used to display information about sensors, monitored events, network topology, and network usage. This research provided a platform capable of meeting the Water Framework Directive's monitoring requirements (WFD). Temperature, phosphate, dissolved oxygen, conductivity, pH, turbidity, and water level are among the parameters being investigated.

MACHINE LEARNING

Machine Learning (ML) methodologies use a set of data to complete a task. The information in ML has a collection of example data. Normally, an entity of data is explained by a set of variables or attributes. Attributes can be categorized as having nominal, binary, or numeric values. The specific task of the ML model is calculated based on the performance metrics, which are enhanced with experience over time. ML algorithms are estimated with the support of statistical and mathematical models.

Figure 2. Block Diagram for Machine Learning

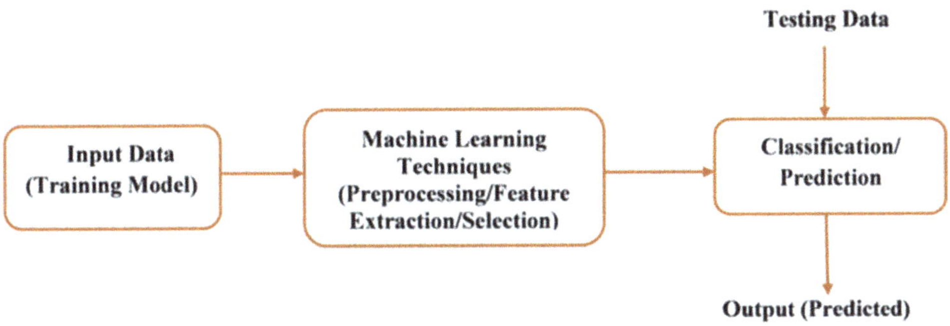

Figure 3. Classifications of Machine Learning Algorithms

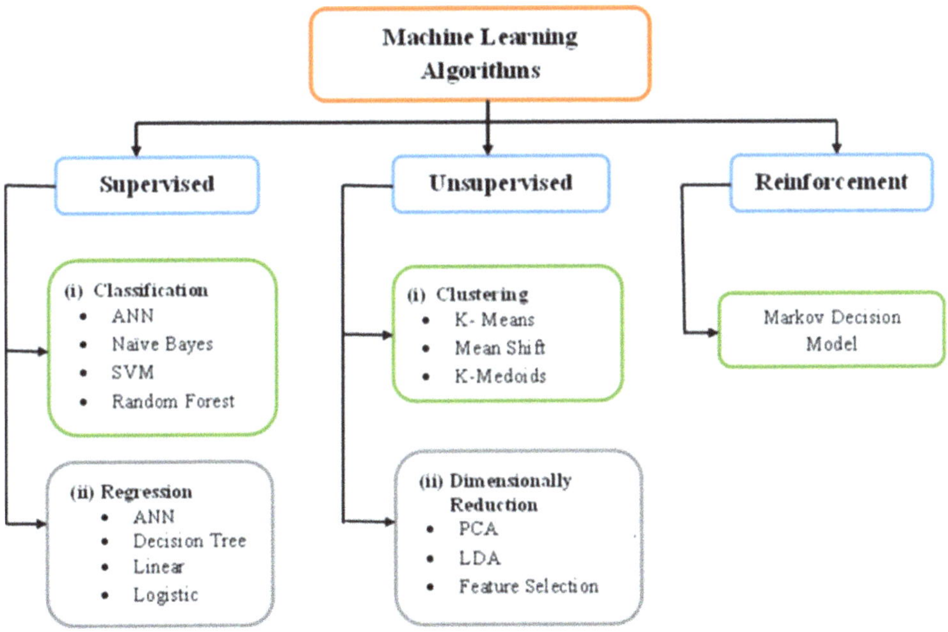

Figure 2 depicts the general idea of a machine learning system. .The raw input data is derived into an appropriate state by preprocessing, which includes data cleaning by removing inconsistent data or noise, data integration, and data transformation (Coumou & Rahmstorf, 2012). The feature extraction and selection create the informative subset of features for the implementation of the learning model (Haghiabi et al., 2018). Overall learning performance is improved by the feedback system. The unnoticed testing data is applied to the trained model in the form of feature vectors. Based on the output, an appropriate decision is obtained through the classification/prediction rule.

Algorithms for Machine Learning

ML is the artificial intelligence domain that gives information to machines and has the capability to learn statistical data instead of programme coding (Anagnostis et al., 2020). ML is categorized into three types, namely, supervised learning, unsupervised learning, and reinforcement learning, as shown in figure 3. Supervised learning is referred to as a single supervisor being used to train the data through training examples. It supports labelling datasets for algorithm training, data classification, and data prediction accurately. Examples of supervised learning are Artificial Neural Networks, Bayesian Networks, Decision Trees, Support Vector Machines (SVM), ID3, k-Nearest Neighbor, Markov Model, etc. Unsupervised learning is referred to as providing an infinite quantity of data to the programme for pattern findings and pattern relationships. By using unsupervised learning, hidden patterns in the datasets can be discovered. For example, k-Nearest Neighbor, k-Means Clustering, etc.

Reinforcement learning is one of the methods of supervised and unsupervised learning that creates a sequence of decisions by training machine learning models. Reinforcement learning-based robots are used in the industry for performing a number of applications. Due to the increase in an enormous amount of data in the agriculture field, ML techniques are used for agricultural production to obtain precise crop predictions.

ROLE OF AI IN WATER MANAGEMENT

The AI approach has proven to make efficient water distribution and water saving systems in water resource management. According to UN statistics, about 1.2 billion people across the world are suffering from water scarcity, while 500 million people are unaware of the situation. In addition, 1.6 billion people are experiencing economic water shortage scenarios like lack of infrastructure and clean and potable water facilities. Water is an essential part of our daily lives. Due to the sudden increase in the occurrences of droughts and water scarcity in various parts of the world in recent years, researchers have been looking for better, more robust techniques to implement water resource management. There is a need to opt for a system for designing and maintaining the existing water management systems. AI Technologies in Water Resource Management build a sustainable water supply for people around the world. AI based Water Management System is shown in Figure 4.

Figure 4. AI based Water Management System

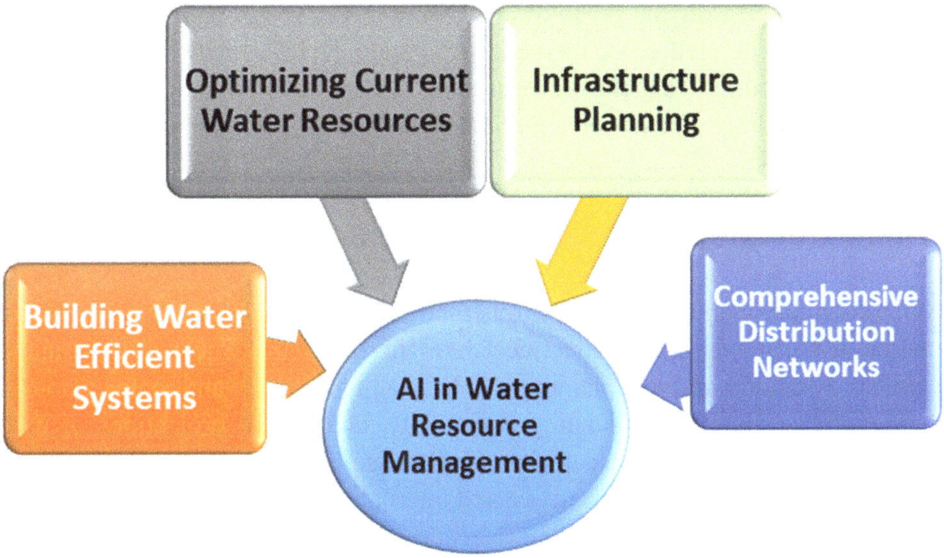

1. Building Efficient Water Systems

In order to build an efficient system for water management, the water managing body has two main tasks. The first task is to seek new sources of water due to the volatility of the resource, and the second is to manage the existing water reserves and system sustainability. Applying various algorithms and programmes to build water plants gives updated statistics of water resources and support in building models. Water operations can become more dynamic and intelligent with the appropriate application of AI techniques such as sensor programs, Neural Networks strategizing, prediction models for future water use, robotic sensors in water disposal plants, and Blockchain for operating financial transactions (Zheng & Casari 2018).

2. AI Optimizing Current Water Resources

The primary goal of United Nations statistics is to maximize current water resources. Optimization and automation are obtained through the decision-making abilities of AI. AI-driven planning allows government agencies and water departments to understand real-time water loss and misuse, design and implement comprehensive distribution networks, and maximize revenue. In order to forecast the groundwater reserves and to measure the groundwater reserve levels, vector regression methods and other artificial intelligence methods are applied. Hybrid Artificial Intelligence Models help in improving the water systems across the world. Researchers and analysts are working together on a global level to contribute to the sustainable use of water resources.

3. Infrastructure Planning for Water Structure

Water distribution and wastewater infrastructure are critical for human health and environmental protection; they must meet the highest standards. Substantial planning is necessary in infrastructure planning due to its aging. Climate and socioeconomic changes create significant planning uncertainty, and simple projections of past developments are no longer acceptable. Various scenarios can stimulate creativity when generating decision alternatives, and internal consistency is ensured using a strategy generation table. Structured decision-making (SDM) procedures can be adapted to inform decisions about sustainable water infrastructures. Due to short-term variation in water availability and demand, long-term uncertainty in climate and population growth, and differing perspectives on the value of water, planners are facing challenges in water infrastructure development and decisions. AI classifies multiple uncertainties and develops a decision framework that combines simulation for probabilistic uncertainty, scenario analysis for deep uncertainty, and multistage decision analysis for uncertainties reduced over time. Scenario analysis for deep uncertainties underlines the significant impact of assumptions about the future and cost of water scarcity in evaluating infrastructure performance.

4. Comprehensive Distributive Network in AI

The WAMAN (WAter MANagement) system seeks to address a number of problems facing recent water utility managers and offers a framework for aiding water utility managers even in the absence of specialized employees. In the field of water management, some important applications of Expert Systems may include EXPLORE Expert System treated water distribution, Network Management System for Water Distribution Systems, Intelligent Control System for a Municipal Water Distribution

Network, A review of the literature, reveals that presently available expert systems have an extremely limited scope. Current work aims to create an expert system (named WAMAN) with a dynamic knowledge acquisition system that is powered by simulated runs of a hydraulic model that has been appropriately calibrated and validated for the specified water utility (Samuel 2000). The Expert System includes a dynamic knowledge acquisition system that is powered by hydraulic model simulation runs that have been properly calibrated and validated for the specified water utility. The Expert System uses CLIPS (an Artificial Intelligence (AI) tool developed by NASA's Software Technology Branch) and integrates computational platforms such as MATLAB, open-source GIS, and a common user interface is used by an RDBMS (Relational Database Management System).

5. Promises and Challenges of AI in Water Management Systems

With the increase in the global population and a decline in the quality of water resources, there is a need for sustainable technologies to ensure a reliable supply of clean water. AI systems estimate the composition of contaminants in the water and raise the efficiency of the water treatment plant. According to researchers, standard controllers can save up to 40% on energy by employing neural networks for optimization. The Jal Jeevan Mission (Urban), which will link 2.86 billion homes to safe water through taps, was announced in the Union Budget for 2021–22. This is consistent with the start of the Center's rural water supply project in 2019. Artificial intelligence in clean tech offers exactly this promise.

Water is one of the four major areas where artificial intelligence can produce $5.2 Tn, according to a 2019 PwC report titled "Artificial Intelligence and the Fate of Planet Earth." A UNICEF and WHO report claims that 1 in 3 people worldwide lack access to clean drinking water. If we do not handle the situation properly, it will get worse in the upcoming years. It might not seem like a big surprise, but AI in water management has the potential to transform how we treat and manage the water sources all around us.AI can impact the water sector around the world(Sandeep and Rakesh 2011).

6. AI in Water Treatment

Globally, when it comes to accessing clean water, one in three people does not have the resources. In most of the areas contaminated by sewage water and other toxic effluents, municipal wastewater treatment is a key support. AI is used to reduce pollutants in the water, which in turn decreases water contamination and the scarcity of clean water. AI can be used to determine the quantity and make-up of harmful pollutants. Real-time data collection is achieved through artificial intelligence, machine learning, and big data for more effective water treatment processes that detect early allocation of efforts while neural networks and IoT reduce energy costs.

AI SYSTEMS SUPPORTS

- Increased Efficiency Of Waste Management Systems
- Protection from Unknown Contaminants
- Reducing Operational Costs

AI in Wastewater Management

An India Today report states that it is estimated that around 40% of piped water in India is lost to leakage. A US EPA analysis claims that an average family can lose 180 gallons of water each week, or 9400 gallons yearly, due to leaks in the home. This quantity is equal to the water required to wash more than 300 loads of laundry. Lots of water is wasted through leaks, burst pipes, etc. This wastage is assisted by AI and IoT. Implementation of AI to analyze real-time water loss and automate pipes to shut off whenever there is a leak improves the amount of water wastage. AI can predict leaks in storage tanks and support mending them before it is too late. Devices connected through the IoT can communicate better and integrate various systems(Samuel 2000).

AI in Smart Irrigation

Agriculture currently uses 70% of all freshwater, with an efficiency rate of less than 50%. By using less water for irrigation, AI can simplify the usage of water in agriculture. Agriculture is the major water-using sector, and most lands use a quality portion of groundwater for the purpose of irrigation. AI systems will leverage Smart Irrigation to minimize water usage and optimize water resources without wastage. AI systems can predict agricultural demands to balance water use by directing sprinkler systems while also detecting the levels of groundwater. Emerging precision-based AI systems predict the weather conditions, climate, and humidity to enable enhanced management of agriculture. Smart farming minimizes leakages and analyses the soil to resolve the plants' condition and water needs through AI sensors. AI creates the digitally-empowered farm and provides solutions that utilize the minimum amount of water.

AI-based solutions include:

- 'Precision-based 'AI systems consist of dissimilar sets of data like satellite images, climate, temperature, humidity, and weather predictions for the best crop management decisions.
- Soil and light sensors share data with an AI solution that determines the best time to water crops and use fertilizers, among others.

AI in Digital Water

Digital technology in AI makes the water management process easier with regression models, data analytics, and algorithms. These cutting-edge technologies help in building efficient water systems and networks. It supports building water plants and acquiring the status of water resources. Nowadays, government sectors and other water resource bodies apply AI to build a smart water system. It builds an efficient infrastructure for water management systems, which can be easily adapted to the varying conditions. AI provides cost-effective and sustainable results that can optimize all water management solutions and predict potential reimbursement.

AI in Water Conservation System

Currently, smart home water meters identify water leakages, cracks in pipelines, and pipe bursts in apartment buildings, leading to more efficient systems. The government sector and other water corporations are implementing the data analytics software to manage and analyze sensor data installed on water systems, such as flow rate, level, volume, pressure and water quality. To address the water utilities' issues of growing demand, the IoT platform is used to provide solutions in the water sector which reduce pilferage and wastage of unaccounted for water. Systems with AI capabilities can more effectively manage water distribution by forecasting water demand. According to research, the number of smart meters installed by water utilities worldwide will increase by 28% by 2026, reaching approximately 400 million units. Every solution is customized, depending on the hardness, chlorine, or other contaminants present in the water. AI offers immense possibilities for clean tech with cost effectiveness (Govindaraju 2000).

CHALLENGES OF ARTIFICIAL INTELLIGENCE

Clean water, sanitation, and hygiene are key issues that lead to soil contamination, and water pollution. Artificial Intelligence Technologies play a critical role in achieving Sustainable Water Resource Development. As per UN statistics, more than 40% of the global population is affected by water scarcity, and nearly 2.2 billion people lack access to clean drinking water. The data on sanitation and hygiene is disgusting, with 4.5 billion people lacking access to a safe sanitation system. Approximately 2.4 billion people don't have access to basic toilets and latrines, and 892 million people still practice open defecation. Almost one thousand children die due to preventable water and sanitation related diarrheal diseases every day.

The unremitting growth of the population globally and escalating environmental degradation, together with climate change, pose considerable challenges to water security. Overall, 70 percent of the deaths occurred in natural disasters like floods and other water-related disasters. More than 80% of wastewater generated by human activities is discharged untreated into rivers or the sea, contributing to water pollution, while approximately 70% of water extracted from rivers, lakes, and aquifers is used for irrigation. To ensure the availability and sustainable management of water and sanitation within the context of sustainable developmental goal (SDG) 6 is inserted.

The SDG 6 target encourages widespread and equitable access to safe and affordable drinking water for all; improving water quality; increasing water utilization efficiency; protecting and restoring water-related ecosystems; and reinforcing neighborhood community participation in civilizing water and sanitation management (Feng et al., 2008). It pays special attention to the needs of women, girls, and people living in exposed situations. Every year, with increasing losses in biodiversity, millions of people are dying due to fewer infrastructure facilities. The current scenario is reshaped through AI power to revolutionize water-use efficiency and sanitation management. Engineering, mapping, and forecasting extensively humanize the sustainability and scalability of water and sanitation services. AI in a data-driven approach provides an opportunity for good governance and reliable information.

Clean Water AI can run continuously in real time, monitors and the analysis is accomplished with a digital microscope connected to a computer operation with the Ubuntu operating system. At present, about 2 billion people use a drinking water source contaminated with feces, AI prototype devices apply pattern recognition and machine learning to monitor and inspect the quality of water. Also, AI can identify dangerous bacteria and harmful particles in the water. Artificial intelligence and machine learning enable more effective water treatment processes, detect problems, assure an early allocation of efforts, and provide more accurate, timely information to save money and water.

EMAGIN's software offers precise, well-timed information on water pollution treatment and manages the system to prevent overflows. According to the ITU report, a 40% gap in water resources and usage is expected to hit us by 2030. The growing global population has an impact on this gap, which also affects the population's cleanliness, food security, and health. Early AI applications in the water sector have demonstrated both promise and challenges. Data-driven water management can use artificial intelligence to fundamentally improve the use of water and its waste in agriculture, toilets, and water-use efficiency.

CONCLUSION

The development of artificial intelligence in water management has the potential to change and manage water resources. The use of AI and cutting-edge technologies such as data analytics, regression models, and algorithms simplifies the water management process. AI-based systems will be sustainable, cost-effective, and capable of predicting potential damage in order to optimize water management solutions.

REFERENCES

Anagnostis, A., Papageorgiou, E., & Bochtis, D. (2020). Application of artificial neural networks for natural gas consumption forecasting. *Sustain.*, *12*(16), 6409. doi:10.3390u12166409

Coumou, D., & Rahmstorf, S. (2012). A decade of weather extremes. *Nature Climate Change*, *2*(7), 491–496. doi:10.1038/nclimate1452

Feng, S., Kang, S., Huo, Z., Chen, S., & Mao, X. (2008). Neural networks to simulate regional ground water levels affected by human activities. *Ground Water*, *46*(1), 80–90. doi:10.1111/j.1745-6584.2007.00366.x PMID:18181867

Govindaraju, R. S. (2000). Artificial Neural Networks in Hydrology. II: Hydrologic Applications. *Journal of Hydrologic Engineering*, *5*(2), 124–137. doi:10.1061/(ASCE)1084-0699(2000)5:2(124)

Haghiabi, A. H., Nasrolahi, A. H., & Parsaie, A. (2018). Water quality prediction using machine learning methods. *Water Qual. Res. J.*, *53*(1), 3–13. doi:10.2166/wqrj.2018.025

Hassan, Z., Hossain, G. J., & Islam, M. M. (2020). Internet of Things (IoT) based water quality monitoring system. *Educ. Res*, *2*(4), 168–180.

Samuel, A. L. (2000). Some studies in machine learning using the game of checkers. *IBM Journal of Research and Development*, *44*(1–2), 207–219. doi:10.1147/rd.441.0206

Sandeep, K., & Rakesh, K. (2011). CLIPS based decision support system for water distribution networks. *Drinking Water Engineering and Science*, *4*(1), 37–50. doi:10.5194/dwes-4-37-2011

Zheng, A., & Casari, A. (2018) *Feature Engineering: Principle and Techniques for Data Scienctists*. O'Reilly Media, Inc. https://www.oreilly.com/library/view/feature-engineering-for/9781491953235/

Chapter 11
New Strategies in Treatment and Enzymatic Processes:
Ethanol Production From Sugarcane Bagasse

T. Vennila
Sri Sairam Engineering College, India

M. S. Karuna
https://orcid.org/0000-0002-6115-6055
MJP Rohilkhand University, India

Bipin Kumar Srivastava
Galgotias College of Engineering and Technology, India

Jayaprakash Venugopal
https://orcid.org/0000-0003-1273-2981
Sathyabama Institute of Science and Technology, India

Raviteja Surakasi
https://orcid.org/0000-0002-0786-0105
Lendi Institute of Engineering and Technology, India

Sampath B.
Mythayammal Engineering College, India

ABSTRACT

Ethanol can be made from lignocellulosic biomass, a promising replacement for fossil fuels. The bioethanol production process is divided into three basic unit operations. Conventional and well-established corn-to-ethanol technology can be cost competitive with the cellulosic ethanol process. For the pre-treatment of bagasse, steam explosion without a catalyst and a mixture of sulfuric and oxalic acids were used. The slab was submitted to enzymatic hydrolysis in the first scheme without any additional treatments like washing or solid-liquid separation. According to a study, post-treating bagasse with a pressure filtering process improved the amount

DOI: 10.4018/978-1-6684-4118-3.ch011

of sugar that might be produced for ethanol. The total cellulosic ethanol process can be made successful. The effect of varying concentrations of sodium sulphite and its effect on both process configurations were also studied. The study provided an 8-10% increase in overall enzymatic efficiency as compared to the control substrate.

INTRODUCTION

In this study, we investigate the biochemical production of fuel ethanol and other vital commodities biochemicals from lignocellulosic (LC) biomass. In order to produce suitable sugars that may then be fermented to produce ethanol, the majority of large-scale LC ethanol production procedures utilise high solids pre-treatment followed by enzymatic hydrolysis. The pre-treatment and enzymatic hydrolysis processes both have a key role in the overall process economics and energy efficiency during the generation of ethanol. This is due to the fact that efficient pre-treatment techniques get beyond the biomass' refractory nature and offer a substrate that is suitable to the enzyme-based hydrolytic processes. In the aqueous phase, these processes highly solubilize xylose in monomeric, oligomeric, or a combination of the two forms. Enzymatic hydrolysis is performed after high-solid (>15 percent) pilot scale pre-treatment in the DA or SE process. Following pre-treatment, the slurry is typically filtered to produce a liquid phase and a solid phase (lignin and cellulose) (xylose or xylo oligomers). Enzymatic hydrolysis of cellulose to sugars and subsequent fermentation to ethanol are two methods for turning it into alcohol. To obtain adequate yields of sugar and ethanol during enzymatic hydrolysis, solid loading of 5–10% is utilised. In the grand scheme of bioethanol production, each of these procedures calls for more process water, lowering ethanol concentration and requiring more energy for distillation(Chandel et al., 2007).

Lower energy requirements for ethanol distillation arise from greater ethanol concentrations (4-5 w/w%) produced by higher sugar concentrations. Using washed and unwashed sugarcane bagasse, only a small number of experiments have been reported on the various process scenarios for cellulose and hemicellulose hydrolysis. The purpose of the study is to investigate alternative post-treatment techniques to improve bagasse's enzymatic hydrolysis efficiency. This study's major objective is to determine whether washing or other adjustments, like solid-liquid separation and washing, may be eliminated. The streams needed for this investigation came from pilot scale treatments at various combined severity factor (CSF) values, which is dependent on acid dose, temperature, and retention duration. Due to its high acidity, low cost, and well-known effectiveness in the majority of pre-treatments, sulfuric acid is preferred. For the hydrolysis of xylan, oxalic acid is a potent dicarboxylic acid that is largely non-corrosive (Canilha et al., 2010).

Sugarcane bagasse (SB) is a fiber-based agricultural residue left after the juice has been extracted from the sugarcane. SB is considered one of the abundant and low-cost renewable feedstocks for production of biofuels and biochemicals. Biomass biomass (SB) can be converted into sugars for further fermentation to ethanol or biochemicals in a number of ways, such as through the use of hydrolysis and enzymatic processes. Several pre-treatment techniques have been researched in the recent decades for the synthesis of bioethanol and biochemicals from SB. Important characteristics of diluted acid (DA) pre-treatment include a straightforward technique, a short residence period, and inexpensive costs. Pentose and hexose are two monomeric sugars that are produced from lignocellulose with DA treatment. The co-utilization of pentose sugars for the full optimization of biofuel production is made possible by the application of DA. Using cellulase enzymes, the pre-treated solid's cellulose is further degraded into glucose sugars, which are subsequently transformed into electricity. There are several literature studies on the use of a single diluted acid for the conversion of hemicelluloses to their monomeric sugar components, indicating that DA treatment is particularly promising for the generation of bioethanol. The majority of research have used mineral acids such hydrochloric, nitric, sulfuric, and phosphoric acid for the processing of lignocellulosic biomass. Organic acids like maleic and oxalic acids, as well as more expensive alternatives to sulfuric acid, are used in addition to minerals acid (Canilha et al., 2010).

At the pilot-scale level, there are no research on the use of mixed acids for the treatment of sugarcane bagasse. The majority of DA pre-treatment research have experimented with varied sulfuric acid concentrations, retention times, and temperatures. The production of xylose, which primarily produces phenolics as a by-product, can be significantly influenced by all variables. The disadvantages of using a single acid, such as the partial substitution of sulfuric acid with other acids, may be eliminated by a pre-treatment process that makes use of the advantages of each acid by combining them with the other acids. In terms of gypsum disposal and construction material, it may be advantageous. For the pre-treatment of organophosphorus derivatives such tri-chloroethylc-arburosulphuric acid (TCA) and trichloroethyl sulfuric acid (TES), binary acid pre-treatment can provide a larger severity range(Da Silveira Dos Santos et al., 2010). The research of mixing different types of acids is fascinating because it can show how different types of acids affect pre-treatment and the subsequent enzymatic hydrolysis. The conditions for the reactions are selected so that a variety of combined severity factors are investigated. For this investigation, mixed acids were used to examine how acid treatment affects the synthesis of glucose and xylose. Due to its high acidity, low cost, and widespread use in the majority of the DA pre-treatment investigations, sulfuric acid was one mineral acid that was used in all mixed acid strategies. A cheap Lewis acid catalyst, ferric chloride attacks the C-O-C and C-H bonds in cellulose and other cell wall polymer networks with specificity. It has been used to partially decrystallize cellulose, speeding up the process and improving the effectiveness of enzyme hydrolysis.

PRE-TREATMENT AND ENZYME-MEDIATED DIGESTION OF SUGAR CANE BAGASSE

(a) Experimental Methods

Materials: Sugar cane bagasse (SB) extraction was place in the nearby area. This SB was used and stored in ambient conditions without washing and had a moisture level of 30–35%. Before being added to the pilot-scale reactor, bagasse was shredded in a hammer mill to a thickness of between 15 and 25 mm.

Compositional Analysis Methods: The total solids, carbohydrate content, and acid-insoluble lignin content of untreated bagasse as well as the solid fraction left over after different pre-treatments have been identified by researchers at the National Renewable Energy Laboratory (NREL) of the US Department of Energy in Boulder, Colorado. In order to determine the monomers hydrolyzed from the oligomer using HPLC, samples of hydrolyzate were brought to a final sulfuric acid concentration of 4 percent weight-to-weight, autoclaved at 121°C for an hour, and then centrifuged. They used a Bio-Rad ion exchange column to quantify glucose, xylose, arabinose, acetic acid, HMF, and furfural (Jackson de Moraes Rocha et al., 2011).

(b) pre-treatment Methods

Continuous pre-treatment: A horizontal screw-type reactor on a pilot size was used for the pre-treatments. This reactor has a maximum total solid handling capacity of 22% (w/w). SB was first reduced in size to between 15 and 25 mm, then supplied to the pre-treatment reactor. The reactor can be run at temperatures between 140 and 200 °C, pressures between 4 and 20 bar, and retention times between 15 and 30 minutes. Steel alloy SS 316 is used to construct the pre-treatment reactor. For a 15-minute holding period, various diluted mixed acid concentrations were dosed onto dry biomass at 150–170 °C. The ultimate concentration of solids after pre-treatment was 20–22% w/w due to the addition of steam and acid.

Pre-treatment procedure: To ascertain the pre-treatment conditions for bagasse and other biomass, a total of 29 tests were carried out at the pilot size. Temperatures ranged from 150–170 °C and acid dosages ranged from 1-2.5 percent for mixed acid doses to 2-3 percent for single acids on a dry biomass basis (Macrelli et al., 2012).

(c) Enzymatic Hydrolysis

The experiment was conducted using commercial Novozymes enzymes at 14–15% total solids and 13–14% total insoluble solids in an anchor impeller reactor. The samples were cultured for 120 hours at 50°C with the pH kept between 5.0 and 5.2. Samples were taken every 24 hours for the HPLC examination of the by-products and sugar. The ratio of the glucose yield attained after 120 hours to the theoretically feasible glucose yield following pre-treatment in the solid fraction was used to evaluate the efficiency of enzyme hydrolysis.

OUTCOMES OF PRE-TREATMENT AND ENZYMATIC DIGESTIBILITY

(a) Effect of different mixed acid pre-treatments on xylose yield and inhibitors

It was discovered that mixed acid pre-treatments at different combined severity factors (CSFs) affected the xylose hydrolysis of cellulose with the aid of enzymes. SB was discovered to contain cellulose, xylan, arabinan, lignin, proteins, water extractives, and ethanol extractives in its dry form (Amores et al., 2013).

SA+OA Pre-treatment: The production of xylose monomer from the dilute SA pre-treated slurry was 83 percent, which was 10% more than the yield from the dilute OA alone treatment (Table 3.1, run no. 2). This is because oxalic acid has a lower pKa1 than H2SO4 (-3) does. Because oxalic acid is more expensive than mineral acids, which are less expensive, using higher doses of it could not be cost-effective. The yield of by-products like acetic acid and furfural is significantly higher in this pre-treatment, despite the fact that sulfuric acid treatment yields C5 sugars more effectively. The pH of slurry treated with SA is 0.9, which is the lowest of all the documented pre-treatments. This low pH could not be good for reactors because it might induce corrosion, which shortens their lifespan and raises capital expenditure. At a pH of 1.2, run no. 9 with inhibitors (0.36 percent acetic acid, 0.01 percent HMF, and furfural) produces the highest yields of monomeric xylose (83 percent). Higher dosages of mixed acid don't show this correlation (runs 9–11). A higher pH (1.5-2.5) is obtained at lower acid dosages, which is advantageous for the pre-treatment reactor. Lesser xylose yields are achieved at lower acid dosages, though. It is better to have a lower sulfuric acid content during the pre-treatment because less gypsum will form.

SA+FeCl3 pre-treatment: Xylose monomeric efficiency of 70.1 percent was provided by diluted FeCl3 pre-treatment (1.5 percent by weight). Higher dosages of ferric chloride have been found to have lower efficiencies. This may be a result of pentose sugars degrading under harsh circumstances. In comparison to metal chloride pre-treatment, the addition of metal ions produces more inhibitors in the liquid phase. The efficacy of monomeric xylose for SA+FeCl3 treatment and SA+OA system pre-treatment is depicted in the figure as a function of CSF. The efficiency obtained for the identical pre-treatment conditions is less than that of the SA+.OA system. More severe pentose sugar breakdown may be to blame for the decline.

SA+PA pre-treatment: A 2 percent diluted PA treatment on its own resulted in a 32 percent xylose monomeric efficiency. Lower efficiency has been noted at lower dosages of phosphoric acid, similar to other acids (excluding SA). In order to obtain comparable conversion efficiencies, higher doses could be necessary. Beyond 160 °C in SA+PA treatment reduces xylose yield and produces more inhibitory compounds (mostly from C5), which is reflected in the composition of the liquid phase. The highest xylan-xylose conversion of the three pre-treatments examined is provided by the sulfuric acid and phosphoric acid combination, which also minimises the combined acid dose. In addition to the CSF, the breakdown of pentose sugars at greater combined severity factor (CSF) levels reduces xylose efficiency. For a single acid treatment, higher acid dosages are needed to achieve competitive conversion (oxalic acid and phosphoric acid). The running costs can eventually go up as a result(Mesa et al., 2014)(Mesa et al., 2014).

Cellulose hydrolysis in pre-treatment: The synthesis of glucose, which is present in small amounts in the resulting liquors, results from the acid pre-treatment of cellulose with water. During the acid pre-treatment, some of the amorphous cellulose is solubilized and turns into glucose. The porosity of the residual cellulose in the solid phase is also increased by acid pre-treatment. Compared to the other pre-treatments examined here, SA+PA had greater glucose concentrations in the liquid phase. Higher glucose yields are obtained during the pre-treatment with an increase in temperature. At 170 °C, 1.5 percent SA plus 1 percent PA, the maximal concentration of free glucose (1.02 percent) is attained after 15 minutes.

(b) Effect of Combined Acids Pre-treatment on Enzymatic Hydrolysis Efficiencies

The enzyme hydrolysis of all pre-treated solid phases was investigated at a total solids content of 15% using a dose of 30 mg/g glucan Novozymes enzyme (equal to 8 FPU/g glucan). The next section discusses the findings of these studies. From a scale-up standpoint, the findings of this research are very helpful since they demonstrate how mixed acid pre-treatment can result in larger amounts of glucose for subsequent fermentation to produce ethanol. There isn't any research for the various mixed acid treatments, despite the fact that there are numerous studies on the use of greater solid concentrations for enzymatic hydrolysis(Wong & Sanggari, 2014).

SA+OA pre-treatment: The glucose production from the enzymatic hydrolysis of the pre-treated particles is used to evaluate the impact of various pre-treatment conditions. The two greatest glucose conversions were found to be 58 and 60 percent. The latter, however, exhibits a reduction in xylose output as a result of pentose sugar breakdown at high temperatures. As the CSF rises, cellulose digestibility rises as well, reaching its peak at a CSF of 2.03. The decreased conversions may result from the pre-lower treatment's monomeric xylose conversion as well as from the release of xylooligomers.

SA+FeCl3 pre-treatment: 53 percent represents the lowest glucose conversion while 55.8 percent represents the highest. This might be because pre-treatment produced more breakdown products than usual. The efficiency of enzymatic hydrolysis is adversely affected by the inhibition of cellulase enzymes in the presence of ferric ions.

SA+PA pre-treatment: The study's three pre-treatment approaches yield the maximum glucose conversion, at 67.1 percent. This is regarded as the ideal situation since it offers a respectable level of xylose and glucose hydrolysis. As the temperature rises under the identical pre-treatment circumstance, the glucose conversion rises.

Overall enzymatic hydrolysis: The effectiveness of the enzymatic hydrolysis seen here is equivalent to that seen in earlier experiments on bagasse treated with a single acid. The nature of the substrate and pre-treatment are just two examples of the many variables that affect how effective an enzyme is. Due to the possibility of conditions varying from study to study, it will be challenging to compare all of these characteristics here. Oxalic acid pre-treated bagasse at 20 FPU/g glucan and steam-exploded sugarcane bagasse that had been impregnated with phosphoric acid yielded an enzymatic efficiency of 67%. The procedure becomes unprofitable if the enzyme dose is increased further since it does not account for the observed incremental efficiency and raises manufacturing costs.

NEW STRATEGY IN PRE-TREATMENT AND ENZYMATIC PROCESSES: SUGAR PRODUCTION FROM SUGARCANE BAGASSE

(a) Pre-treatment Methods

Continuous pre-treatment: A horizontal screw-type reactor on a pilot scale was used for pre-treatment. This reactor has a total solid handling capacity of up to 22% w/w. This reactor was a component of a reactor feeder blow tank system that had an hourly feed rate of 30 kg dry weight. The pre-treatment reactor is constructed of SS 316 steel and can be used between the ranges of 140 and 200 °C, 4 and 20 bar of pressure, and 15 and 30 minutes of retention time. A metering-style dosing pump was subsequently used to pump this DA solution under active steam pressure.

Monomeric pre-treatment (high-severity mixed-acid treatment): The pre-treatment digester received a batch of 15 kg of size-reduced dry bagasse, or 21.4 kg on a weight basis, fed at a rate of 30 kg per hour. The digester was filled with 30 kg of feed, 20 kg of which had been heated to 160 °C for 15 minutes while being dissolved in an acid solution. The output solid's concentration increased to up to 20% w/w as a result of the pre-injection treatments of steam and an acid solution (Camargo et al., 2016).

Mild acid pre-treatment (intermediate severity mixed-acid treatment): Using combinations of 0.5 percent sulfuric acid and 0.5 percent oxalic acid on a dry bagasse base for 15 minutes, mixed-acid pre-treatment was carried out at 180 °C.

SE pre-treatment (low-severity pre-treatment): The SE pre-treatment operation was carried out at 180 °C for 15 minutes.

(b) Post-treatment Methods and Enzymatic Hydrolysis

Whole slurry hydrolysis Method (Scheme 1): The flow chart of various processes for Whole slurry hydrolysis Method is illustrated in Figure 1. In a 10 L reactor with an anchor impeller and 13–14% total insoluble solids (TIS), hydrolysis experiments were conducted and incubated for 120 hours at 50 °C. In order to bring a batch of 7 kg of pre-treated slurry within the pH range of 5.2-5.5, 500 mL (monomeric), 260 mL (mild acid), and 180 mL (steam explosion) of 40 percent (w/v) NaOH were used. The enzymatic slurry's pH was kept within the range of 5.0-5.2. Every 24 hours, samples were taken for the HPLC examination of sugar and by-products. The proportion of actual glucose yield to hypothetical glucose yield in the solid fraction indicates how effectively glucose is hydrolyzed by enzymes.

Reslurried hydrolysis Method (Scheme 2): The high-solid pre-treatment slurry was mixed to 30 kg of xylose or xylooligomer stream (aqueous) after the initial filtration step to create TIS. As a result, there was a decrease in the concentration of insoluble solids, allowing for efficient solid-liquid separation. Using a hydraulic cylinder and an 11-plate filter press, filtration pressures of 8–9 bar (g) were reached. With a film thickness of 10 mm, a cake thickness of 30 mm and a filter area of 3 m^2 were accomplished. Then, at 13–14 percent TIS, hydrolysis experiments were carried out in a 10L reactor fitted with an anchor impeller. Using 500 mL (monomeric) and 255 mL, the enzymatic slurry was incubated at 50 °C for 120 hours while keeping the pH between 5.0 and 5.2. The flow chart of various processes for reslurried hydrolysis methods is illustrated in Figure 2.

Figure 1. Processes of whole slurry hydrolysis Method (Scheme 1)

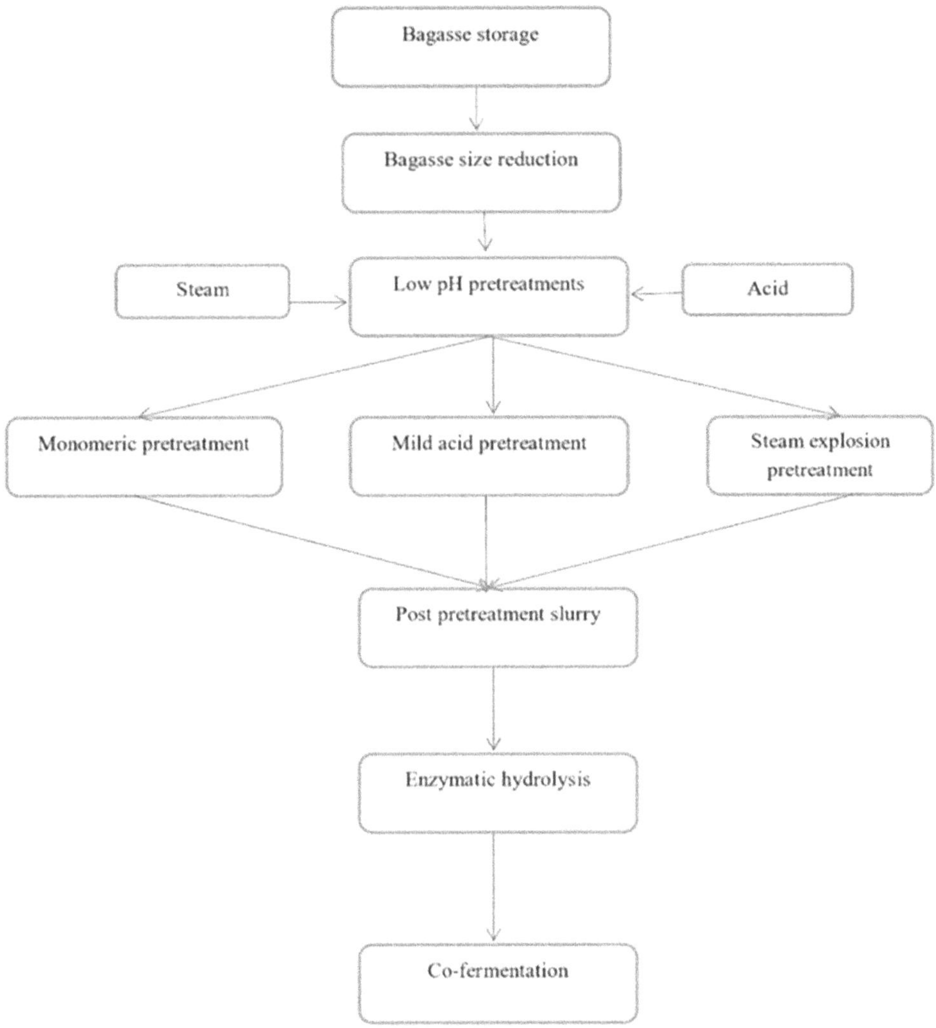

Wet cake hydrolysis (Scheme 3): Using the commercial Novozymes enzyme with the same enzyme loading as previously described, all experiments were carried out in a 10L reactor. After separating the pre-treated slurry and 37 kg of fresh water to a consistency of 16–17% total solids and 14–15% w/w TIS, 25 kg of solid phase were recovered. For the first time, a sugar, water, and other hydrolysis products enzymatic slurry has been created and is ready to be employed in a lab for the

Figure 2. Processes of Reslurried hydrolysis Methods (Scheme 2)

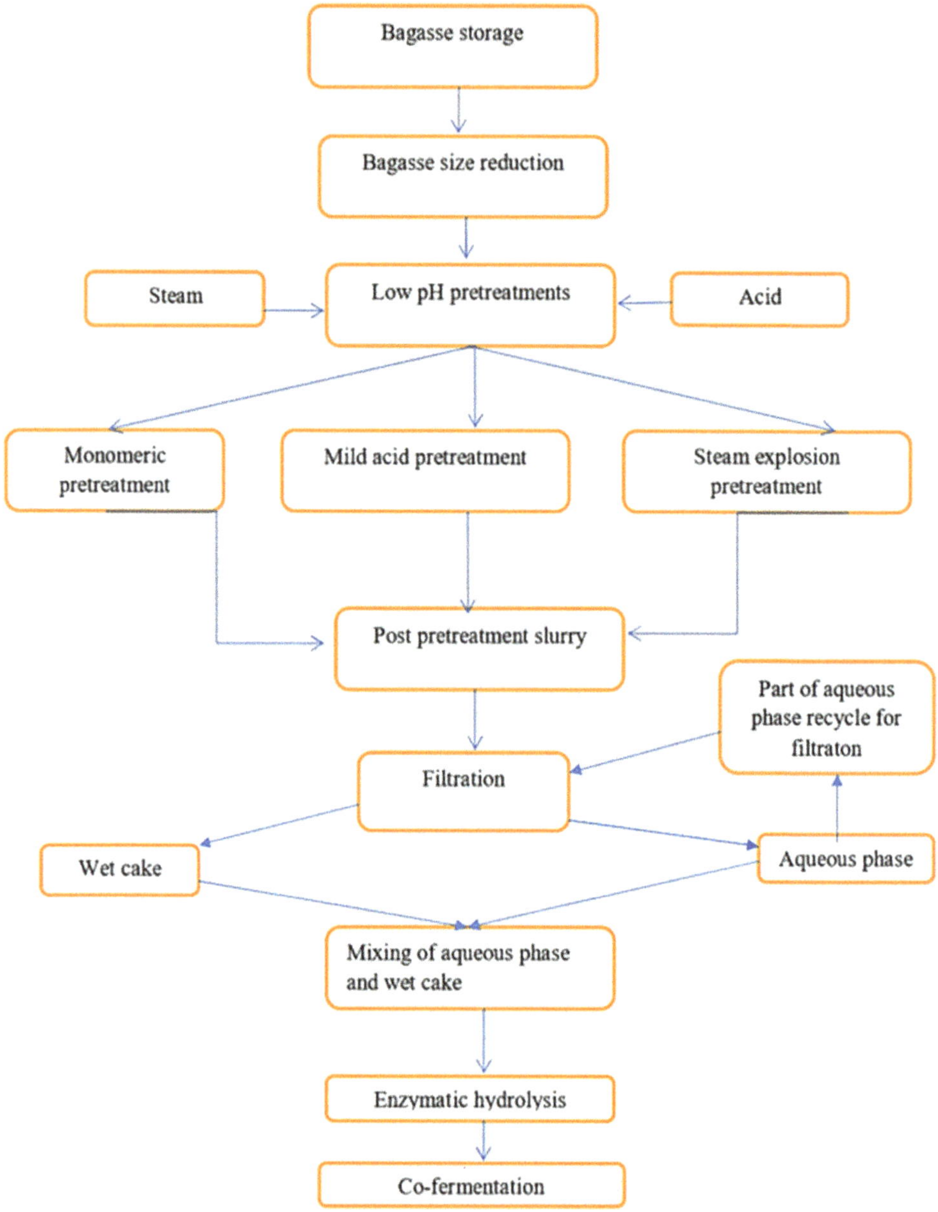

examination of sugar and by-products. The flow chart of various processes for Wet cake hydrolysis method is illustrated in Figure 3.

Co-Fermentation: In a broth containing yeast extract, peptone, and glucose, yeast extract peptone dextrose (YPD) was cultivated. The mixture was sterilised for

Figure 3. Processes of Wet cake hydrolysis (Scheme 3)

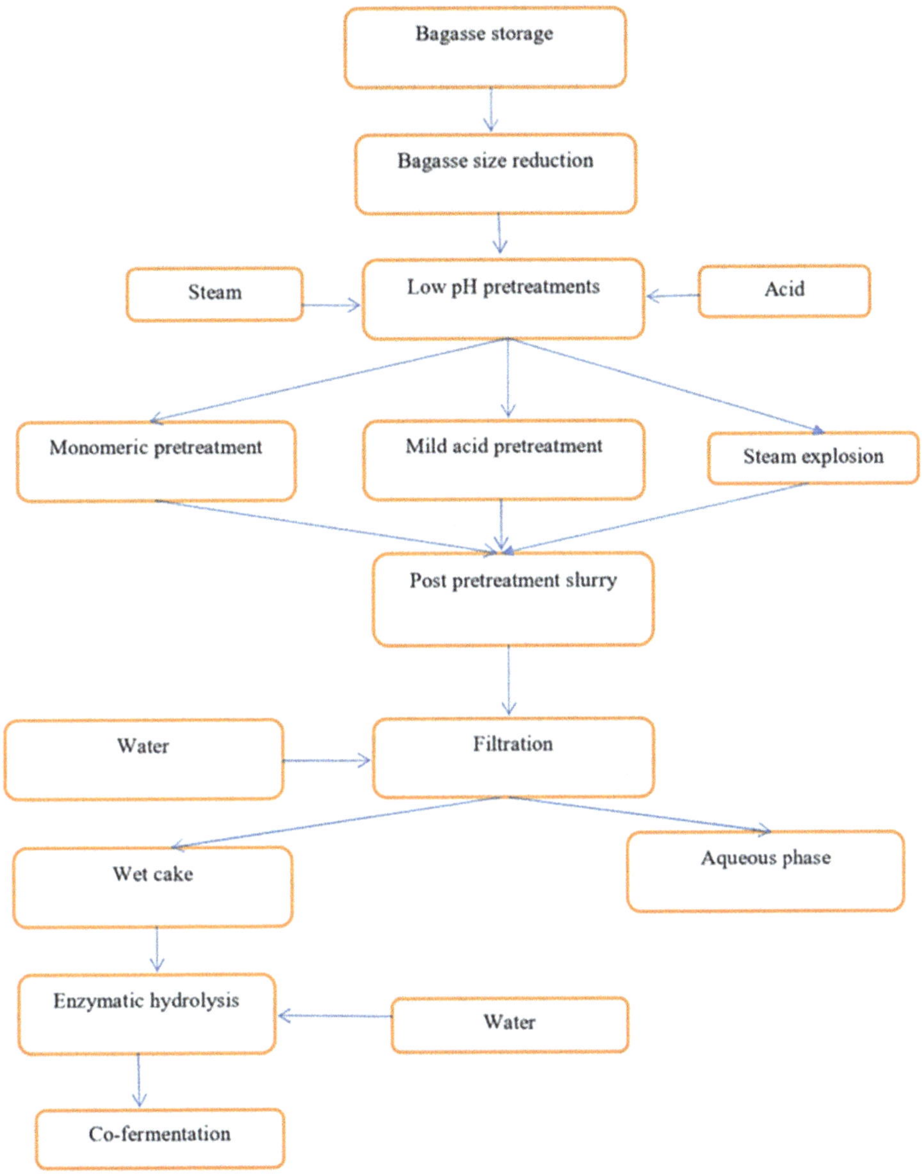

20 minutes at 121 °C. After that, a yeast culture preserved in glycerol stock kept at -80 °C was added to the broth as an inoculant. The cells were cultured in 500 mL Erlenmeyer flasks with 100 mL of media at 32 °C and 150 rpm in an orbital shaker. The inoculum was sterilised at 121°C for 20 minutes after having its pH

adjusted to 5.5 using 40 percent w/v NaOH. Stage 1 inoculum was injected into 2L Erlenmeyer flasks with 1L medium at 32°C and 150 rpm in an orbital shaker. Using a 40 percent w/v NaOH solution in a 10-L reactor with 9L of enzymatic hydrolyzate, the pH of the inoculum was brought down to 5.5. Using a pitched-blade impeller, the temperature was 32°C, and the inoculum volume was 10% of the fermentation medium. The fermenting medium was supplemented with additional nutrients, 1 g/L of urea, and 1 g/L of diammonium phosphate (Cheng et al., 2019).

Viscosity Measurement: A Brookfield rheometer (RHE02000V2.7) with a shear rate of 50 s1 was used to test the viscosity of monomeric, mildly acidic, and SE treated bagasse samples at room temperature. For measuring viscosity, a total solids content of around 7-8% (w/w) was utilised. To determine viscosity in the gigantic reed study, a slurry with total solids of 20% w/w was diluted with fresh water to a solid consistency of 7-8% w/w. Similar findings have been made for enormous cassowaries, and they probably apply to other species as well.

Dynamic Light Scattering: All pre-treated slurries derived from various pre-treatment methods were subjected to DLS tests. The upper 1% of the dispersion was taken and diluted 100 times for the DLS tests after samples (1 percent w/w) were sonicated into water.

EFFECT OF IMPROVED PRETREATMENT AND POST-PRETREATMENT PROCESS ON ENZYMATIC HYDROLYSIS EFFICIENCIES

It is discussed how different plans affect the concentration of insoluble particles and the production of ethanol (Shen et al., 2022).

(a) Whole-slurry hydrolysis scheme (Scheme 1)

According to a study published in the Journal of Clinical Pharmacology and Biomarkers, the monomeric treatment achieved glucan hydrolysis efficiency of 54.5 percent and overall xylose efficiency of 85.1 percent for moderate acid treatment during enzymatic hydrolysis. Acetic acid concentration was 0.52 percent at the beginning of the enzyme hydrolysis for high-severity monomeric treatment, 0.28 percent for moderate acid treatment, and 0.18 percent for SE pre-treatment. The pre-treatment process utilised has a significant impact on the by-products that are produced.

For the mild acid treatment, the acetic acid concentration rises from 0.28 percent to 0.53 percent. Pretreatment with low pH typically results in some hemicellulosic sugar degradation. The transformation of xylo-oligomers into xylose monomers can account for this. A conversion of glucan hydrolysis is seen with the SE pre-treatment. Other SE treatments have no impact on biomass aside from eliminating hemicellulose as xylo-oligomers. During enzyme hydrolysis, glucan efficiency is 48.5 percent, while xylose monomer conversion ranges from 15 to 83.1 percent.

Enzyme activity can be suppressed by pre-treatment with steam, acid, or heat. It is necessary to add more extra enzymes in order to increase glucan conversion. For SE treatment, the solubilization of acid-soluble lignin results in the greatest release of phenolic chemicals. Hormone-like substances, such as HMF and furfural, prevent the hydrolysis of enzymes and have an impact on the cellulase enzymes. The fibrous and solid-like characteristics of the slurry can also be credited with the enzyme efficiency of SE pre-treatment. As a result, the reaction sites will be overrun with enzymes, preventing the process from working to its full capacity.

Due to the higher viscosity of the bagasse slurry produced by SE pre-treatment compared to that produced by the other treatments, enzymatic efficiency is reduced, and production costs are raised. Since bagasse has a lengthy fibre structure, measuring viscosity can be challenging, especially for bagasse with significant solid loadings. of the steam-exploded bagasse slurry has been compared to that of Scheme 2 while that of the monomeric-treated slurry for Scheme 1 is 0.038 Pa.s. Particle size impacts the enzymatic process and controls viscosity.

(c) Re-slurried hydrolysis scheme (Scheme 2)

According to Scheme 2, for monomeric treatment, the glucan and xylan hydrolysis efficiencies are 65.9 percent and 85.2 percent, respectively, after enzyme treatment. For mild acid treatment, comparable enzymatic efficiencies have also been attained. At the beginning and the end of the enzyme hydrolysis, sugar yields, inhibitors, and glucan and xylan hydrolysis efficiencies are measured. Similar to Scheme 1, mild acid and SE treatments result in an increase in the content of xylose and acetic acid. This probably means that reaching better enzyme hydrolytic efficiency requires physical rearrangement of the bagasse, which takes place at the mesoscopic level in response to Scheme 2. The character of the bagasse appears to change after pressure filtration of the pre-treated slurry. According to DLS data, the particle size for monomeric pre-treatment using Scheme 2 is 1647.2 nm as opposed to 5222.1 nm for those using Scheme 1. The pressure filtration procedure utilised in this investigation increases the amount of new, internally accessible surface area, which improves the enzymes' ability to penetrate the cellulose. The accessibility of cellulose enzymes to the substrate is increased due to post-pre-treatment milling, which results in smaller particle sizes and better enzyme conversion efficiencies. In enzyme hydrolysis, the amount of xylose released into the slurry increased from

15% to 84.0%, matching the outcome from Scheme 1. The conversion of cellulose to glucose has been successful for all three pre-treatment methods.

(d) Wet cake hydrolysis (Scheme 3)

The majority of publications published employ Scheme 3, which is the standard routine method. These washing procedures prevent the loss of sugars while removing the greatest number of soluble inhibitors attached to the substrate. Furthermore, due to its high water need, washing of the pre-treated biomass will be challenging to repeat on a commercial scale. As a result of removing soluble inhibitors attached to the substrate, there is an increase of 5-7 percent in glucan hydrolysis efficiency as compared to Scheme 2. Acetic acid, furan aldehydes, and phenol concentrations are substantially lower in Scheme 3 because the inhibitors released during the hemicellulose hydrolytic treatment are washed away. After pre-treatment, hemicellulose is still present in the solid fraction. Another explanation is that the pre-treatment liquid contains substances (such as solubilized lignin derivatives) that, upon chilling the pre-treated biomass, can re-precipitate and re-condense, blocking the cellulase enzymes. The journal Biomaterials has published a study on the enzyme hydrolysis of xylan to xylose for mild acid and steam pre-treatments. The aqueous stream in Scheme 3 that is filtered apart will produce relatively little ethanol following fermentation since it is so diluted. Xylo-oligomers will need to be changed into xylose monomers using an enzyme or an acid in a subsequent step.

(e) Effect of the different schemes on sugar yield

This study has considered pressure filtration, solid-liquid separation, and solid-liquid separation for ethanol production. For mass balance calculations, the mass balance of sugars and each of its constituent parts has been considered, with an emphasis on cellulose/glucose, xylan/xylose, and total solids. This demonstrates that the various enzymatic hydrolysis strategies utilised in this investigation should be effective regardless of the pre-treatment type. Scheme 2 outperforms Scheme 1 in terms of sugar concentration and mixed fermentable sugar yield for the monomeric treatment. This discrepancy is primarily caused by the reduced enzyme hydrolysis efficiency obtained with Scheme 1 at greater solids. In Scheme 2, the mild acid and steam pre-treatments exhibit a similar pattern. Scheme 3 has the best enzymatic hydrolysis efficiency, notwithstanding

(f) Co-fermentation of bagasse enzymatic hydrolyzate generated from different schemes:

A genetically altered co-fermenting S. cerevisiae strain with high sugar levels and inhibitor tolerance was used in the co-fermentation experiments. The initial mixed sugar content for Schemes 1 and 2 was in the range of 80–90 g/L, of which 45–55 g/L was glucose and the remaining was xylose. Consumption of xylose didn't start until all of the glucose had been gone. The decreased retention time needed for glucose eating indicates that the rate of glucose consumption was significantly higher than the rate of xylose consumption. Using Scheme 2 for monomeric treatment, the maximum ethanol generation (5.2 percent v/v) matched the maximum amount of total sugar liberated (89.5 g/L). After fermentation, Scheme 2 for monomeric treatment yields a maximum of 5.1–5.2 percent v/v ethanol. This shows that the co-fermenting strain can produce significant ethanol conversion rates with high sugar and inhibitor tolerance levels. In Scheme 3, glucose serves as the main initial sugar source whereas xylose plays a less significant role. Due to a lower initial xylose concentration at the start of co-fermentation in Scheme 3 than in other schemes, a shorter retention period is needed for ethanol conversion. The overall retention time for the manufacturing of ethanol was 48 hours, of which 24 hours were spent using glucose and xylosphate.

POST-PRETREATMENT METHODS USING SODIUM SULFITE

(a) Sodium sulphite treatment on whole unwashed slurry

The Post-pre-treatment Methods Using Sodium Sulphite processes are demonstrated in Figure 4. The slurries that were produced after the diluted mixed acid pre-treatment discussed in the previous section underwent sodium sulphite treatments. First, the process for treating sulphite is described below. Then, the enzymatic hydrolysis of entire slurries and wet cakes is covered in the sections that follow. A 15-liter reactor with an anchor-type impeller was used to treat the sodium sulphite solution at 300 RPM. Each batch of 3 kg of process water was added to a batch size of 7 kg pre-treatment slurry (20 percent w/w total solids) to create a solids consistency of 15 percent w/w total solids (TS) and 10 percent w/w total insoluble solids (TIS).

(1) The slurry was treated with sodium sulphite at 70°C after being chilled to 50°C with cooling water circulated in a jacket.
(2) At 70°C, sodium sulphite (1 percent w/w) was added to the slurry, which was subsequently cooled to 50°C using cooling water circulated inside a jacket.

Figure 4. Processes of Post-pretreatment Methods Using Sodium Sulphite

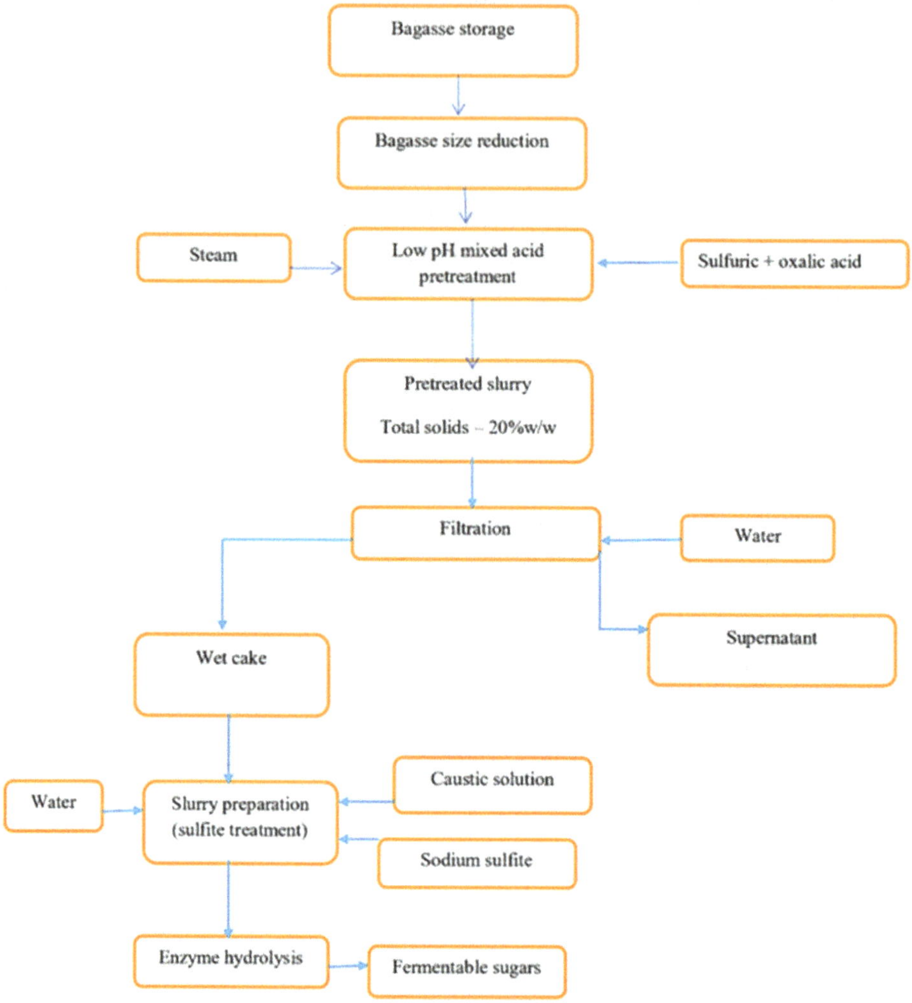

Using 475 mL of 40 percent (w/v) NaOH, the pH of the slurry was adjusted to 4.3–4.5 before being treated with an alkaline solution.

(3) The slurry was then maintained at 70°C while being cooled to 50°C using cooling water circulated in a jacket.

(4) 450 mL of 40 percent (w/v) NaOH were used to pre-treat diluted slurry to a pH range of 3.8-4.0. At 70 °C, sodium sulphite (4% w/w) was added with a 30-minute retention period. After that, the slurry was cooled using cooling water that circulated inside the jacket to 50°C.

(b) Sodium sulphite treatment of wet cake

It was possible to create a filter press with 11 plates, a surface area of 3 m^2, and a cake thickness of 30 mm. Before being introduced to water for the generation of hydrocarbons, dilution slurry is first treated with 0.5 percent, 1 percent, 2 percent, and 4 percent w/w sodium sulphite. A 15-liter reactor with an anchor-type impeller was used to treat the sodium sulphite solution at 300 rpm.

Four different methodologies as explained earlier were used here also.

(1) At a temperature of 70°C and a retention duration of 30 minutes, 225 mL of 40 percent (w/v) NaOH and 0.5 percent sodium sulphite were added to create a dilute slurry. The addition of sulphite resulted in an increase in the ultimate pH of the slurry to 5.0–5.2. The cooling water contained in the water jacket was then used to reduce the temperature to 50°C.

(2) The slurry was made with 202 mL of 40 percent (w/v) sodium hydroxide and 1 percent (w/w) sodium sulphite, and it had a pH between 4.3 and 4.5. It was also heated to 70°C for 30 minutes. Slurry was brought further for enzymatic hydrolysis after being further cooled to 50°C.

(3) At a temperature of 70 °C and a pH of 3.8-3.9 for 30 minutes, sodium sulphite was added to a solution of sodium hydroxide and water to create the slurry.

(4) The pH was stabilised using sodium sulphite, and the temperature was maintained at 70°C for the duration of the 30-minute retention period. The resulting slurry was then chilled using cooling water circulated in a jacket to 50°C. The pH was raised to 5.0–5.2 after treatment, and the temperature was held at 60°C until the slurry was suitable for processing.

(c) Enzymatic Hydrolysis

Whole slurry enzymatic hydrolysis (sulphite treated): In a reactor with an anchor impeller (250rpm) and 13–14% total solids—of which 9–10% w/w insoluble solids—a 10 kg enzyme hydrolysis experiment was run. The slurry was incubated at 50°C for 120 hours while commercial Novozymes enzyme was introduced in varying concentrations of 20, 30, 40, 60, and 90 mg/g of glucan. The slurry's pH was kept within the range of 5.0 and 5.2. Every 24 hours, samples were taken for the HPLC analysis of sugar(Boopathi et al., 2023; Boopathi S et al., 2022; Kumar et al., 2023; Sampath et al., 2023).

Whole slurry enzymatic hydrolysis of control (without sulphite): In a 15L reactor with an anchor impeller (250 RPM) and 13–14% w/w total particles, including 9–10% w/w insoluble solids, hydrolysis tests were carried out. A batch of 7 kg of complete pre-treatment slurry was diluted with 3 kg of process water and

500 mL of 40 percent (w/v) NaOH to bring the pH level to a range of 5.2-5.5. A slurry of sodium sulphite (sodium sulphite) and glucan (glucan) was prepared for the investigation and incubated for 120 hours at 50°C. The enzymatic slurry's pH was kept within the range of 5.0-5.2. According to the materials and techniques, samples were taken every 24 hours for the HPLC analysis of sugar and by-products.

Wet cake enzymatic hydrolysis (Sulfite addition): Studies on enzyme hydrolysis employed the wet cake sodium sulphite treated slurry as is. A batch of 10 kg of slurry, consisting of TIS and TS, was processed in a 15 L reactor with an anchor impeller running at 250 RPM. The slurry was treated with commercial Novozymes enzyme at doses of 20, 30, 40, 60, 80, and 90 mg/g. Utilizing sodium sulphite in a variety of dosages, SB's enzymatic hydrolysis efficiency was improved. Treatment for process intensification was applied to the wet cake and entire slurry of dilute mixed acid pre-treated SB. Every 24 hours, samples were taken for the HPLC examination of sugar and by-products. Following the addition of 4 percent w/w sodium sulphite at 70°C for 30 minutes, a hydrolysis efficiency of 57 percent for whole slurry was obtained at an enzyme concentration of 30 mg/g glucan. The efficiencies listed above are higher than they would be with no sulphite addition. The morphology of the control sample and the sulphite-treated samples differed, as seen in SEM pictures.

RESULTS OF PRE-TREATMENT WITH SODIUM SULFITE

A dosage response study was conducted using enzyme doses ranging from 20 to 90 mg/g glucan in order to better understand how enzyme dosages affect the conversion of cellulose to glucose. This study examined several sodium sulphite dosages along with a number of enzyme loadings. The lowest dose of addition in this investigation that produced the best efficiency in ethanol hydrolysis was sodium sulphite. Process optimization studies later in the process's lifespan may benefit more from the findings of this study on various enzyme doses.

Both the control and sodium sulphite treated samples exhibit an increase in cellulose to glucose conversion as the enzyme dose increases. The comparative impact of enzyme dosage (5–10 g/l) on the efficiency of glucose hydrolysis in the control sample (4 percent w/w). Glucan hydrolysis is positively impacted by the addition of sodium sulphite at enzyme doses ranging from 20 to 90 mg/g glucan. The ideal enzyme dosage, which is a compromise between effectiveness and price, may range from 30 to 40 mg per glucan. To speed up the enzymatic hydrolysis of different pre-treated biomasses, polymers like PVP (polyvinyl pyrrolidinone), PEG (polyethylene glycol), and surfactants Tween 80 and Tween 20 are frequently added. In this study, quick experiments were run to compare the results of sodium sulphite and polymer additions.

When PEG 8000 was added, wet cake hydrolysis achieved an enzyme efficiency of 62.3 percent, which is comparable to the addition of sodium sulphite. This demonstrates that sodium sulphite exhibits a similar strong effect on polymer additives. To ensure a fair comparison, the cellulases enzyme loading was also kept constant at 30 mg/g glucan. Regarding elements like the reduction of enzyme-lignin binding and the mitigation of inhibitors, the mechanisms of action of polymers and anionic compounds may be comparable. As an anionic additive, sodium sulphite, for instance, includes electrostatic interactions, whereas interactions of surfactants with biomass substrates have been mostly characterised as hydrophobic. Deeper features of the mechanism of action of these substances may be revealed by a thorough investigation involving several physical and chemical characterizations. From a techno-economic perspective, sodium sulphite post-pre-treatment treatment might offer superior value in the whole process because it is readily available and reasonably priced.

CONCLUSION

After a high-intensity, diluted mixed-acid treatment, the most xylose sugars were produced. Larger sugar concentrations, a higher yield of fermentable mixed sugars, and higher ethanol concentrations were produced via pressure filtering and the remixing of liquid with filtered solid. Enzymatic hydrolysis is used in the first two processes, while intensive washing and a solid-liquid separation step are used in the third. They discovered that applying a high-solid model led to an increase in ethanol concentration and a decrease in the capital and energy costs related to reactors and distillation, respectively.

A promising replacement for fossil fuels, ethanol can be made from lignocellulosic biomass. The complete process of producing bioethanol is divided into three basic unit operations. Recent studies demonstrate that the conventional and well-established corn-to-ethanol technology can be cost competitive with the cellulosic ethanol process. It is crucial to pre-treat lignocellulosic biomass, which is an expensive process in terms of money and energy. The goal of the research is to increase the existing efficiency of hemicellulose-to-xylose conversion and cellulose-to-glucose enzymatic conversion. For the pre-treatment of bagasse, steam explosion (SE) without a catalyst and a mixture of sulfuric and oxalic acids at low and high loadings were used. Three distinct post-pre-treatment process schemes were looked into in order to improve the effectiveness of the enzymatic hydrolysis of the pre-treated bagasse. The pre-treated slurry was made in three distinct methods for this study. The entire pre-treated slab was submitted to enzymatic hydrolysis in the first scheme (Scheme 1) without any additional treatments like washing or solid-liquid separation. Following filtering for further processing, the liquid phase underwent pressure filtration to produce a solid and liquid phase. According to a study, post-treating bagasse with a pressure filtering process improved the amount of sugars that might be produced for ethanol. The total cellulosic ethanol process can be made successful, and possibly cost-efficient as well, using several process modification techniques described.

Two distinct post-pre-treatment process configurations were explored for the sodium sulphite treatment procedure. The pre-treatment slurry was filtered and rinsed with more process water to separate it into a solid and liquid phase. The solid phase was then treated with sulphite and then hydrolyzed with enzymes. It was also investigated how different sodium sulphite concentrations affected both process setups. Comparing the study's experimental substrate to the control substrate without the addition of sulphite, it showed an 8–10% increase in overall enzymatic efficiency.

REFERENCES

Amores, I., Ballesteros, I., Manzanares, P., Sáez, F., Michelena, G., & Ballesteros, M. (2013). Ethanol Production from Sugarcane Bagasse Pretreated by Steam Explosion. *Electronic Journal of Energy and Environment*, *1*(1), 25–36. doi:10.7770/ejee-V1N1-art519

Boopathi, S., Saranya, A., Raghuraman, S., & Revanth, R. (2022). Design and Fabrication of Low Cost Electric Bicycle. *International Research Journal of Engineering and Technology, 5*(July), 5–7. www.irjet.net

Boopathi, S., Venkatesan, G., & Anton Savio Lewise, K. (2023). Mechanical Properties Analysis of Kenaf–Grewia–Hair Fiber-Reinforced Composite. In *Lecture Notes in Mechanical Engineering* (pp. 101–110). Springer. doi:10.1007/978-981-16-9057-0_11

Camargo, L. A., Pereira, S. C., Correa, A. C., Farinas, C. S., Marconcini, J. M., & Mattoso, L. H. C. (2016). Feasibility of Manufacturing Cellulose Nanocrystals from the Solid Residues of Second-Generation Ethanol Production from Sugarcane Bagasse. *BioEnergy Research*, *9*(3), 894–906. doi:10.100712155-016-9744-0

Canilha, L., Carvalho, W., De Almeida Felipe, M. D. G., De Almeida E., Silva, J. B., & Giulietti, M. (2010). Ethanol production from sugarcane bagasse hydrolysate using Pichia stipitis. *Applied Biochemistry and Biotechnology, 161*(1–8), 84–92. doi:10.1007/s12010-009-8792-8

Chandel, A. K., Kapoor, R. K., Singh, A., & Kuhad, R. C. (2007). Detoxification of sugarcane bagasse hydrolysate improves ethanol production by Candida shehatae NCIM 3501. *Bioresource Technology*, *98*(10), 1947–1950. doi:10.1016/j.biortech.2006.07.047 PMID:17011776

Cheng, M. H., Wang, Z., Dien, B. S., Slininger, P. J. W., & Singh, V. (2019). Economic analysis of cellulosic ethanol production from sugarcane bagasse using a sequential deacetylation, hot water and disk-refining pretreatment. *Processes (Basel, Switzerland)*, *7*(10), 642. doi:10.3390/pr7100642

Da Silveira Dos Santos, D., Camelo, A. C., Rodrigues, K. C. P., Carlos, L. C., & Pereira, N. (2010). Ethanol production from sugarcane bagasse by Zymomonas mobilis using simultaneous saccharification and fermentation (SSF) Process. *Applied Biochemistry and Biotechnology, 161*(1–8), 93–105. doi:10.1007/s12010-009-8810-x

Jackson de Moraes Rocha, G., Martin, C., Soares, I. B., Souto Maior, A. M., Baudel, H. M., & Moraes de Abreu, C. A. (2011). Dilute mixed-acid pretreatment of sugarcane bagasse for ethanol production. *Biomass and Bioenergy*, *35*(1), 663–670. doi:10.1016/j.biombioe.2010.10.018

Kumar, P., Sampath, B., Kumar, S., Babu, B. H., & Ahalya, N. (2023). Hydroponics, Aeroponics, and Aquaponics Technologies in Modern Agricultural Cultivation. In Trends, Paradigms, and Advances in Mechatronics Engineering (pp. 223–241). IGI Global.

Macrelli, S., Zacchi, G., & Mogensen, J. (2012). Techno-economic evaluation of 2nd generation bioethanol production from sugar cane bagasse and leaves integrated with the sugar-based ethanol process. *Biotechnology for Biofuels*, *5*(1), 152. . doi:10.1186/1754-6834-5-22 PMID:22502801

Mesa, L., Morales, M., González, E., Cara, C., Romero, I., Castro, E., & Mussatto, S. I. (2014). Restructuring the processes for furfural and xylose production from sugarcane bagasse in a biorefinery concept for ethanol production. *Chemical Engineering and Processing: Process Intensification*, *85*, 196–202. doi:10.1016/j.cep.2014.07.012

Sampath, B., Sasikumar, C., & Myilsamy, S. (2023). Application of TOPSIS Optimization Technique in the Micro-Machining Process. In Trends, Paradigms, and Advances in Mechatronics Engineering (pp. 162–187). IGI Global.

Shen, G., Yuan, X., Chen, S., Liu, S., & Jin, M. (2022). High titer cellulosic ethanol production from sugarcane bagasse via DLCA pretreatment and process development without washing/detoxifying pretreated biomass. *Renewable Energy*, *186*(January), 904–913. doi:10.1016/j.renene.2022.01.062

Wong, Y. C., & Sanggari, V. (2014). Bioethanol production from sugarcane bagasse using fermentation process. *Oriental Journal of Chemistry*, *30*(2), 507–513. doi:10.13005/ojc/300214

Chapter 12
Bioethanol Production From Biomass Extracted From Sunn Hemp Seed:
A Case Study

Ashok Kumar Koshariya
School of Agriculture, Lovely Professional University, India

J. Madhusudhanan
Anand Institute of Higher Technology, India

M. Jinnah Sheik Mohamed
Nellai College of Engineering, India

S. S. Kerur
KLE Technological University, India

Srinivas Viswanth Valeti
Aditya Engineering College, India

V. Haribalaji
Narasu's Sarathy Institute of Technology, India

ABSTRACT

In this chapter, the general procedures for the biomass and bio-ethanal manufacturing processes are explained. The various physical and chemical pre-treating methods for biomass are elaborated. The various acid hydrolysis methods, enzyme immobilisation methods, fermentation methods, and optimising the parameters and factors for hydrolysis have been exemplified. The stage-by-stage bioethanol production processes from Sunn hemp seed has been illustrated. Biomass preparation, pre-treatment processes, hydrolysis, synthesis of ferrites, and study of bioethanol characterizations are graphically exemplified. The microstructure of untreated and treated hemp biomass has been displayed. The variations in bioethanol concentration with increasing treatment time have been discussed. The incorporation of optimization parameters for producing quality bioethanol is also interpreted.

DOI: 10.4018/978-1-6684-4118-3.ch012

Figure 1. Various sources for the biomass production

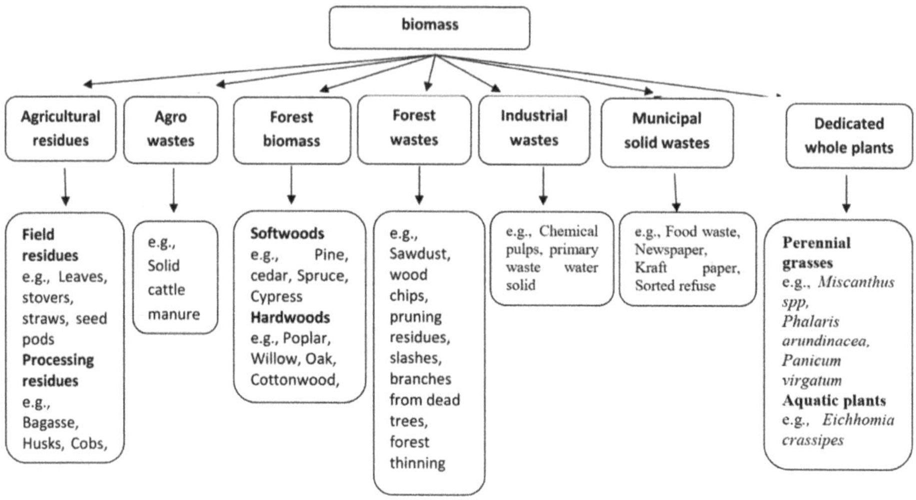

INTRODUCTION

Coal, oil, and gas are the main types of conventional fossil fuels used for primary energy today. Increasing energy demand, greenhouse gas emissions, and global climate change have propelled a number of experts to create new solutions. Around 80% of the energy generated by global renewable energy carriers comes from biomass. When needed, it can be stored and used to generate energy, fuel, and heat. These are referred to as bioenergy, which is defined as fuels derived from biological sources that are solid, liquid, or gaseous. Examples of bioenergy produced from first-generation feedstock include bio-alcohol made from corn, wheat, sugar beets, and sugarcane. Syngas, which may then be processed into biofuels and power, can be created from biomass (Khoo, Wen, Tang, & Chen, 2020). Biogas has been commercially created via anaerobic digestion. The heat and electricity produced by the biogas have been used for cooking and lighting. The various sources of biomass production are illustrated in Figure 1.

Despite significant scientific progress, it still takes technical advancements to make bioenergy production competitive with fossil fuels. For instance, a considerable challenge still exists in the large-scale cultivation of algae for the production of microalgal biofuels. Due to a lack of adequate infrastructure for the production process and high production costs, there are obvious technological limitations in the production of biofuels. Before beginning the processes for producing biofuel, pre-treatment techniques must be devised to remove fermentable sugars from lignocellulosic biomass.

The core idea behind nanoscience and nanotechnology applications is nanomaterials. It is possible to increase the energy and financial efficiency of bioenergy production. To obtain excellent quality and a high yield, pre-treatment, enzymes, and fermentation processes can be investigated. Materials considered to be nanoscale are those whose minimum dimension is less than or equal to 100 nm. This incredibly small size results in high surface area-to-volume ratio. These nanoparticles can also display many morphologies, which has expanded the range of applications for them (Alonso et al., 2010). Nanomaterials are excellent candidates in various biofuel systems because of their vast surface areas and special qualities like high catalytic activity, crystallinity, durability, storage, stability, and adsorption capacity. Nanostructured materials can react with other molecules more quickly than large particles. Energy crops are used to produce bioethanol, which is used as an alternative to petroleum. Many different feedstocks, including first-, second-, and third-generation corn, have been used to make bioethanol. Additionally, efforts are being made to create gasoline ethanol using municipal solid wastes. The bioethanol sector requires a steady flow of trustworthy raw materials. Softwoods and hardwoods are included in forest biomass, whereas sawdust, pruning, bark thinning residues, and wood chips are included in forest waste. Softwoods, which have lower density and rapid growth, include hemlock, redwood, spruce, pine, cedar, and fir. Municipal solid waste.

PRE-TREATMENT OF LIGNOCELLULOSIC BIOMASS

Yield is limited by hydrolysis of lignocellulosic biomass without pre-treatment. Pre-treatment is required to improve enzyme hydrolysis. There have been adopted a number of physical, chemical, physiochemical, and biological pre-treatment methods. The purpose of pre-treatment is to make more cellulose available for hydrolysis. The development of degradation products that prevent the enzymatic hydrolysis and fermentation of biomass is constrained by an optimum pre-treatment technique (Jha et al., 2022). To minimise the complex matrix of lignocellulosic biomass, various pre-treatment techniques are needed, depending on the kind of biomass. Pre-treatment techniques ought to be affordable and efficient on a variety of substrates with little preparation. Pre-treatment methods for Biomass production is shown in Figure 2.

Physical Methods

(a) **Mechanical size reduction method:** By reducing biomass into smaller particles, increasing the specific surface area, and reducing the degree of polymerization, it alters the crystallinity of cellulose. Mechanical pre-treatment has the drawbacks of being time-consuming, costly, and energy-intensive.

(b) **Pyrolysis:** One of the thermochemical pre-treatments employed to transform lignocellulosic biomass into fermentable sugars is pyrolysis. Hemicelluloses decompose between 250 and 300 degrees Celsius, cellulose between 325 and 400 degrees Celsius, while lignin decomposes between 300 and 5500 degrees Celsius.

(c) **Steam Explosion:** The lignocellulose structure is altered by the steam explosion treatment carried out at high temperature and pressure, followed by a quick decompression. Part of the lignin in biomass that has been heated to elevated temperatures (150–250 °C) is solubilized, and the hemicelluloses are degraded. The temperature, pressure, and duration of steam exposure all affect the structural changes that occur in biomass during steam pre-treatment. Eighty percent of the hemicellulose and eighty-five percent of the pectin component were solubilized during the autohydrolysis of sunflower seed hulls that had been prepared for five minutes at 200 °C before being subjected to explosive defibration. When rice straw was steam pre-treated at 3.53 MPa for 2 minutes, the pore size increased and the highest amount of glucose that could be produced by enzymatic hydrolysis was 83 percent. However, rice straw that had been processed for 4.02 MPa had less fibre porosity. Higher temperatures and longer residence durations during pre-treatment of bagasse enhanced the total conversion of biomass into simple sugars. The best potential sugar yield of 70% was obtained by adding auto-hydrolysis to coastal Bermuda grass that had been prepared at 150 °C for 60 minutes. H_2SO_4 or SO_2 in a steam explosion can reduce duration and temperature.

Chemical Methods

By eliminating lignin and hemicelluloses, lowering the degree of polymerization, and raising the crystallinity of lignocellulosic biomass, chemical pre-treatment has been widely employed to increase the biodegradability of cellulose. The paper and pulp industries made considerable use of chemical preparation(Ha et al., 2020). According to reports, chemicals significantly affect how lignocellulosic biomass is structured. Pore size, surface area, and lignin and hemicelluloses that can be dissolved.

Figure 2. Pre-treatment methods for Biomass production

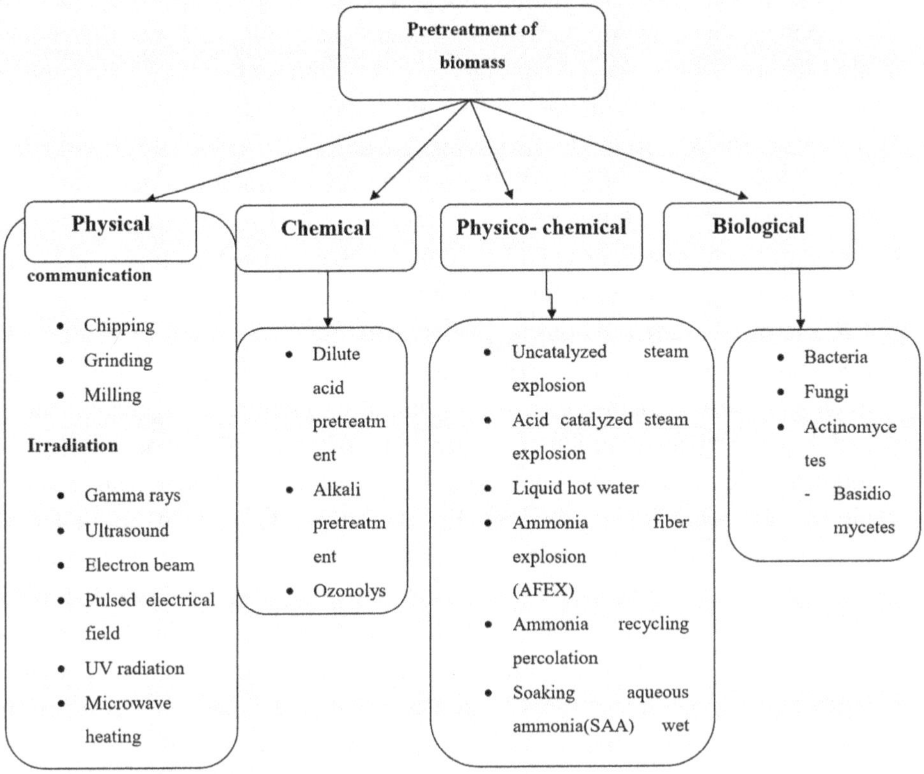

(a) **Acid pre-treatment:** Numerous inhibitory substances, including furan derivatives, aliphatic acids, and other phenolic compounds, are produced during the acid hydrolysis of lignocellulosic biomass. In order to get rid of the acid before sugar fermentation, more washing is necessary. Chemical, physical, and biological detoxification techniques are employed. Ion exchange resins, neutralisation, and adsorption are all used in the process of chemical detoxification. High temperatures and pressures are used for a dilute acid pre-treatment to transform xylan into xylose, which can then be fermented by microorganisms to produce ethanol. 55 percent of the cellulose in wood can be turned into simple sugar using batch-by-batch dilute acid hydrolysis. Although the sugar production is higher when using the high-temperature continuous mode, the sugar stream is also more diluted.

(b) **Alkali Pre-treatment:** Chemicals like NaOH, Na_2CO_3, $Ca(OH)_2$, and ammonia, which are non-corrosive and non-polluting, are used in the alkaline pre-treatment process. Lignin and other uranic acid substitutions on hemicelluloses are removed because they hinder the accessibility of enzymes. Hemicelluloses and lignin interact during the sodium hydroxide pre-treatment process. Each component's level of polymerization changes as a result of this reaction, which also affects surface area, crystallinity, porosity, and surface area. The removal of amorphous areas from the biomass alters the crystallinity index. The ideal pre-treatment conditions for sugarcane bagasse, corn cobs, and wheat straw were 0.5 percent NaOH at 121°C for 90, 60, and 30 minutes, respectively. The best conditions for delignification were demonstrated in water hyacinth treated with 1 percent caustic soda and 1 kg/cm2 steam pressure for 1 hour. Mesocarp from oil palm(McKendry, 2002).

(c) **Ultrasound assisted Alkaline pre-treatment:** In contrast to acidic pre-treatment, which must be carried out under harsh conditions, alkaline pre-treatment often uses non-corrosive, non-polluting alkali. Less sugar degradation and furan derivative production are seen. Although time-consuming and requiring a significant amount of alkali, alkaline pre-treatment is an effective method for removing lignin from lignocellulose. To intensify the alkaline pre-treating of oligosaccharides, a suitable physical or chemical method is necessary. Due to the possibility that the connections in the lignin and xylan networks could be broken by free radicals produced as a result of cavitation-induced breakdown of water molecules, ultrasound has been utilised to delignify lignocellulose. The homolyses of lignin-carbohydrate linkages that could result from the ultrasonic irradiation could result in the creation of a new type of cellulose.

HYDROLYSIS

The process of hydrolysis turns lignocellulosic biomass's carbohydrate polymers into straightforward fermentable sugars. Lignin and hemicellulose removal reduce crystallinity, and pre-treatment methods that promote porosity can significantly enhance cellulose hydrolysis. Lignosaccharides are typically broken down using one of two techniques: either diluted and concentrated acids, or an amalgam of acidic and non-aqueous solutions(Luo et al., 2014).

Acid Hydrolysis: The process of acid hydrolysis involves exposing the cellulose to an acid for a predetermined amount of time at a predetermined temperature. This causes the cellulose and hemicellulose polymers to break down into simple sugar molecules.

(a) **Dilute acid hydrolysis:** Dilute acid hydrolysis is used to produce reaction times in seconds or minutes, which permits continuous processing. It is carried out under high temperature (120-210 °C) and pressure. It has also been suggested that diluted acids like H_2SO_4, HCL, nitric acid, and phosphoric acid can increase cellulose synthesis. High reaction rates can be achieved and cellulose hydrolysis is improved with pre-treatment with diluted H_2SO_4. High temperature (T > 160 °C) and a continuous-flow process are the two main types of dilute-acid pre-treatment techniques that are frequently employed (5-10 solid loadings). A heterogeneous reaction is hydrolysis in the presence of weak acids. Cellulosic materials are turned to sugar in the first reaction, while sugars are transformed into other secondary compounds, such as furfural, in the second process. In order to hydrolyze hemicellulose into xylose and other sugars, biomass is treated with dilute H_2SO_4.

(b) **Concentrated Acid Hydrolysis:** Lignocellulosic biomass is hydrolyzed using concentrated acids. With less breakdown, they quickly transform cellulose and hemicelluloses into glucose and xylose. At first, the 70% concentrated acid dissolves the cellulose's crystalline structure. The single, isolated cellulose chains undergo a hydrolytic reaction in the second stage as a result of hydrolysis. Enzymatic hydrolysis at lower operating temperatures and air pressures depends on effective pre-treatment. Compared to the dilute acid hydrolysis technique, this requires a longer retention period and yields more ethanol.

(c) **Inhibitory products in hydrolysate:** In a first-order procedure known as hydrolysis, sugars made from hemicelluloses, such as glucose, mannose, and galactose, are broken down into furan derivatives, primarily hydroxyl methyl furfural (HMF). Released sugars are transformed into other chemical compounds that are harmful to fermentation microbes through subsequent reactions. The hydrolysate contains acetic acid due to the acetyl group in hemicelluloses. The survival of the cell is adversely affected by acetic acid, a weak acid with a high thermodynamic dissociation constant. Cell death and decreased biomass output are both caused by high acid concentrations. Hydroxylmethylfurfural compounds, which are created during the hydrolysis of cellulose and hemicelluloses, are the strongest inhibitor substances found in the hydrolysate. These two reactions are first order rate reactions that are controlled by temperature and acid concentration. Furfural diminishes ethanol production, biomass yield, and specific growth rate when it is present in the medium. When phenolic chemicals are present in a medium, they may partition into biological membranes, resulting in a loss of integrity and altering the activity of enzymes. Phenolic chemicals produced by the breakdown of lignin have the potential to promote microbial development and the development of biofilm.

(d) **Enzymatic Hydrolysis:** Enzyme-catalyzed hydrolysis is the process by which polysaccharides are changed into straightforward fermentable sugars. The prospect of higher conversion, less byproduct creation, a low energy need, and minimal operating conditions are all provided by avoiding the use of poisonous and corrosive chemicals. By targeting the glycosidic bonds, cellulase enzymes catalyse the breakdown of cellulose into glucose. Bioethanol is produced by further fermenting this glucose with the aid of microorganisms. Commercial cellulase enzyme synthesis is carried out by fungi. For the generation of ethanol, lignocellulosic biomass is hydrolyzed by enzymes in a mild environment with no corrosion. Other wood-eating insects, such as termites, are being researched as possible cellulase enzyme manufacturers. Complete hydrolysis of eucalyptus chips was achieved with a 10 FPU/g enzyme dose. When the substrate concentration was less than 5% and the reaction time was between 12 and 72 hours, the rate of hydrolysis increased as the enzyme concentration (1-4 IU/mL) rose. After the saccharification of sunflower hulls, the effects of treatment with NaOH were examined. After being treated with 0.5 percent NaOH at 121°C for 1.5 hours, the maximal hydrolysis rate was attained. A 50 FPU/g enzyme dosage enhanced hydrolysis yield. Using 5% hydrogen peroxide and 2% NaOH to treat sugarcane bagasse resulted in saccharification rates of 63.3 and 54.7 percent, respectively. An 86 percent saccharification rate with 40 FPU/g of enzyme concentration, a pH of 4.5, and 2.5 percent substrate loading were obtained after pre-treating the biomass of Antigonum leptopus with water.

Factors Affecting Hydrolysis

Numerous factors, including the source of the enzyme, the pre-treatment technique used to prepare the substrate, the quantities of the enzyme and substrate, mass transfer, and intrinsic kinetics, affect the rate of hydrolysis. Additionally, crucial are product inhibition, product stability, and enzyme stability at various temperatures (Osman et al., 2021). pH, temperature, agitation, substrate porosity, product inhibition, aeration, and dilution rate are a few of them. Focus must be placed on increasing cellulase activity and perfecting the hydrolytic process in order to increase efficiency. Enzymes fluctuate between their active and inactive states during the hydrolysis process as a result of modifications and the generation of residual charges at the active site. According to research on Trichoderma reesei, pH variations at 50 °C can also cause structural alterations.

Figure 3. Chemical Components of from Lignocellulosic Materials

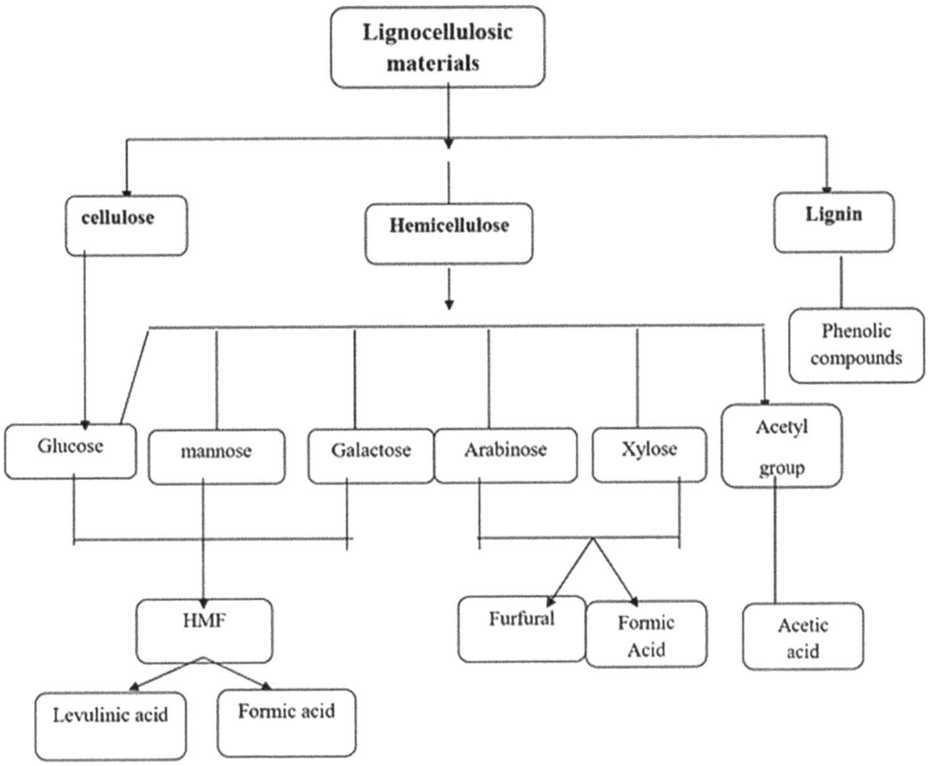

ENZYME IMMOBILIZATION

The structure of lignocellulosic biomass presents a number of obstacles that can be overcome by immobilising cellulase on solid materials. The activity of various immobilisation methods for various enzymes has been found to be higher than that of free enzymes. Enzyme attachment to a magnetic nanoparticle results in increased surface area, strong covalent cross-linking, and improved separation from the reaction mixture. Based on its magnetic behaviour, super magnetism controllability, and electromagnetic properties, ferrite-based materials have long been studied. Due to its unique characteristics, including chemical stability, thermal stability, and particle size dependent magnetic properties, zinc ferrite has long been the focus of research. Due to its distinct qualities, glutaraldehyde has been employed as a crosslinker and is one of the more efficient ones. The various chemical components of from Lignocellulosic materials are displayed in Figure 3.

OPTIMIZING THE PARAMETERS

An important area of research is modelling and optimising biochemical and biotechnological processes using statistical experimental designs. Some benefits of statistical design approaches include reduced experiment numbers, the search for factor relativity, applicability for multiple factor experiments, locating the most suited settings, and response predicting. RSM is a collection of statistical and mathematical methods for generating empirical models. This statistical technique sought to maximise an outcome (response variable) that was affected by a number of independent variables (input variables). Box and Wilson (1951) and central composite designs are the two methods for process analysis and modelling that are most frequently employed (CCD). In the 1950s, ridge analysis was touted as a fantastic method for studying the response surface (Boopathi, 2019, 2021; Boopathi & Myilsamy, 2021; Sampath et al., 2022; Sampath & Myilsamy, 2021).

FERMENTATION

A wide range of biomass resources have the potential to offer inexpensive feedstock for the manufacture of bioethanol since it can lower greenhouse gas emissions and contribute to environmental protection. Sugars including glucose, fructose, and sucrose undergo a biological process called ethanol fermentation in which ethanol and carbon dioxide are produced as by-products. The cost-effective commercialization of ethanol production from lignocellulosic biomass depends largely on the accessibility of adequate energy sources. The hydrolysate after initial pre-treatment includes simple sugars in addition to a number of inhibitory substances that prevent microbial fermentation. Hexoses are simple for S. cerevisiae to ferment, but pentoses are more challenging.

(a) **Microorganisms used in fermentation:** Aeurobasidium pullulans, Fusarium avenaceum, Fusarium graminearum, Fusarium lycopersici, oxysporum, sambucium, solani, tricinetum, and Paecilomyces are among the fungi that may ferment D-xylose. S. cerevisiae is a popular fermenting yeast used to produce ethanol. But it can only use hexose; pentose sugars cannot be fermented by it. The ability of Candida, Clavispora, Kluveromyces, Pachysolen, Pichia, and Schizosaccharomyces to metabolise xylose has been tested.

(b) **Fermentation of Lignocellulosic Hydrolysate:** It has been discovered that enzymatic hydrolysis of beet pulp results in increased fermentable sugar yields and better ethanol production conditions. In under 24 hours, the pulp can be fermented, yielding 8 to 9 percent (V/V) ethanol. The ability of four different strains of Saccharomyces cerevisiae to select xylose as a substrate from a variety of substrates has been investigated. In the presence of used sulphite liquor, it has been claimed that this strain produced 1.2 g/l of ethanol (Ong et al., 2020). When S. cerevisiae was used to ferment wheat straw hydrolysate, 40 grams of ethanol were generated for every game of total sugar. The simultaneous alcoholic fermentation of glucose and xylose generated from lignocellulosic biomass was investigated using separate and co-culture techniques.

After fermentation with Saccharomyces cerevisiae and Pachysolen tannophillus, rice straw prepared with 4 percent NaOH and enzyme hydrolysate containing a mixture of pentoses and hexoses yielded 24.5 percent ethanol. Eichhornia crassipes (Water hyacinth) acid hydrolysate fermented by Candida shehatae NCIM 3501 produced 0.35 g/g ethanol. The effects of inhibitory substances produced by steam pretreatment on enzymatic hydrolysis and fermentation were investigated by Palmqvist et al. The best production of ethanol from a mixture of glucose and xylose (0.5 g/g) and largest benefit of ethanol (3.75 g/L in 40 h) were produced by cultures of P. stipitis CBS 5773 with a respiratory impaired. It has been researched how to turn aspen wood hydrolysate into ethanol.

(c) **Yeast immobilization for bioethanol production:** A variety of biomasses have been utilised to produce the cost-effective lignocellulosic ethanol that is in increasing demand. Yeast is commonly used and is known to be a highly tolerant strain in the fermentation medium. The procedure of yeast cell immobilisation has been proven to be effective in achieving increased cell density in a bioreactor. Utilizing immobilised yeast has the benefit of being re-used, making product recovery simple, and avoiding product and substrate inhibition. Alginate's high compatibility, simplicity of manufacture, and inexpensive cost made it a successful supportive substance.

(d) **Distillation:** The best recovery method for the more flammable fermentation products is distillation. It has been successfully proven that some distillation techniques can inexpensively recover diluted volatile goods from contaminants. Along with the commercialization of biomass hydrolysis and fermentation technologies, there is an increasing need for innovative technologies. The recovery yield in the majority of industrial bioreactors is around 4% (w/w). High-quality anhydrous ethanol is gathered using a molecular sieve with three different pore diameters throughout the distillation procedure.

(e) **Sunn hemp as an energy source:** The best recovery method for the more flammable fermentation products is distillation. It has been successfully proven that some distillation techniques can inexpensively recover diluted volatile goods from contaminants. Along with the commercialization of

Figure 4. Sunn hemp dry and pre-treated biomass

biomass hydrolysis and fermentation technologies, there is an increasing need for innovative technologies. The recovery yield in the majority of industrial bioreactors is around 4% (w/w). High-quality anhydrous ethanol is gathered using a molecular sieve with three different pore diameters throughout the distillation procedure.

CASE STUDY: BIOETHANOL FROM SUNN HEMP SEED

(1) Biomass Preparation

Sunn hemp seeds were grown in Warangal, Telangana, and were harvested 9 to 12 weeks after planting. After subjecting 1 g of biomass sample to a temperature of 550 °C for two hours, the percentage of ash was calculated. To eliminate dirt remnants, the gathered biomass was rinsed. The biomass was crushed to fit through a 1.5-mm screen after being air dried(Akia et al., 2014). Weight loss following drying was used to calculate the moisture content. The sunn hemp dry and pre-treated biomass are shown in Figure 4.

(2) Pre-treatment

The process and biomass conversions by pre-treating methods are illustrated in Figure 5.

Figure 5. Process of pre-treating bio-mass

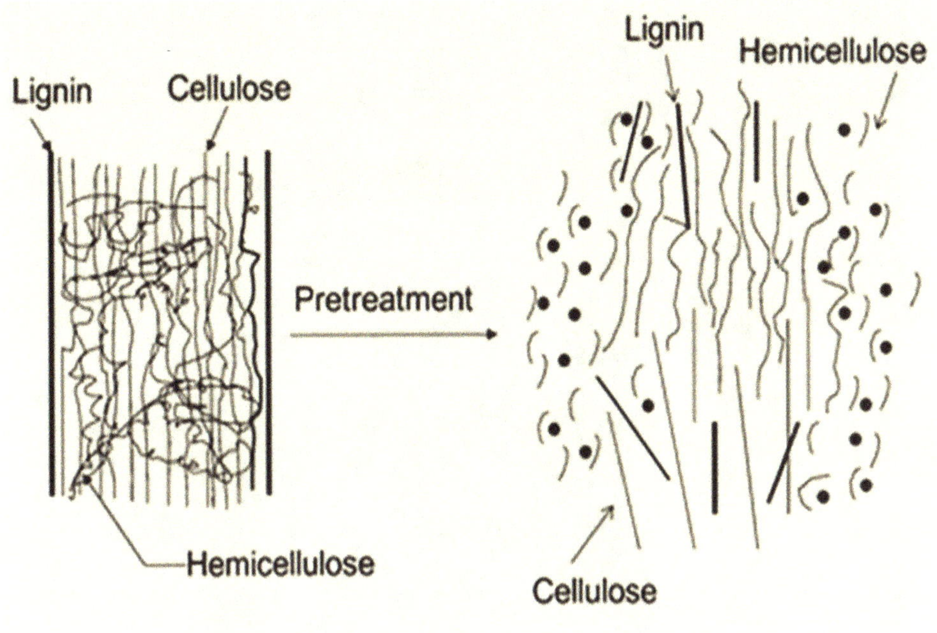

(a) **Alkaline Pre-treatment:** One gram of sunn hemp biomass was submerged for 20 minutes at 121°C in solutions of NaOH with the required concentrations (0.5, 1, and 1.5). Following pre-treatment, filtering was used to recover the solid residues. Solid wastes were dried to a consistent weight and kept at room temperature in a Ziploc bag for later compositional analysis.

(b) **Ultrasound-alkaline pre-treatment:** The Sonic Vibra Cell Model: VCX500 sonicator, which has a 0.5-cm probe with a 20 kHz frequency and a 500 W ultrasonic processor, was used to perform ultrasound-assisted alkaline pretreatment. Sunn hemp biomass weighing one gramme was dissolved in 10 ml of a 1 percent NaOH solution.

(c) **Proximate compositional analysis:** The carbohydrate composition of dried Sunn hemp biomass and dried biomass that had undergone ultrasonic treatment with 1 percent NaOH was analysed. The forage analysis method was used to compute acid detergent lignin (ADL) and neutral detergent fibre (NDF). In order to determine the cellulose content, ADL and ADF were subtracted.

(d) **Enzymatic hydrolysis:** Hydrolysis tests were conducted in a 250 mL Erlenmeyer flask at a solid loading of 1% in 0.1 M sodium acetate buffer at pH 4 with an enzyme loading of 25 IU/g of biomass. In a thermal incubator, the flask was shaken for 60 hours at 50°C and 160 RPM (Scigenics Biotech). Following

Figure 6. Ferrite particles

hydrolysis, samples were filtered through a 0.22-m nylon membrane driven by a syringe, and their glucose and xylose yields were determined using an HPLC system (LC-20AD, Rezex column, refractive index detector RID-10A, Shimadzu, Japan) with a Rezex column at an ambient temperature of 80 °C. HPLC grade water was used as the mobile phase at a flow rate of 0.6 mL/min.

(e) **Analytical methods:** Sunn hemp biomass' X-ray diffraction (XRD) peaks were captured using a Bruker D8 Advanced X-ray Diffractometer. Utilizing the peak height approach, the crystallinity index (ICR) was computed from PXRD data.

(f) **Synthesis of Ferrite particles:** The ferrite particles are shown in Figure 6. Under hydrothermal conditions, ferrite particles were created using the co-precipitation process with alkaline solution. Under steady stirring, a sodium hydroxide aqueous solution was added to neutralise the pH. To get rid of any remaining water, the precipitate was further cleaned with ethanol.

(g) **Immobilization of cellulase on the activated ferrite nanoparticle:** The ferrite particles were sonicated for 1 hour at a concentration of 4 mg/ml while suspended in deionized water. The suspension was then centrifuged for 25 minutes at 10,000 rpm after being sonicated. To get rid of any unbound glutaraldehyde, the activated support particles were washed in 0.1 M sodium acetate buffer(Alonso et al., 2010). By determining the ratio of the total amount of protein bound to the total amount of protein provided for immobilisation, the Bradford test at 595 nm with comassive brilliant blue G250 was used to evaluate the immobilised enzyme binding efficiency. To remove unattached protein, the immobilised enzyme on the ferrite particle support rinsed with buffer.

(h) Characterization

XRD Analysis: Cu K-alpha radiation, which has a wavelength of 0.154 nm and an angular range of 10 to 80°, was used to produce the X-rays. Using a Bruker D8 Advanced X-ray Diffractometer, the crystalline structure of the ferrite particles was characterised and recorded.

Particle size analyser: Researchers at the University of Bristol have examined ferrite particles using a high-performance nanomaterials platform (Nano S90 version 7.02). Using the dynamic light scattering method, ferrite particles are created by dissolving them in various solvents such water, ethanol, and isopropanol.

(i) **Enzyme assay:** When free and immobilised enzymes react with carboxymethyl cellulose as a substrate, the amount of glucose produced is measured to assess the enzyme activity. A sample mixture of the enzyme and substrate was used in the reaction, which was carried out at 50°C and 60°C and incubated for 30 min. Researchers from the University of Bristol in the UK took measurements at 540 nm of the amounts of reducing sugars generated during enzymatic hydrolysis. By adding 3,5-Dinitrosalicylic acid reagent and holding the mixture in a boiling water bath for 10 minutes, the enzyme substrate reaction was paused to allow for colour production.

(j) **Thermo stability study:** At various temperatures (30, 40, 50, 60, 70, and 80 C), the thermostability of an immobilised enzyme was investigated, and values of sugar concentration were established.

(k) **Reusability assay:** Enzyme activity experiments were used to determine the immobilised enzyme's reusability. as shown by the hydrolysis reaction with substrate solution using 0.1 M sodium acetate buffer at 60 °C for 30 min (pH 5). Particles were re-suspended in buffer and substrate solutions after each cycle of the experiment until the activity was lower than 10%.

(l) **Enzymatic Hydrolysis of Biomass using free and immobilized enzyme:** By measuring enzyme activity, the immobilised enzyme's reusability was determined. shown by the hydrolysis reaction with the substrate solution at 60 °C for 30 min while using 0.1 M sodium acetate buffer (pH 5). Until the activity fell below 10% after each cycle of the experiment, particles were re-suspended in buffer and substrate solutions. Cellulose Hydrolysis is calculated by equation (1).

$$\text{Cellulose Hydrolysis}\,(\%) = \frac{\text{Volume reduction of sugar concentartion} \times 0.9 \times 100}{\text{Additon of celluose}} \tag{1}$$

(m) Fermentation

Microorganisms and maintenance: The pentose yeast Pichia stipites 3507 and Saccharomyces cerevisiae were provided by NCL Pune in India. On YEPX agar media, pentose yeast was kept alive, and on agar 20 g L-1, pH 5.0, glucose was developed. A single-cell yeast colony was inoculated in YEPD and YEPX broth and incubated in a 100-ml Erlenmeyer flask for 24 hours for hexose yeast and 48 hours for pentose yeast at 30°C. By inoculating 1% of the primary culture cell concentration in 1000 ml of yeastrophenyl (YEPX) under analogous circumstances, secondary culture preparation was completed. 1000 ml of Sunn hemp hydrolysate were mixed with yeast extract, yeast nitrogen base, ammonium sulphate, and magnesium sulphate to create ethanol. Secondary yeast cultures were centrifuged at 8000 rpm for 10 minutes after 24 and 48 hours.

Ethanol fermentation by co-culture: At a pH of 5.25, hemp hydrolysates were taken along with fermentation ingredients and sterilised for 20 minutes at 121 °C. A secondary culture of yeast made from hexose and pentose was taken, centrifuged for 10 minutes at 8000 rpm, and then washed with sterile water.

Immobilization of yeast cells by sodium alginate: Sodium alginate solutions were combined with yeast, S. cerevisiae, and P. stipitiscell suspensions at a 1:1 ratio. The cellulose-alginate mixture was added dropwise with a hypodermic syringe to a cold, sterile 2.0 percent $CaCl_2$ solution that was kept on a magnetic stirrer with moderate agitation.

(n) Estimation of ethanol using solvent extraction followed by potassium dichromate method

Tri-n-butyl phosphate was added to one millilitre of material, which was then violently vortexed for one hour. A new test tube was filled with 750 litre of the solvent phase (top layer) following phase separation. The 750 litre of dichromate reagent was added and vortexed vigorously for 30 minutes.

RESULTS AND INTERPRETATION

(a) Sunn hemp biomass preparation

Effect of NaOH pre-treatment: Untreated Sunn hemp biomass has a total cellulose, hemicellulose, and lignin content of 83 percent. The Van Soest technique results revealed reduced hemi thrombosis levels and a greater cellulose content of 55 to 56 percent in biomass processed at higher NaOH concentrations. When alkaline samples were treated to sonication, the cellulose content rose from 55% to 63%. A research by the University of Bristol found that 10 to 15 minutes of sonication resulted in the degradation of 50% of hemicellulose and 64% of lignin. In a study conducted by researchers at the University of Bristol, it was discovered that ultrasound was more effective than alkaline pre-treatment at removing hemicellulose and lignin(Abualigah et al., 2022). The ultrasound-assisted alkaline procedure is particularly appealing as a method because it selectively removes lignin and hemicellulose without destroying cellulose.

X-Ray Diffraction studies: Sono aided alkaline pre-treatment, alkaline, and untreated Sunn hemp samples' X-ray diffraction spectra were investigated. The percentage of crystalline cellulose in the entire solid is determined by the crystallinity index. 39 percent, 44 percent, and 53 percent were the crystallinity indices. The crystallinity index would rise if the amorphous lignin and hemicelluloses were largely eliminated. Untreated Sunn hemp was found to have a crystallinity size that was 19.4 nm larger than that of the alkaline and sono aided alkali treated samples.

SEM Analysis: The untreated biomass has a hard, tightly packed structure. After pre-treatment, cavitation's effects and the release of lignin and hemicelluloses cause the structure to break. The untreated, and alkaline treated biomass is shown in Figure 7. The sono-assisted alkaline treated biomass is illustrated in Figure 8.

Figure 7. Untreated and Alkaline Treated biomass

(b) Optimization of enzyme loading

The variation of bioethanol concentration with respect to treating time is illustrated in Figure 9. The final product of bio-ethanol is displayed in Figure 10. The effectiveness of biomass hydrolysis depends greatly on the enzyme to substrate ratio. After 48 hours, the reducing sugar output rose from 7 to 9 mg/mL as the enzyme loading increased from 25 to 30 IU/g. When the enzyme loading was increased from 25 to 30 IU/gm, the reducing sugar yield first increased but then gradually stabilized (Khan et al., 2022). The result of hydrolysis may result in feedback inhibition for ongoing hydrolytic activities by increasing enzyme concentration and enzyme loading.

(c) Optimization of enzymatic hydrolysis by Design expert

It is possible to model and improve biochemical and biotechnological processes using the response surface methodology. By combining mathematical and statistical methodologies, empirical models can be created. For optimum reducing sugar yield, the optimal feasible hydrolysis condition was identified and optimised using a central composite design. Results of a two-way ANOVA to identify the impact of specific variables on the generation of glucose and xylose in a quadratic response surface model(Jha et al., 2022; Khan et al., 2022).

(d) Model Validation

According to a study, the response surface methodology's optimised variables for enzymatic hydrolysis were fairly close to experimental investigations that

Figure 8. Sono-assisted alkaline treated biomass

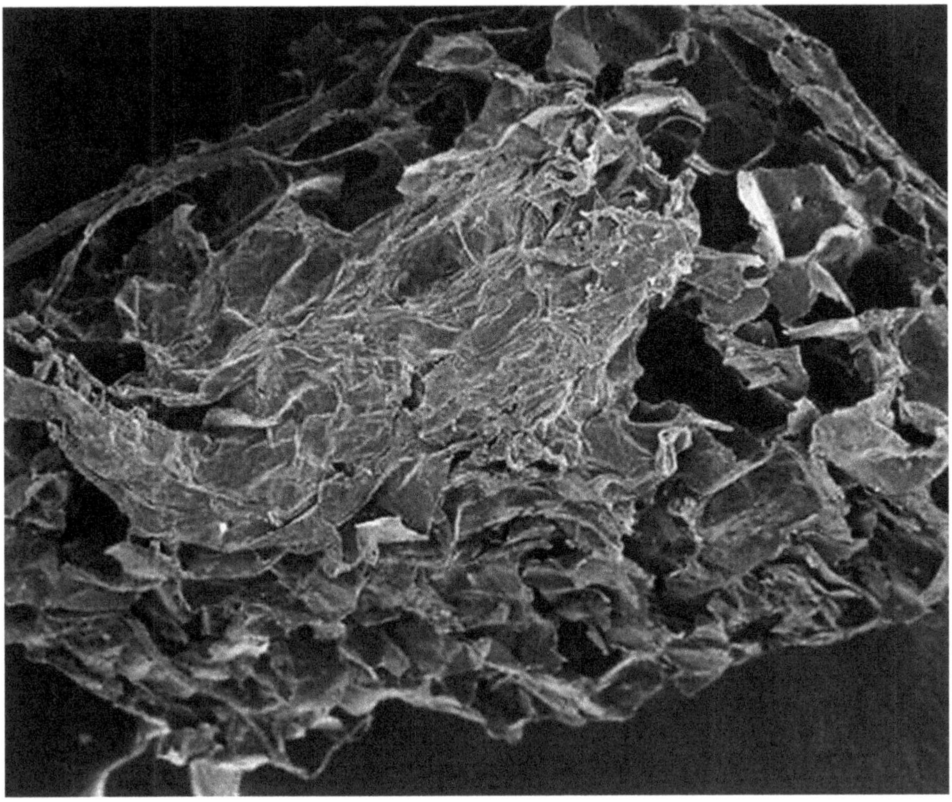

demonstrated a release of 5.17 mg/ml of glucose. The researchers offered thirty potential strategies for lowering sugar release for the input variables. The largest amount of xylose released in an experiment was 2.24 mg/ml, which was quite similar to the experimental study's 2.08 mg/ml. The values obtained by the response surface methodology and the experimental results closely match each other. This demonstrates how the process variables can be optimised and enzymatic hydrolysis studied using statistical experiment design. A number of 3D surface plots for creating reduced sugar have been created by researchers at the University of Bristol in the UK. The plotted reaction seems.

(e) Inferences

121°C for 20 minutes with ultrasound assisted alkali pre-treatment revealed favourable cellulose content. The largest amount of glucose released from 25 units of enzyme loading after 48 hours was 5.43 mg/mL. The trials were performed with optimised

Figure 9. Variations of bioethanol concentration versus time

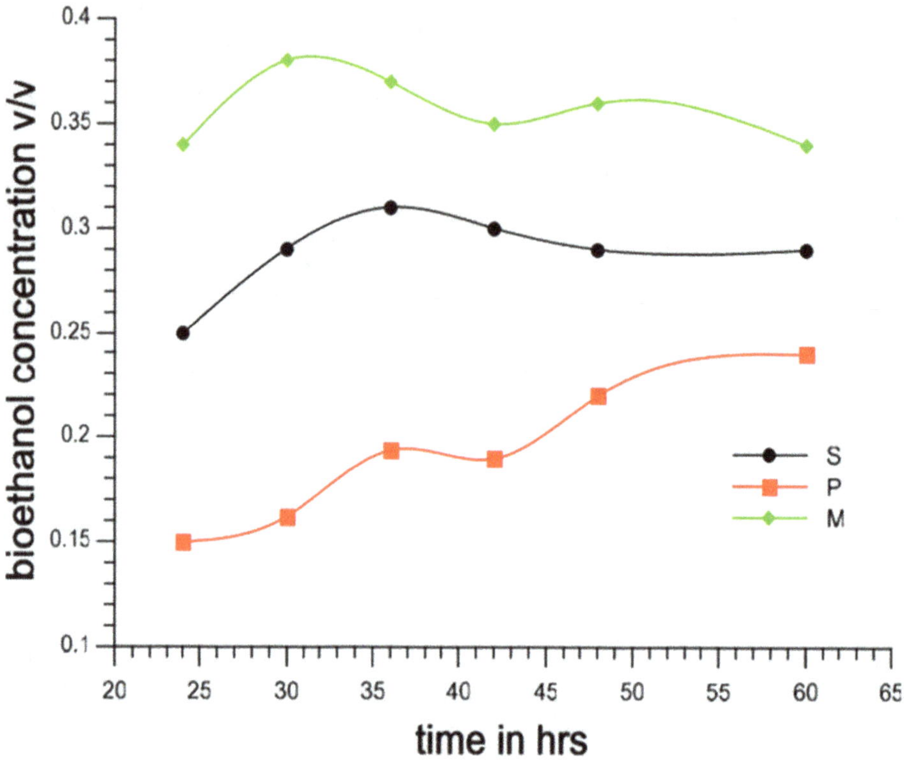

variables, and the outcomes matched those of the optimal circumstances. When compared to axenic culture, mixed fermentation of Saccharomyces cervasiae and Pichia stipilis produced more ethanol. Adapted microorganisms were used to ferment the sugars hexose and pentose. At a ferrite loading dosage of 4 mg/ml and an enzyme concentration of 20 units, approximately 74% of binding was accomplished. It is important to research the properties of fuel, including their calorific value, cetane number, octane number, and oxygen content. It is possible to conduct scale-up experiments to create bioethanol on an industrial scale. Large-scale production requires and preferably undergoes a techno-economic review.

SUMMARY

The general procedures for the production of bioethanol and biomass are described in this chapter. Examples of the numerous methods for acid hydrolysis, enzyme

Figure 10. Final product of bioethanol

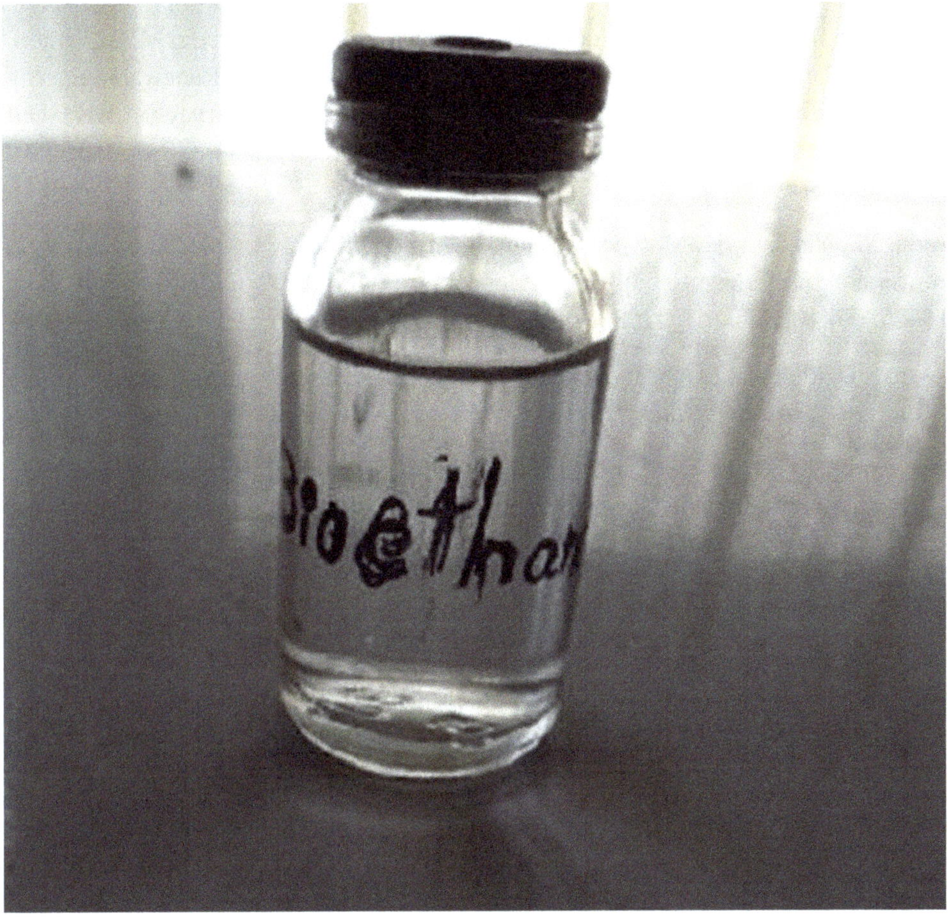

immobilisation, fermentation, and optimising the parameters and variables for hydrolysis may be found. The detailed steps for producing bioethanol from hemp seed have been illustrated. The Sunn hemp seeds were grown in Warangal, Telangana, and harvested 9 to 12 weeks after planting. The biomass was crushed to fit through a 1.5-mm screen after being air dried. Weight loss following drying was used to calculate the moisture content. More ethanol was produced by mixed fermentation of Saccharomyces cervasiae and Pichia stipilis than by axenic culture. A techno-economic evaluation is required for large-scale production and should ideally be conducted. Scale-up experiments can be performed to produce bioethanol on a large scale.

REFERENCES

Abualigah, L., Zitar, R. A., Almotairi, K. H., Hussein, A. M., Elaziz, M. A., Nikoo, M. R., & Gandomi, A. H. (2022). Wind, Solar, and Photovoltaic Renewable Energy Systems with and without Energy Storage Optimization: A Survey of Advanced Machine Learning and Deep Learning Techniques. *Energies*, *15*(2), 578. doi:10.3390/en15020578

Akia, M., Yazdani, F., Motaee, E., Han, D., & Arandiyan, H. (2014). A review on conversion of biomass to biofuel by nanocatalysts. *Biofuel Research Journal*, *1*(1), 16–25. doi:10.18331/BRJ2015.1.1.5

Alonso, D. M., Bond, J. Q., & Dumesic, J. A. (2010). Catalytic conversion of biomass to biofuels. *Green Chemistry*, *12*(9), 1493–1513. doi:10.1039/c004654j

Boopathi, S. (2019). Experimental investigation and parameter analysis of LPG refrigeration system using Taguchi method. In SN Applied Sciences (Vol. 1, Issue 8). doi:10.100742452-019-0925-2

Boopathi, S. (2021). Improving of Green Sand-Mould Quality using Taguchi Technique. *Journal of Engineering Research*. doi:10.36909/jer.14079

Boopathi, S., and Myilsamy, S. (2021). Material removal rate and surface roughness study on Near-dry wire electrical discharge Machining process. *Materials Today: Proceedings*, *45*, 8149–8156. doi:10.1016/j.matpr.2021.02.267

Ha, G. S., El-Dalatony, M. M., Kurade, M. B., Salama, E. S., Basak, B., Kang, D., Roh, H. S., Lim, H., & Jeon, B. H. (2020). Energy-efficient pretreatments for the enhanced conversion of microalgal biomass to biofuels. *Bioresource Technology*, *309*(January), 123333. doi:10.1016/j.biortech.2020.123333 PMID:32305842

Jha, S., Nanda, S., Acharya, B., & Dalai, A. K. (2022). A Review of Thermochemical Conversion of Waste Biomass to Biofuels. *Energies*, *15*(17), 1–23. doi:10.3390/en15176352

Khan, A. A., Minai, A. F., Pachauri, R. K., & Malik, H. (2022). Optimal Sizing, Control, and Management Strategies for Hybrid Renewable Energy Systems: A Comprehensive Review. *Energies*, *15*(17), 6249. Advance online publication. doi:10.3390/en15176249

Khoo, K. S., Wen, Y. C., Doris, Y. Y. T., & Pau, L. S. C. (2020). Nanomaterials Utilization in Biomass for Biofuel and Bioenergy Production. *Energies*, *13*(4), 892. doi:10.3390/en13040892

Luo, J., Fang, Z., & Smith, R. L. Jr. (2014). Ultrasound-enhanced conversion of biomass to biofuels. *Progress in Energy and Combustion Science*, *41*(1), 56–93. doi:10.1016/j.pecs.2013.11.001

McKendry, P. (2002). Energy production from biomass (part 1): Overview of biomass. *Bioresource Technology*, *83*(1), 37–46. doi:10.1016/S0960-8524(01)00118-3 PMID:12058829

Ong, H. C., Chen, W. H., Singh, Y., Gan, Y. Y., Chen, C. Y., and Show, P. L. (2020). A state-of-the-art review on thermochemical conversion of biomass for biofuel production: A TG-FTIR approach. *Energy Conversion and Management, 209*(December 2019). doi:10.1016/j.enconman.2020.112634

Osman, A. I., Mehta, N., Elgarahy, A. M., Al-Hinai, A., Al-Muhtaseb, A. H., & Rooney, D. W. (2021). Conversion of biomass to biofuels and life cycle assessment: a review. In Environmental Chemistry Letters (Vol. 19, Issue 6). Springer International Publishing. doi:10.100710311-021-01273-0

Sampath, B., & Myilsamy, S. (2021). Experimental investigation of a cryogenically cooled oxygen-mist near-dry wire-cut electrical discharge machining process. *Strojniski Vestnik. Jixie Gongcheng Xuebao*, *67*(6), 322–330. doi:10.5545v-jme.2021.7161

Sampath, B., Pandian, M., Deepa, D., & Subbiah, R. (2022). Operating parameters prediction of liquefied petroleum gas refrigerator using simulated annealing algorithm. *AIP Conference Proceedings*, *2460*(1), 070003. doi:10.1063/5.0095601

Chapter 13
A Glance at Trends Manufacturing Systems Modeling and Optimization

Gilberto Pérez-Lechuga
https://orcid.org/0000-0003-1222-5031
Universidad Autónoma del Estado de Hidalgo, Mexico

José Francisco Martínez-Sánchez
Universidad Autónoma del Estado de Hidalgo, Mexico

Tomás Gómez-Rodríguez
Universidad Autónoma del Estado de Hidaldo, Mexico

ABSTRACT

In this chapter, the authors develop an analysis of the main optimization methods used in flexible manufacturing systems (FMS) problems. The results are obtained from a random sample of more than 100 documents published from 1986 to 2020 dedicated to the optimization of FMS. These are classified by their applications in the most significant fields of this area in analytical methods and heuristic methods. The analysis also addresses some aspects of computational complexity found in ordinary optimization as well as the most common NP-hard problems of the FMS. The results obtained are presented, and the virtual future of the FMS is projected as well as the new challenges and opportunities to venture into this important field of human activity. These are highly representative since they suggest that, despite the continuous growth in the complexity of the models treated, the majority of authors studied here continue using traditional techniques based on scientific optimization methods, leaving aside the emerging techniques of stochastic optimization based on metaheuristics.

DOI: 10.4018/978-1-6684-4118-3.ch013

INTRODUCTION

A flexible manufacturing system (FMS) is defined as the integration of manufacturing or assembly processes, material flow and computer communications and control. The objective is to let the production floor respond rapidly and economically to changes in system operations. The concept of flexibility in FMS has attained significant transcendence in meeting the challenges for a variety of products at shorter lead-times, together with higher productivity and quality (Oracle ERP).

A Manufacturing System is composed of the following elements: a) Machines controlled numerically, b) Automated handling of materials b) moving conveyors, c) automated guided vehicles (agv), d) Automatic tool changers and robots, e) Pallets and fixtures, f) Control computers, (Buzacot, J.A. & Shantikumar, J.G, 1993). A manufacturing line is formed from a continuous flow of materials, storage areas and processes where the piece or product to be made visit each station only once. Usually, at the beginning of each assembly line there is a raw material warehouse and at the end of it, a finished product warehouse. The permanence times of the product in each station are almost always random regardless of whether each station is operated automatically. Source randomness comes from events such as power failures, repairs, stoppages, materials, operator fatigue and more. In some cases, the line also handles temporary warehouses or temporary storage areas for decoupling the sequenced operation between machines (Oracle ERP). Manufacturing can be discrete and continuous. In the first, a number of separate operations is involved, independent in a sense that these operations can be separated arbitrarily in time and space (location). In the second, the process involves a continuous stream of manufactured product flowing through chemical, mechanical processes in a continuous movement.

Stochastic manufacturing is the analysis of the manufacturing systems from a randomly approach. This means that the design, modeling, analysis and / or optimization of an FMS under this approach must include the uncertainty (in the form of probability) inherent in manufacturing systems with little or no information. The factors of uncertainty that affect production are: a) The Environmental uncertainty and b) System uncertainty. The first concept includes the uncertainty of customer demands and the uncertainty of suppliers. The second is related with the uncertainties to the production process including operation yield uncertainty, production lead time uncertainty, quality uncertainty, machine failure and changes to product structure, demand, inventories level and more (Pérez Lechuga, G. et al., 2015).

Today, manufacturing plays a preponderant role in the economy of nations and its wealth is based on the transformation of raw materials for the production of goods. High tech manufacturing deserves special mention. From the economic and financial point of view, the influence of manufacturing goes far beyond the direct contribution to the national product and employment. The manufacturing industry is a global business that can sustain all the economic activity of a country. Spending on goods represents more than half of all consumer spending, while the manufacture of goods represents about two thirds of all exports from most developed countries. The manufacturing industry is the global foundation of the economic power of the richest and most powerful nations in the world. These are the ones that control most of the world production of manufacturing technology. That is, it is not enough to have factories and produce more goods, but you have to know how to build the machinery that makes merchandise. The key is to know how to build the ``means of production" (Pérez, L.G. et al., 2017), (Six Reasons Manufacturing is Central to the Economy).

The manufacturing industry is the most important cause of economic growth in developed countries. The growth of production of manufacturing machinery and technological improvements in that machinery are the main engines of economic growth. And machinery industries depend on sustained long-term economic growth.

World trade is based on goods, not services. A country cannot negotiate services for most of its assets. According to the World Trade Organization (WTO), 80% of world trade between regions is trade in goods - that is, only 20% of world trade is in services. The economic importance of their labor creates 4% of the world's gross domestic product and 6% of the world exports (Pérez, L.G. et al., 2015). In addition, most of the jobs that are created directly or indirectly depend on the manufacturing industry and the consolidation of the sector could provide tens of millions of new jobs (Pérez, L.G. et al., 2017), (Six Reasons Manufacturing is Central to the Economy).

The importance of manufacturing in the world of finance implies the redirection of the surplus resources of the non-financial sector and that the economy produces, which means that indirectly, even finances are dependent on manufacturing. Manufacturing has a great importance and the OECD has considered that is one of the emerging fields that requires immediate attention for its growth and strengthening through future technology that will have a great impact in the next 15 to 20 years () (for example the additive manufacturing Future. Technology Trends, 2016).

With the recent development of Industry 4.0, new areas of opportunity have been identified for the application of optimization and analysis of systems from a quantitative perspective. In general, names such as Hybrid Optimization (HO), Artificial Intelligence (AI) and more emerging techniques that outline a new field of development since its initial German conceptualization in 2011. There are various definitions in this regard, considering it as synonyms and variants of the same trend, for example "manufacturing smart", "digital transformation " "fourth industrial revolution "and "manufacturing digital transformation" are part of the slang that is used today in the middle of this important branch, (Culot, et al., 2020).

This document discusses the fields of manufacturing and related areas that, due to their importance, have received great attention in terms of mathematical modeling and optimization. First, the traditional fields of this specialty are presented where there are greater opportunities to develop mathematics and apply them in the modeling, simulation and optimization of these. To obtain an overview of the techniques applied in their solution, a sampling was made of the most important methodologies used from 1996 to 2020 in order to evaluate the most usual methods and procedures in the solution of the proposed instances. In particular, it is sought to make a contrast between the traditional methods of optimization with the methods of heuristics in order to find new fields of application of these, included the techniques of recent advances in the optimization of stochastic and large-scale models. This proposal is not intended to be a review of the state of the art in terms of mathematical modeling of manufacturing systems, rather, it presents a general overview of the methods used to address these problems from a quantitative perspective for engineers, practitioners, entrepreneurs, mathematicians and operations researchers.

A CLASSIFICATION OF PROBLEMS IN MANUFACTURING

In practice there are hundreds of classifications and sub-classifications of manufacturing systems. Most of them focus on the processes of physical-chemical transformation that materials undergo to be transformed into a finished product by adding value to the process. Others classify them according to the type of FMS, its operation, and the nature of the demands imposed on the system, the programming environment and the system's response capacity to disturbances and more, see for example (Chamadugu, R., Stecke, K. E., 1994).

In line with the objectives of this study, our classification is similar to that proposed in (Madhav, P. A., 1986) and is as follows:

1. Design problems of the operation:
    a. Design and planning
    b. Scheduling and control
    c. Cellular manufacturing
    d. Automatic systems
2. Modern FMS problems
    a. Additive manufacturing
    b. Smart manufacturing or 4.0 Industry
3. Planning, control, logistics and operation problems
    a. Enterprise resource planning (ERP)
    b. Material and requirement planning (MRP)

c. Manufacturing resource planning (MRP-II)
d. Lean manufacturing (Just in Time manufacturing)
e. Agile manufacturing
f. Inventory management
g. Planning, control and integration of production
h. Supply chain management

The related problems in all the fields are faced using statistical and mathematical models and with simulation techniques of discrete and continuous systems for example, in (Buzacot, J.A. & Shantikumar, J.G, 1993) an extraordinary material is developed related to the modeling and optimization of flexible manufacturing systems. For a more recent review of models related to the organization and logistics of the operation in FMS, it is suggested to see (Yadav, A. & Jayswal, S.C., 2017).

To solve instances of manufacturing problems, mathematical models of the following types are constructed: a) analytical models, b) simulation models.

The analytical models are mathematical models that have a closed form solution, i.e., the solution to the equations used to describe changes in a system can be expressed as a mathematical analytic function (Analytical Models). This type of models makes assumptions about the behavior of the real model and from the perspective of these, a solution is built that is valid only under the hypotheses of work made. The solution of the analytical models frequently requires the use of complicated numerical methods or the integration of two or more algorithms to consolidate a global solution to it, see for example (Pérez-Lechuga, G., 2018). In most cases, it requires the use of more than two algorithms or Hybrid Optimization (HO) that interact with each other, extracting the best of each one and / or complementing each other to solve complex instances through the use of AI. A HO aims to combine two or more algorithms that solve the same problem, either by choosing one (based on the data) or by switching between them during the course of the optimization process. The combination through the use of AI of the best properties of the chosen algorithm allows the overall algorithm to be better than the individual components, (Cavazos et al., 2006), (Pérez-Lechuga et al, 2021).

In this category, we can find deterministically formulated models and models with uncertainty. When the model contains uncertainty, different branches of mathematical programming have been used to find approximate solutions in the form of mathematical expectation or average values.

The simulation models make use of probability distributions (known or not) to introduce the uncertainty associated to the problem. Simulation models can express a better approximate reality of a production system. Events such as faults, dead times, lack of material, delays and more can be introduced into the model in the form of deterministic equations or through equations that contain random variables. Here, the associated uncertainty is handled in the form of probability distributions. In this regard there is a large number of computer programs aimed at the simulation of manufacturing systems such as PROMODEL, ARENA, IQMS, Petri Nets, Kanban and more. Simulation models are useful for expressing results in the form of average and/or dispersion values (for example average manufacturing times, required materials, unproductive times, delays, costs and more).

In some cases, a mixture of solutions of analytical models with simulation models could be useful, as well as a study of their sensitivity from statistical models or experimental design (perturbation). Here, as in most simulation models, the biggest disadvantage is the number of runs needed to obtain well-defined patterns of behavior.

One of the most important applications in the modeling of manufacturing systems is the optimization of some of their parameters. Its application requires only conventional model but in most of the time, it needs the use of specialized models of mathematical programming, (integer, linear, non-linear, stochastic and more). In the case of simulation models, some software incorporates routines to optimize certain parameters of the modeled system in order to find better indicators of its efficiency. However, in real problems the common method is to find instances with many variables and large amounts of constraints making the problem very complex regardless of their nature (deterministic, stochastic, linear or non-linear). Here the concepts of dimensionality and computational complexity take on great importance.

A mathematical programming model can be written as follows

$$\min f(x) : g(x) \leq 0, \quad l \leq x \leq u \tag{1}$$

where $f : \mathbb{R}^n \to \mathbb{R}$, $g : \mathbb{R}^n \to \mathbb{R}^m$ and x, l, and u are in $\mathbb{R}^n$.

In turn a stochastic programming needs some adjustments. The objective function could be written in terms of mathematical expectation and the group of constraints may also appear in the form of probabilities or mathematical expectations

$$\min E(f(x_1, \xi_1 \dots x_T, \xi_T)) \tag{2}$$

Or some other function of the decisions and random variables. Constraints containing random coefficients can have one of the following two forms

$$P\{g_i(x,s) > 0\} \leq \epsilon \tag{3}$$

$$E_s\left\{g_i(x,s): g_i(x,s) > 0\right\} \leq h \tag{4}$$

where E is the mathematical expectation operator.

A special case of this structure is the chance-constrained problem

$$\min\{f(x): g_1(x) \leq 0,\ P_\xi\left\{g_2(x,\xi) > 0\right\} \leq \epsilon,\ l \leq x \leq u\} \tag{5}$$

The model defined in Equation (5) is known as multi-stage programming and the most representative problems of the FMS can be written using the group of models from Equations (1) to (5) generating conflict in the times of solution in large instances.

To measure the computational effort required to solve an optimization instance, the concept of computational complexity is often used. The branch of computer science in charge of classifying and comparing the practical difficulty of solving problems on finite combinatorial objects is known by this name. The **P**-type problems (p*olynomial problems*) are those that can be solved by a conventional Turing machine in a number of steps which is proportional to the polynomial function of the size of its input. The **NP**-type problems (or *non-deterministic polynomial time problems*) consist of those problems which can be correctly decided by some computation of a non-deterministic Turing machine in a number of steps which is a polynomial function of the size of its input. The complexity class **EXP** is the set of decision problems that can be solved in a deterministic Turing machine in time $O(2p(n))$, where $p(n)$ is a polynomial function on n (Computational Complexity Theory). Therefore, the selection of the method to optimize a given instance is a delicate process that requires considering the classic factors of convergence in reasonable time and acceptable precision (Computational Complexity Theory). In optimization models associated with SMF, the following complete NP problems are recognized within others: a) Integer programming, b) Mixed integer linear programming, c) The knapsack problem, quadratic knapsack problem, and its variants, d) The traveling salesman problem, e) Bottleneck traveling salesman, f) Vehicle routing problem, g) Flow shop scheduling problem (for example, the problem of sequencing operations), h) Quadratic assignment problem, i) Minimum degree spanning tree, j) k-Chinese postman, k) Shortest total path length spanning tree, l) Minimum spanning tree.

Current methods of programming (for example to solve the Equations 1 to 5) have difficulties with their formulation and solution. Models such as integer programming and mixed integer linear programming can provide an efficient solution but present formulation problems in the accounting of their shared resources, concurrency, routing flexibility, batch size, etc. Similar case occurs in routing systems, minimum spanning tree and more. For this, if a problem can be solved polynomially, no doubt classical optimization techniques are the best option to find optimal solutions. However, if the problem has exponential computational complexity, it is preferable to apply techniques based on heuristics. In these cases, classical optimization can be prohibitive in time and cost, while the heuristic provides acceptable solutions in reasonable time. Moreover, if the problem is non-convex, a heuristic optimization (HO) will probably find a better solution.

Traditional numerical methods are usually based on iterative search algorithms that start with a (deterministic or arbitrary) solution which is iteratively improved according to some deterministic rule. In relation to optimization in manufacturing systems, methods from mathematical programming are often applied because these can solve problems where the constraints contain both equalities and inequalities. These, frequently use some of the following techniques for their solution: Linear Programming, Quadratic and Concave Programming, Dynamic Programming, Stochastic Programming, Nonlinear Programming, Integer Programming, Binary Programming (Heuristic optimization).

The central common feature of all HO methods is that they start off with a more or less arbitrary initial solution, iteratively produce new solutions by some generation rule, evaluate these new solutions, and eventually report the best solution found during the search process. There are many advantages to using heuristics for the solution of complex or random problems, some of them are: a) reaching new and better solutions, b) using an optimal number of search agents, c) limiting search spaces, d) incorporating a priori knowledge, e) having great flexibility for certain types of specific constraints, f) ease of implementation, g) convergence speed, h) reliability, i) low computational complexity.

Some (non-exhaustive list) of modern heuristic methods for optimization are: Ant colony optimization, Artificial bee colony, Artificial immune systems, Artificial neural networks, Automatic computing, Bacterial foraging, Biological computing, Chaos optimization, Cloud computing, Combinatorial optimization, Computational intelligence, Continuous optimization, Cultural algorithms, Differential evolution, Direct search, Evolutionary computing, Fuzzy optimization, Genetic algorithms, Granular computing, Hybrid algorithms, Local and global search, Memetic algorithms, Meta-heuristic methods, Particle swarm optimization, Pattern search, Simulated annealing, Simulation and modeling, Soft computing techniques, Support vector machines, Swarm intelligence, Tabu search, and Variable neighborhood search.

A SAMPLING OF THE TECHNIQUES USED IN THE OPTIMIZATION OF FLEXIBLE MANUFACTURING MODELS

Although the literature is rich in methods and procedures associated with the modeling and optimization of manufacturing systems, especially in line (or mass) production models, most of them have characteristics that allow them to be characterized as conducive to using a certain technique to your analysis. The results of the sampling are shown below. Tables (1 to 3) illustrate the problem associated with MFS, the techniques used to optimize it, and the most important references that approach this technique.

Design problems of the operation:

a) *Planning problems*: The planning of a FMS covers a wide range of activities ranging from the selection of the layout of machines and equipment to the design of a total preventive maintenance system (TPM). In the planning or designing manufacturing, warehousing, logistics and other operational and strategic situations it is common to use advanced software such as FlexSim.
b) *Scheduling and control*: The purpose of scheduling is to disaggregate the general production plan into time phased activities which will optimize criteria.

Minimizing flow times, keeping close conformance to deadlines, and efficient utilization of resources are the three broad classes of decision-making criteria used in scheduling (Cassandras C.G. & Lafortune, S., 2010).

c. *Cellular manufacturing:* The objective of cell formation problem is to group parts with similar design features or processing requirements into part families and the corresponding machines into machine cells (Fantahun, M.D., & Chen, M.A., 2006).
d. *Automatic systems:* An automated manufacturing system is the total integration of software and machinery to create a system that performs manufacturing processes autonomously through computer programming (What are automated manufacturing systems?).

Modern manufacturing problems:

a) *Additive manufacturing*: Additive manufacturing uses data computer-aided-design (CAD) software or 3D object scanners to direct hardware to deposit material, layer upon layer, in precise geometric shapes, (How does additive manufacturing works?.)

Table 1. Design problems of the operation

Planning problems	Reference	Cellular manufacturing	Reference
Empirical Methods, Empirical Proposal Based on Succes Cases.	(De Carlo, F., et al., 2013)	Mathematical Programmming.	(Buzacott. J.A. & Shantikumar, J.G., 1993), (Madavhi, I.A. et al., 2009)
Heuristic Methods, Tabu Search, Simulated Annealing, Genetic Algoritms.	(Ammir Sadrazade, 2012), (Arostegui et al., 2006), (Kusna, 2010)	Genetic Algorithms.	(Madavhi, I.A. et al., 2010)
Non-linear Programming and Simulation Methods.	(Zhenyuan, 2011)	Mixed-Integer Non Linear Programming.	(Kia, R., et al., 2012),
Multiple Attribute Decisions.	(Manilla et al, 2011)	Integer Non-Linear Programming, Simulated Annealing.	(Ah-kioon, 2009)
Swarm Intelligence and Evolutionary Computation.	(Shah et al., 2012)	Ant Colony Optimization, Simulated Annealing, Tabu Search.	(Ghosha, T. et al, 2011), (Six reasons Manufacturing is Central to the Economy), (AriafaraS, et al., 2009)
Hybrid Multi Criteria Decisions.	(Gumus & Gguneri, 2009)	Integer Programming.	(Albadawi, Z., 2005)
Scheduling and Control	**Reference**	**Automatic Systems**	**Reference**
Integer Linear Programming.	(Faizrahnemoon, 2012)	Data Models.	(Andreas Hees, et al., 2015)
Fuzzy Graphical Evaluation and Review Techniques.	(Norouzi,G. et al., 2015)	Meta Express Models.	(López-Ortega et al., 2001)
Control Theory.	(Baruah, S.A., 2015), (Yilmaz, O., Suleiman, T., 2018)	Petri Nets.	(Zhou, M., et al., 1993)
Mixed Integer Programming.	(Andrade, A. et al., 2014)	Automata.	(Cassandras, C.G., 2010)
Genetic Algorithms.	(Zhou,J. et al., 2015), (Yilmaz, O. et al., 2018)	Graph Theory, Automata and Petri Nets.	(Li, Z., et al., 2012)
Bee Based Algorithms.	(Yilmaz, et al., 2018)	Deadlock Control Methods.	(Fanti, M.O. et al, 2004)
Flowe Pollination Algorithms.	(Yilmaz, et al., 2018)	Cuckoo Search.	(Burnhual, S. et al., 2013)
Particle Swarm Optimization.	(Yilmaz, et al., 2018)		
Multi-Agents and Dynamic Systems.	(Yilmaz, et al., 2018)		
Big Data.	(Yilmaz, et al., 2018)		
Cuckoo Search.	(Yilmaz, et al., 2018)		

b) *Operation models:* An operating model includes blueprints, maps, a vendor matrix, people models, decision grids, Supply Chain Operations (SCO), Supply Chain Management (SCM), and other elements, such as a scorecard for evaluating performance, (Muñoz E., Capón-García, 2019), (Iacono, F. et al., 2019), (Ivanov, D. et al., 2016).

Planning, control, logistics and operation problems:

Enterprise Resource Planning (ERP): The acronym ERP stands for enterprise resource planning. It refers to the systems and software packages used by organizations

Table 2. Modern manufacturing problems

Additive manufacturing	Reference
Finite Element Method.	(Rai, J.K., et al., 2008), (Brenner, S.C. et al., 2017), (Jin, J. M., 2014)
Non-Linear Mathematical Programming.	(Rosen, D. W., (2007)
Continuous Simulation.	(Roberts, I.A., 2009)
Differential Equations.	(Manzhirov, A.V., 2015)
Manifolds	(Manzhirov, A.V., 2014)

to manage day-to-day business activities, such as accounting, procurement, project management and manufacturing.

a) *ERP systems* tie together and define a plethora of business processes and enable the flow of data between them. By collecting an organization's shared transactional data from multiple sources, ERP systems eliminate data duplication and provide data integrity with a "single source of truth.", (Pérez, L.G. et al., 2010), (Oracle ERP What is ERP?).
b) *Material and requirement planning (MRP)*: Material requirements planning is a system for calculating the materials and components needed to manufacture a product. It consists of three primary steps: taking inventory of the materials and components on hand, identifying which additional ones are needed and then scheduling their production or purchase (Material Requirement Planning MRP).

c) *Manufacturing resource planning (MRP-II):* Successor to the material requirements planning (MRP), it integrates planning of all aspects (not just production) of a manufacturing firm. MRP-II includes functions such as business planning, production planning and scheduling, capacity requirement planning, job costing, financial management and forecasting, order processing, shop floor control, time and attendance, performance measurement, and sales and operations planning (Manufacturing Resource Planning).

Table 3. Planning, control logistics and operations problems

Planning problems	Reference	Lean manufacturing and Just in Time	Reference
Decision Analysis, Analytic Hierarchy Process.	(Chin, C. et al, 2005)	Multi-Attribute Selection.	(Barla, S.B., 2003)
Analysis of Variance.	(Lenny, Koh, S. C. et al., 2007)	Linear Programming and Simulation.	(Yousefi, M., et al., 2012), (Hao, Q., et al., 2008), (Thapa, G., 2015)
Data Envelopment Analysis	(Lall, V., et al., 2016), (Talluri, S., et al., 2006)	Non-Linear Programming.	(Zhang, W., et al., 2010), (Thapa, G., 2015)
Fuzzy Data	(Ghapanchy, A., et al., 2008)	Non-Linear Integer Programming.	(Cao, D., et al., 2005)
Forecasting, Linear Programming	(Ouimet et al., 2001)	Simulated Annealing.	(Shahabudeen, P., et al., 2003)
Information Systems	(Liao, X., et al., 2007)	Discrete Simulation.	(Savsar, M., et al., 2011)
Material and Requirement problem (MRP)	**Reference**	Minimax Problem's Approach.	(Thapa, G., 2015)
Stochastic optimization.	[66, 67] (Mula, J., etv al., 2006-A), (Mula, J. et al., 2006-B)	**Agile manufacturing**	**Reference**
Fuzzy Linear Programming.	(Karsak, E. et al., 2009), (Rui, L., 2014)	Spatial Point Contact Constraint Theory.	(M. Hong, 1996)
Fuzy Set Theory.	(Wei, C.C., et al., 2004)	Statistical Analysis.	(Al-Tahat, et al., 2012)
Artificial Neural Network.	(Yazgan, H.R. et al., 2009)	Mixed Integer Linear Peogramming.	(Jarimo, T, 2005)
Fuzzy Logic.	(Kilic, H.S., 2014)	Genectic Algorithms and Simulated Annealing.	(Sherif, M., 2013)
Discrete Simulation.	(L. Sun, et al., 2009)	Graphs Theory	(Raj, S.A., 2013)
Mixed Integer Linear Programming.	(Yazıcı, E., 2016)	Petri Nets	(Li, P. et al., 2000)
Manufacturing Resource Planning (MRP II)	**Reference**	**Inventory management**	**Reference**
Linear and Non-Linear Programming.	(Voss, S., et al., 2006)	Analytical Hierarchy Process.	(Braglia, M., 2004)
Electronic Data Interchange.	(Choudhary, S.K., 2016)	Fuzzy Sets.	(Ziukov, S.A., 2015)
Multi-Stage Programming.	(Wazed, M. A., et al, 2011)	Neural Networks, Neuro-Fuzzy Aproximation.	(Gumus, A.T. et al., 2009)
Planning control and integration of production		Mixed Integer Linear Peogramming.	(Ekşioğlu, S. D., et al., 2009), (Fanti, M.P. et al., .2004)
Stochastic optimization.	(Mula, J., et al., 2006 A), (Cárdenas, G.E, et al., 2012), (Pérez-Lechuga, et al., 2011)	Geometric Programming, Multi-Criteria Evaluation, Fuzzy Goal Programming.	(Mandal, N.K., et al., 2005)
Linear Programming, Network Optimization, Non-Linear Mixed Integer Programming.	(Shapiro, J.F., 1993)	Finances	(Chang, C.T., et al, 2008)
Fuzzy Multi-Objective Linear Programming.	(Wang, R.H., et al., 2004)	Chance-Constrained Programming, Linear Programming.	(Thapa, G., 2015)
Mixed Integer Linear Programming.	(Pochet, Y., 2006), (Brah, S., et al., 1991)	Linear Programming.	(Urek, B., et al., 2016)
Genetic Algorithms..	[51] (Li, X. et al., 2017)	Non-Linear Programming.	(Urek, B., et al., 2016)
Supply Chain Management	**Reference**	**Supply Chain Management**	**Reference**
Mixed Integer Linear Programming.	(Sawik, T., 2011), (Tayur, S., 1999), (Pérez- Lechuga, G., 2018)	Simulation, Differential Equations, Probability Theory, Linear Programming, Stochastic Finances, Stochastic Programming.	(Tayur, S., et al., 1999)
Lipschitz Optimization, Linear Programming, Mixed Integer Linear Programming, Non-Linear Programming, Multi-Objective Programming, Stochastic Programming, Genetic Algorithms, Cuckoo Algorithm, Hybrid Optimization.	(Lee, Y.H., 2016), (Mula, J., etval., 2006 B)	Combinatorial Optimization, Forecastings.	(Yazgan, H.R., et al., 2009)

d) *Lean manufacturing and/or Just in Time manufacturing:* Lean Manufacturing is "a philosophy/management system on how to operate a business". Focusing this philosophy/system of tools in the elimination of waste allowing the reduction of the time between the client's order and the shipment of the product and improving quality and reducing costs. Just-in-time (JIT) manufacturing is a production model in which items are created to meet demand, not created in surplus or in advance of need. The purpose of JIT production is to avoid the waste associated with overproduction, waiting and excess inventory. These are three of the seven waste categories defined in the Toyota Production System (known in North America as the lean production model) (Just-In-Time manufacturing (JIT manufacturing)), (Lean manufacturing).

e) *Agile manufacturing:* Agile manufacturing is a term applied to an organization that has created the processes, tools, and training to enable it to respond quickly to customer needs and market changes while still controlling costs and quality (What is agile manufacturing?).

f) *Inventory management*: Inventory management is the management of inventory and stock. As an element of supply chain management, inventory management includes aspects such as controlling and overseeing ordering inventory, storage of inventory, and controlling the amount of product for sale (What is inventory management?).

g) *Planning, control and integration of production:* The administrative process that takes place within a manufacturing business and which involves making sure that sufficient raw materials, staff and other necessary items are procured and ready to create finished products according to the schedule specified. A typical large manufacturing business engaging in production planning will aim to maximize profitability while maintaining a satisfied consumer base (Pochet & Wolsey, 2006).

h) *Supply chain management*: Supply chain management (SCM) is the active management of supply chain activities to maximize customer value and achieve a sustainable competitive advantage. It represents a conscious effort by the supply chain firms to develop and run supply chains in the most effective and efficient ways possible. Supply chain activities cover everything from product development, sourcing, production, and logistics, as well as the information systems needed to coordinate these activities (What is supply chain management? (SCM)).

RESULTS OF THE ANALYSIS

The results obtained show that only 16.94% of the authors use heuristics in the solution of their optimization instances for manufacturing problems. The rest (83.06%) use analytical models. Figures (1) and (2) show the corresponding percentage distribution and Pareto graphs of the sample obtained for both analyses.

The most representative percentages of the analytical methods are: Mixed integer linear programming 10.20%, Fuzzy techniques 8.16%, Nonlinear programming 7.14%, Simulation 7.14%, Stochastic programming 7.14%, Linear programming 5.10%, Finite element method 3.06%, Multicriteria programming 3.06%. The rest is

Figure 1. The percentage distribution and Pareto of analytical models

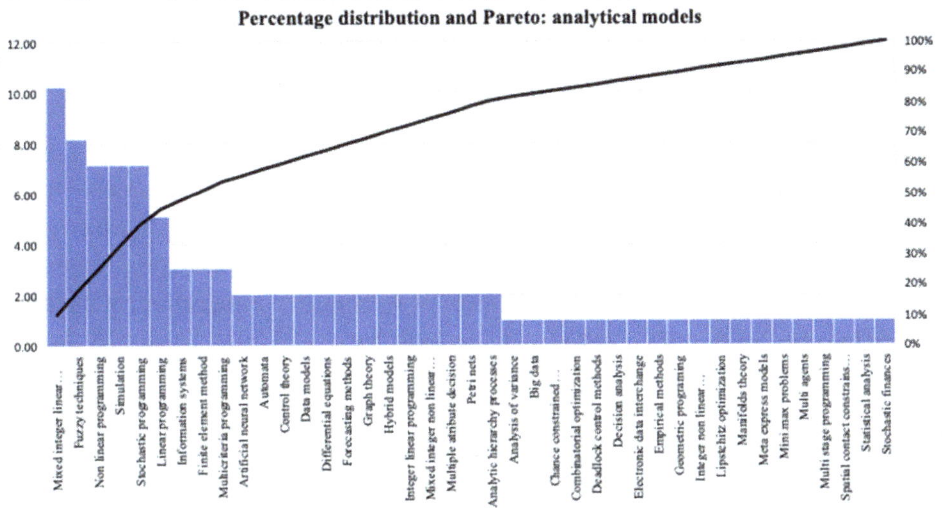

shared among other methods. These suggest that most authors prefer to use optimize traditional optimization methods for the analysis of manufacturing systems.

In the same way, the heuristic methods were distributed as follows: Genetic algorithms 30%, Simulated annealing 25%, Cuckoo search 15%, Tabu search 10%,

Figure 2. The percentage distribution and Pareto of heuristic models

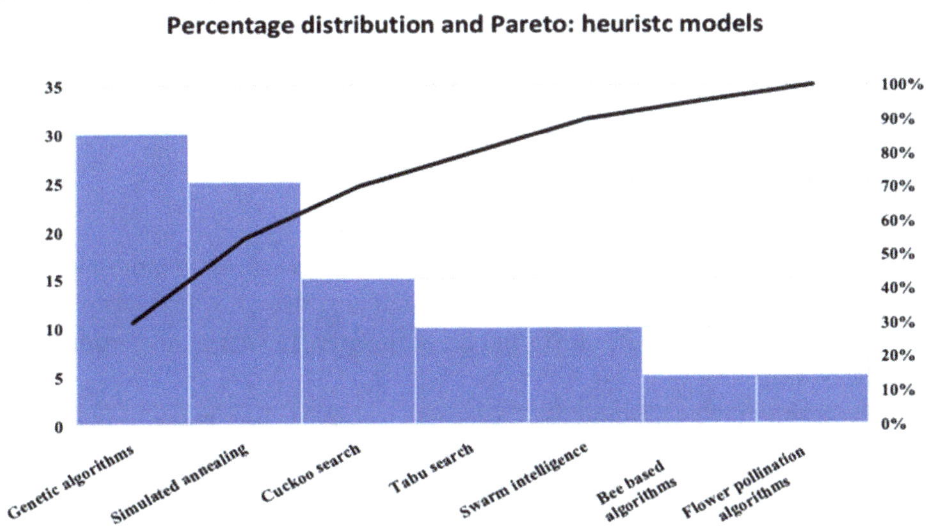

Swarm intelligence 10%, Bee based algorithms 5%, Flower pollination algorithms 5%. In relation to the heuristic, and according to the Pareto of the sample, 80% of the sampled works apply Genetic algorithms, Simulated annealing, Cuckoo search, Tabu search as main tools in the optimization of their models.

In relation to the heuristic, and according to the Pareto of the sample, 80 percent of the sampled works apply Genetic algorithms, Simulated annealing, Cuckoo search, Tabu search as main tools in the optimization of their models. Likewise, the analytical models grouped within the 80 percent interval consider the following methods: Mixed integer linear programming, Fuzzy techniques, Nonlinear programming, Simulation, Stochastic programming, Linear programming, Finite element method, Multicriteria programming, Analytic hierarchy processes, Artificial neural network, Automation, Control theory, Data models, Differential equations, Forecasting methods, Graph theory, Hybrid models, Integer linear programming, Mixed integer nonlinear programming, Multiple attribute decision, Petri nets, Information systems.

The percentage distribution of sampling in planning, control, logistics and operation problems is shown in Figure (3).

In the same way, Figure (4) shows the results of the sampling for the problems related to the design problems of the operation. Finally, figure (5) shows the percentage distribution obtained from the sample for problems related to modern manufacturing.

In relation to the data shown in Fig (3) observe that the methods for the optimization of the supply chain have received little attention from a quantitative perspective, way below inventory management, Just in Time in manufacturing and Material Requirement Planning.

The sampling done also indicates the great dispersion of methods and techniques currently available to solve an instance from different perspectives. Finally, and although there is a great attraction for traditional methods of linear programming and its variants, fuzzy techniques, and traditional approaches to stochastic programming, undoubtedly, heuristic methods are a tool that is here to stay.

CONCLUSION

Over the last few years, the fourth industrial revolution has attracted more and more attentions all around the world. In the current literature, there is still a lack of efforts to systematically review the state of the art of this new industrial revolution wave (Ziukov, S.A., 2015).

These results not only summarize the current research activities in this field (e.g., main research directions, algorithms analytical and heuristic) employed, but also indicate existing deficiencies and potential research directions through proposing

Figure 3. The percentage distribution of planning, control, logistics and operation problems

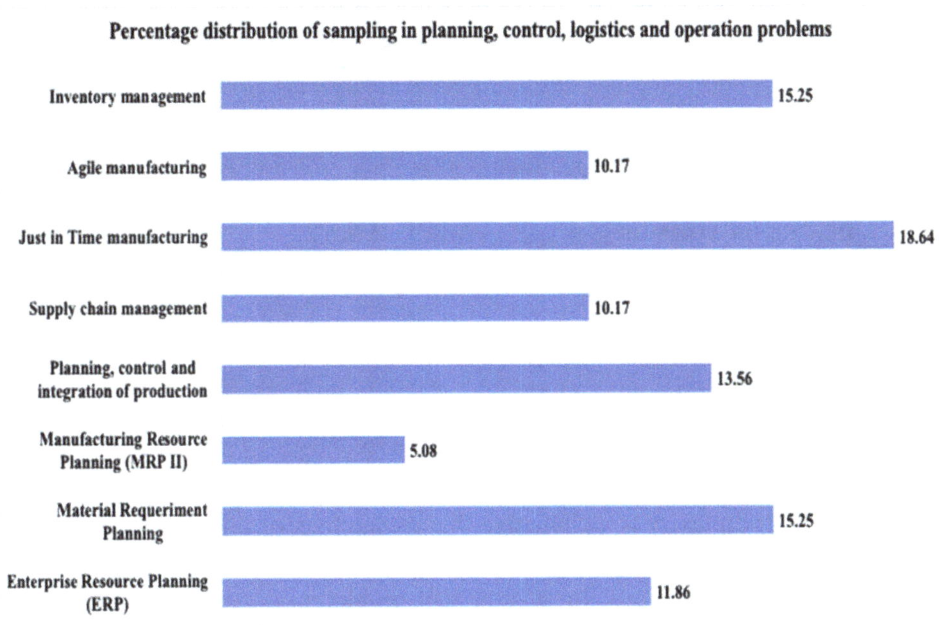

a research agenda. Findings of this analysis can be used as the basis for future research in industry.

The results obtained in this study suggest that even with the great development in terms of heuristics, traditional analytical algorithms are still widely used in the optimization of instances associated with flexible manufacturing problems. The

Figure 4. The percentage distribution of design problems of the operation

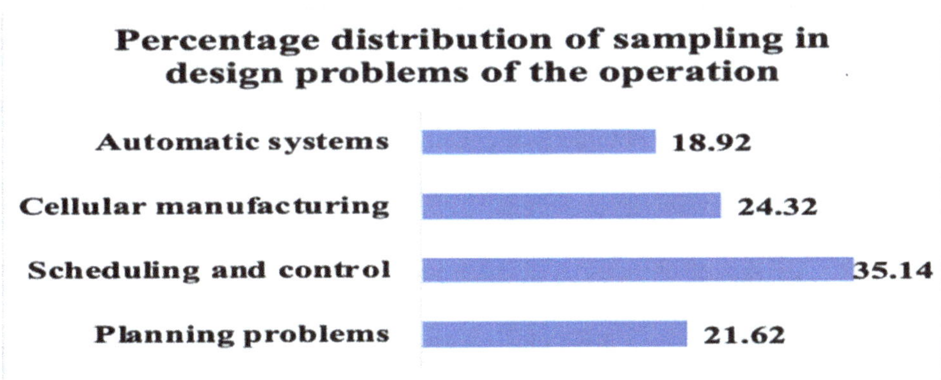

Figure 5. The percentage distribution of design problems of modern manufacturing

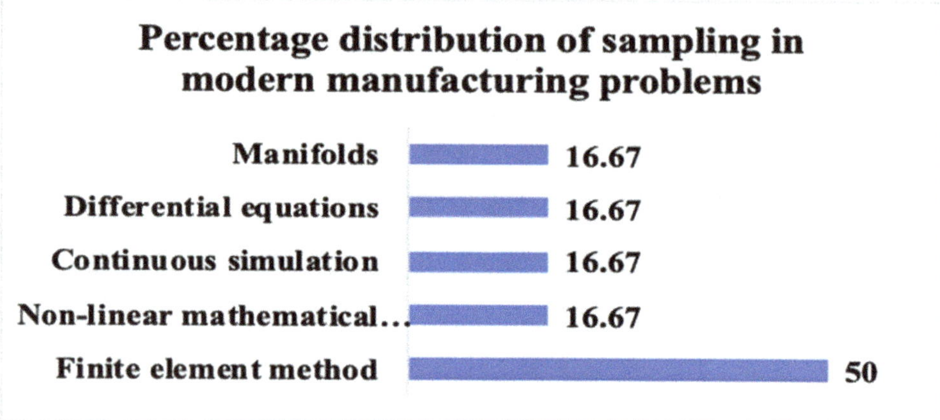

development and application of new heuristic techniques to solve models of great computational complexity is a challenge that researchers of this important branch must assume. It is interesting to note that in none of the sample values obtained is artificial intelligence used. In several real problems associated with routing, plant location, sequencing of operations, logistics and supply chain, it is common to find problems of great size and computational complexity. For example, in the case of traveling salesman problem the use of heuristics as ant's colony and/or greedy algorithms optimally solve combinatorial problems of great scale, hence the importance of its use.

A great variety of problems associated with FMS belong to the group of problems called NP-hard, so we cannot expect to be able to solve instances of practical problems of arbitrary size to optimum. Depending on the size of an instance or depending on the available CPU time, we should often be satisfied with the computation of approximate solutions and it may sometimes be impossible to evaluate the actual quality of the approximate solutions. There is a large number and variety of difficult problems, which arise in practice and need to be solved efficiently, and this has promoted the development of efficient procedures in an attempt to find good solutions, even if they are not optimal. Such methods, in which the speed of the process is as important as the quality of the solution obtained are called heuristic or approximate algorithm.

REFERENCES

Ah-kioon, S., Bulgak, A. A., & Bektas, T. (2009). Integrated cellular manufacturing systems design with production planning and dynamic system reconfiguration. *European Journal of Operational Research, 192*(2), 414–428. doi:10.1016/j.ejor.2007.09.023

Al-Tahat, M. D., & Bataineh, K. M. (2012). *Statistical analyses and modeling of the implementation of agile manufacturing tactics in industrial firms*. Academic Press.

Albadawi, Z., Bashir, H. A., & Chen, M. (2005). A mathematical approach for the formation of manufacturing cells. *Computers & Industrial Engineering, 48*(1), 3–21. doi:10.1016/j.cie.2004.06.008

Algin, V., Ishin, M., Siarhei Paddubka, S., Starzhinsky, V., Shil'ko, S., Rackov, M., & Čavić, M. (2020), Development of information model of power transmissions in the light of Industry 4.0. Mathematical Modelling of Technological Processes and Systems, 4(2), 54-63.

Andrade, A., Santoro, M., & De Tomi, G. (2014). Mathematical model and supporting algorithm to aid the sequencing and scheduling of mining with loading equipment allocation. *Rem: Rev. Esc. Minas, 67*(4).

Andreas Hees, A., & Reinhart, G. (2015). Approach for production planning in reconfigurable manufacturing systems. *9th CIRP Conference on Intelligent Computation in Manufacturing Engineering - CIRP ICME '14.*

Ariafara, S., & Ismail, N. (2009). An improved algorithm for layout design in cellular manufacturing systems. *Journal of Manufacturing Systems, 28*(4), 132–139. doi:10.1016/j.jmsy.2010.06.003

Arostegui, M. Jr, Kadipasaoglu, S., & Khumawala, B. (2006). An empirical comparison of Tabu Search, Simulated Annealing and Genetic Algorithms for facilities location problems. *Journal of Production Economics, 103*(2), 742–754. doi:10.1016/j.ijpe.2005.08.010

Barla, S. B. (2003). A case study of supplier selection for lean supply by using a mathematical model. *Logistics Information Management, 16*(6), 451–459. doi:10.1108/09576050310503420

Baruah, S. (2015). A scheduling model inspired by control theory. *Proceedings of the 6th International Real-Time Scheduling Open Problems Seminar (RTSOPS).*

Braglia, M., Grassi, A., & Montanari, R. (2004). Multi-attribute classification method for spare parts inventory management. *Journal of Quality in Maintenance Engineering*, *10*(1), 55–65. doi:10.1108/13552510410526875

Brah, S., Hunsucker, J., & Shah, J. (1991). Mathematical Modeling of Scheduling Problems. *Journal of Information and Optimization Sciences, 12*.

Brenner, S. C., & Carstensen, C. (2017). *Finite Element Methods*. Wiley Online Library.

Burnhual, S., & Deb, S. (2013). Scheduling optimization of flexible manufacturing systems using cuckoo search-based approach. *The International Journal of Advance Manufacturing Technology, 64(*5-8).

Buzacott, J. A., & Shanthikumar, J. G. (1993). Stochastic Models of Manufacturing Systems. Prentice Hall.

Cao, D., & Chen, M. (2005). A mixed integer programming model for a two-line CONWIP-based production and assembly system. *International Journal of Production Economics*, *95*(3), 317–326. doi:10.1016/j.ijpe.2004.01.002

Cárdenas, G. E., Pérez-Lechuga, G., Tuoh-Mora, J. C., & Medina-Marín, J. (2012). Obtaining the optimal short-term hydrothermal coordination scheduling: A stochastic view point. *Applied Mathematical and Computational Sciences, 3*(4), 361-379.

Cassandras, C. G., & Lafortune, S. (2010). *Introduction to Discrete Event Systems*. Springer.

Cavazos, J., Moss, J. E. B., & O'Boyle, M. F. P. (2006). Hybrid Optimizations: Which Optimization Algorithm to Use. In A. Mycroft & A. Zeller (Eds.), Lecture Notes in Computer Science: Vol. 3923. *Compiler Construction. CC 2006*. Springer. doi:10.1007/11688839_12

Chamadugu, R., & Stecke, K. E. (1994). Classification and review of FMS scheduling procedures. *Production Planning & Control, The Management of Operations, 5*(1).

Chang, C., Teng, J.-T., & Goyal, S. K. (2008). Inventory lot-size models under trade credits: A review. *Asia-Pacific Journal of Operational Research*, *25*(01), 89–112. doi:10.1142/S0217595908001651

Chin, C, Chen, W., Chien, F., & Wang, J.J. (2005). An AHP-based approach to ERP system selection. *International Journal of Production Economic, 96*(1), 47-62.

Choudhary, S. K. (2016), Study about the Types of Information Technology Service for Supply Chain. *Proceedings of the World Congress on Engineering and Computer Science 2016, 2.*

Culot, G., Nassmbeni, G., Orzes, G., & Sartor, M. (2020). Behind the definition of Industry 4.0: Analysis and open questions. *International Journal of Production Economics, 226*, 107617. doi:10.1016/j.ijpe.2020.107617

De Carlo, F., & Arlo, M.A. (2013). Layout Design for a low-capacity manufacturing line: A case study. *International Journal of Engineering Business.*

Ekşioğlu, S. D., Acharya, A., Leightley, L. E., & Arora, S. (2009). Analyzing the design and management of biomass-to-biorefinery supply chain. *Computers & Industrial Engineering, 57*(4), 1342-1352.

Faizrahnemoon, M. (2012). *Mathematical modelling of the scheduling of a production line at SKF* [Master's Thesis]. University of Gothenburg.

Fantahun, M. D., & Chen, M. (2006). A comprehensive mathematical model for the design of cellular manufacturing systems. *International Journal of Production Economics, 103*(2), 767–783. doi:10.1016/j.ijpe.2005.10.008

Fanti, M. P., & Zhou, M. (2004). Deadlock control methods in automatic manufacturing systems. IEEE Transactions on Systems, 34(1).

Future Technology Trends. (2016). https://read.oecd-ilibrary.org/science-and-technology/oecd-science-technology-and-innovation-outlook-2016/future-technology-trends_sti_in_outlook-2016-5-en#page1

Ghapanchi, A., Jafarzadeh, M. H., & Khakbaz, M.-H. (2008). Fuzzy-Data Envelopment Analysis approach to Enterprise Resource Planning system analysis and selection. *International Journal of Information Systems and Change Management, 3*(2), 157. doi:10.1504/IJISCM.2008.020693

Ghosh, I. (2014). An Intelligent Hybrid Multi Criteria Decision Making Technique to Solve a Plant Layout Problem. *International Journal of Industrial Engineering Research and Development, 5*(3), 13–23. doi:10.34218/IJIERD.5.3.2014.002

Ghosha, T., Senguptaa, S., Chattopadhyayb, M., & Dana, P. K. (2011). Meta-heuristics in cellular manufacturing: A state-of-the-art review. *International Journal of Industrial Engineering Computations.*

Gumus, A. T., & Guneri, A. F. (2009). A multi-echelon inventory management framework for stochastic and fuzzy supply chains. *Expert Systems with Applications, 36*(3), 5565-5575.

Hao, Q., & Shen, W. (2008). Implementing a hybrid simulation model for a Kanban-based material handling system. *Manufacturing*, *24*(5), 635–646. doi:10.1016/j.rcim.2007.09.012

Heuristic Optimization. (n.d.). Recovered from file:///Users/gilbertoperezlechuga/Downloads/9780387258522-c2.pdf

Hong, M.M., Payandeh, S., & Gruver, W.A. (1996). Modeling and analysis of flexible fixturing systems for agile manufacturing. *IEEE International Conference on Systems, Man and Cybernetics. Information Intelligence and Systems.*

How does additive manufacturing work? (n.d.). Recovered from https://www.ge.com/additive/additive-manufacturing

Iacono, F., Ferretti, S., Mezzadra, A., Magni, L., & Toffanin, Ch. (2019). Industry 4.0: Mathematical model for monitoring sterilization processes. *2019 IEEE* International Conference on Systems, Man and Cybernetics (SMC), 610-615. 10.1109/SMC.2019.8914206

Ivanov, D., Dolgui, A., & Sokolov, B. (2016). A dynamic model and an algorithm for short-term supply chain scheduling in the smart factory industry 4.0. *International Journal of Production Research*, *54*(2), 386-402. doi:10.1080/00207543.2014.999958

Jarimo, T., & Pulkkinen, U. (2005). A multi-criteria mathematical programming model for agile virtual organization creation. In *Collaborative Networks and Their Breeding Environments*. Springer. https://link.springer.com/content/pdf/10.1007%2F0-387-29360-4_13.pdf

Jin, J. M. (2014). *The Finite Element Method in Electromagnetic* (3rd ed.). Wiley & Sons.

Just-in-Time manufacturing. (n.d.). https://translate.google.com/?hl=es#en/es/justtime%20manufacturing%20(JIT%20manufacturing)

Karsak, E. E., & Özogul, C. O. (2009). An integrated decision-making approach for ERP system selection. *Expert Systems with Applications*, *36*(1), 660–667. doi:10.1016/j.eswa.2007.09.016

Khusna, D., Zawiah, S., Jamasri, J., & Aoyama, H. (2010). A Proposed Study on Facility Planning and Design in Manufacturing Process. *Proceedings of the International Multiconference of Engineers and Computers Scientists 2010*, 3.

Kia, R., Baboli, A., Javadian, N., Tavakkoli-Moghaddam, R., Kazemi, M., & Khorrami, J. (2012). Solving a group layout design model of a dynamic cellular manufacturing system with alternative process routings, lot splitting and flexible reconfiguration by simulated annealing. *Computers & Operations Research, 39*(11), 2642–2658. doi:10.1016/j.cor.2012.01.012

Kilic, H. S., Zaim, S. M., & Delen, D. (2014). Development of a hybrid methodology for ERP system selection: The case of Turkish Airlines. *Decision Support Systems, 66*, 82–92. doi:10.1016/j.dss.2014.06.011

Lall, V., & Teyarachakul, S. (2016). Enterprise Resource Planning (ERP) System Selection: A Data Envelopment Anaysis (DEA) Approach. *Journal of Computer Information Systems, 47*(1).

Lean Manufacturing. (n.d.). http://www.leansolutions.co/wp-content/uploads/2011/07/Indicadores-Lean-manufacturing.png

Lee, Y. H., Golinska, P., & Wu, J. Z. (2016). Mathematical Models for Supply Chain Management. *Mathematical Problems in Engineering*.

Lenny Koh, S. C., & Mike Simpson, M. (2007). Could enterprise resource planning create a competitive advantage for small businesses? *Benchmarking, 14*(1), 59–76. doi:10.1108/14635770710730937

Li, P., Shi, K., & Zhang, J. (2000). *Modeling Agile Manufacturing Cell using Object-Oriented Timed Petri Net*. Recovered from: http://keshi.ubiwna.org/paper/Modeling%20Agile%20Manufacturing%20Cell%20using%20Object-Oriented%20Timed%20Petri%20net.pdf

Li, X., Guo, S., Liu, Y., Du, B., & Wang, L. (2017). A production planning model for make-to-order foundry flow shop with capacity constraint. *Mathematical Problems in Engineering, 2017*, 6315613. doi:10.1155/2017/6315613

Li, Z., & Wu, N. (2012). Deadlock control of automatic manufacturing systems based on Petri Nets- A literature review. IEEE Transactions on Systems, 42(4).

Liao, X., Li, Y., & Lu, B. (2007). A model for selecting an ERP system based on linguistic information processing. *Information Systems, 32*(7), 1005–1017. doi:10.1016/j.is.2006.10.005

Liao, Y., Deschamps, F., Loures, E.F., & Ramos, L.F.P., (2017). Past, present and future of Industry 4.0- a systematic literature review and research agenda proposal. *International Journal of Production Research, 55*(12).

López-Ortega, O., & Pérez-Lechuga, G. (2001). An express meta model for sharing and exchanging flexible manufacturing resources. In *World Congress Manufacturing*. Rochester Institute of Technology.

Madavhi, I. (2009). Genetic algorithm approach for solving a cell formation problem in cellular manufacturing. *Expert Systems with Applications, 36*(3), 6598–6604. doi:10.1016/j.eswa.2008.07.054

Madavhi, I., Aalaei, A., Paydar, M. M., & Solymanpur, M. (2010). *Designing a mathematical model for dynamic cellular manufacturing systems considering production planning and worker assignment*. Academic Press.

Madhav, P. A. (1986). *Mathematical Modelling of Flexible Manufacturing Systems*. University of Windsor.

Mandal, N. K., Roy, T. K., & Maitic, M. (2005). Multi-objective fuzzy inventory model with three constraints: A geometric programming approach. *Fuzzy Sets and Systems, 150*(1), 87–106. doi:10.1016/j.fss.2004.07.020

Maniya, K., & Bhatt, M. G. (2011). An alternative multiple attribute decision making methodology for solving optimal facility layout design selection problems. *Computers & Industrial Engineering, 61*(3), 542–549. doi:10.1016/j.cie.2011.04.009

Manufacturing resource planning (MRP-II). (n.d.). http://www.businessdictionary.com/definition/manufacturing-resource-planning-MRP-II.html

Manzhirov, A. V. (2015). Mechanical design of viscoelastic parts fabricated using additive manufacturing technologies. *Proceedings of the World Congress on Engineering 2015, 2*.

Manzhirov, A. V., & Lychev, S. A. (2014). Mathematical Modeling of Additive Manufacturing Technologies, *Proceedings of the World Congress on Engineering 2014, 2*.

Material requirements planning (MRP). (n.d.). https://searcherp.techtarget.com/definition/material-requirements-planning-MRP

Mula, J., Peidro, D., Diaz-Madroñero, M., & Vicens, E. (2010). Mathematical programming models for supply chain production and transport planning. *European Journal of Operational Research, 204*(3), 377–390. doi:10.1016/j.ejor.2009.09.008

Mula, J., Poler, R., & García, J. P. (2006a). MRP with flexible constraints: A fuzzy mathematical programming approach. *Fuzzy Sets and Systems, 157*(1), 74–97. doi:10.1016/j.fss.2005.05.045

Mula, J., Poler, R., & García, J. P. (2006b). Models for production planning under uncertainty: A review. *International Journal of Production Economics, 103*(1), 271–285. doi:10.1016/j.ijpe.2005.09.001

Muñoz, E., & Capón-García, E. (2019). Intelligent Mathematical Modelling Agent for Supporting Decision-Making at Industry 4.0. In J. Mejia, M. Muñoz, Á. Rocha, A. Peña, & M. Pérez-Cisneros (Eds.), *Trends and Applications in Software Engineering. CIMPS 2018. Advances in Intelligent Systems and Computing* (Vol. 865). Springer.

Norouzi, G., Heydari, M., Noory, S., & Bagherpour, M. (2015). Developing a mathematical model for scheduling and determining success probability of research projects considering complex-fuzzy networks. *Journal of Applied Mathematics, 2015*, 809216. doi:10.1155/2015/809216

Omar, M., Sarker, R., & Othman, W. A. M. (2013). A just-in-time three-level integrated manufacturing system for linearly time-varying demand process. *Applied Mathematical Modelling, 37*(3), 1275–1281. doi:10.1016/j.apm.2012.03.041

Oracle, E. R. P. (n.d.). What Is ERP? https://translate.google.com/?hl=es#en/es/Oracle%20ERP%0AWhat%20Is%20ERP%3F

Ouimet, Charu, & Chaubal. (2001). *No. 6308163 B1*. United States Patent. https://translate.google.com/?hl=es#es/en/Method%20for%20controlled%20optimization%20of%20enterprise%20planning%20models

Pérez, L. G., Bujari, A. A., & Venegas, M. F. (2017). Economic aspects of the production planning in lines of manufacturing under conditions of uncertainty. In Modeling of Economic and Financial Phenomena: A Contemporary Vision. Editorial Castdel.

Pérez, L. G., Gress, E. S. H., Karelyn, A., & Orán, M. G. M. (2010). Some efficiency measures in the operation of flexible manufacturing systems: A Stochastic Approach. *Proceedings of the 2009 SIAM Conference on Mathematics for Industry*, 31-36.

Pérez, L. G., Pérez, R. E., & Venegas, M. F. (2015). Stochastic Optimization of Manufacture Systems by Using Markov Decision Processes. In Handbook of Research on Modern Optimization Algorithms and Applications in Engineering and Economics. IGI Global.

Pérez-Lechuga, G. (2018). Optimal logistics strategy to distribute medicines in clinics and hospitals. *Journal of Mathematics in Industry, Springer*, *8*(1), 2. doi:10.118613362-018-0044-5

Pérez-Lechuga, G., Flores-Rivera, C., Karelyn, A., & Tarasenko, A. (2011). A multi-period stochastic model to link the supply chain production planning in a manufacturing organization. *Advances in Science and Statistics Applications, 6*(4), 255-283.

Pérez-Lechuga, G., Madrid F. K. N., & Fiore, H. (2021). Hybrid Optimization and Artificial Intelligence Applied to Energy Systems: A review. In *Advances of Artificial Intelligence in a Green Energy Environment*. IGI Global.

Pochet, Y., & Wolsey, L. A. (2006). *Production Planning by Mixed Integer Programming*. Springer.

Production planning. (n.d.). Recovered from http://www.businessdictionary.com/definition/production-planning.html

Rai, J. K., & Xirouchakis, P. (2008). Finite element method-based machining simulation environment for analyzing part errors induced during milling of thin-walled components. *International Journal of Machine Tools & Manufacture*, *48*(6), 629–643. doi:10.1016/j.ijmachtools.2007.11.004

Raj, S. A., Sudheer, A., Vinodh, S., & Anand, G. (2013). A mathematical model to evaluate the role of agility enablers and criteria in a manufacturing environment. *International Journal of Production Research*, *51*(19).

Roberts, I. A., Wang, C. J., Esterlein, R., Stanford, M., & Mynors, D. J. (2009). A three-dimensional finite element analysis of the temperature field during laser melting of metal powders in additive layer manufacturing. *International Journal of Machine Tools & Manufacture*, *49*(12-13), 916–923. doi:10.1016/j.ijmachtools.2009.07.004

Rosen, D. W. (2007). *Design for additive manufacturing: a method to explore unexplored regions of the design*. Recovered from: https://pdfs.semanticscholar.org/101a/8f03ed20725ae71d30537e90f783a398e86f.pdf

Rui, L., Xiaofei, Z., & Guiwu, W. (2014). Models for selecting an ERP system with hesitant fuzzy linguistic information. *Journal of Intelligent & Fuzzy Systems*, *26*(5), 2155–2165. doi:10.3233/IFS-130890

Sadrazadeh, A. (2012). A genetic algorithm with the heuristic procedure to solve the multi-line layout problem. *Computers & Industrial Engineering*, *62*(4), 1055–1064. doi:10.1016/j.cie.2011.12.033

Savsar, M., & Jawini, A. N. (1995). Simulation analysis of just-in-time production systems. *International Journal of Production Economics*, *42*(1), 67–78. doi:10.1016/0925-5273(95)00166-2

Sawik, T. (2011). *Scheduling in Supply Chains Using Mixed Integer Programming*. Wiley.

Shah, K., Ripon, N., & Glette, K. (2012). Adaptive variable neighborhood search for solving multi- objective facility layout problem with unequal area facilities. *Swarm and Evolutionary Computation*.

Shahabudeen, P., Krishnaiah, K., & Narayanan, M. T. (2003). Design of a Two-Card Dynamic Kanban System Using a Simulated Annealing Algorithm. *International Journal of Advanced Manufacturing Technology*, *21*(10–11), 754–759. doi:10.100700170-002-1387-y

Shapiro, J. F. (1993). Chapter 8 Mathematical programming models and methods for production planning and scheduling. *Management Science*, *4*, 371–443.

Sherif, M. (2013). *Agile manufacturing system scheduling using genetic algorithms and simulated* [Master Degree Thesis]. American University in Cairo. http://dar.aucegypt.edu/handle/10526/3397

Six Reasons Manufacturing is Central to the Economy. (2011). Recovered from http://rooseveltinstitute.org/six-reasons-manufacturing-central-economy/

Solymanpur, M., Vrat, P., & Shankar, R. (2004). Ant colony optimization algorithm to the inter-cell layout problem in cellular manufacturing. *European Journal of Operational Research*, *157*(3), 592–606. doi:10.1016/S0377-2217(03)00248-0

Sun, L.L., Heragu, S.S., Chen, L., Spearman, M.L. (2009). Simulation analysis of a multi-item MRP system based on factorial design. *Winter Simulation Conference (WSC), Proceedings of the 2009*, 2107–2114.

Talluri, S., Narasimhan, R., & Nair, A. (2006). A., Vendor performance with supply risk: A chance-constrained DEA approach. *International Journal of Production Economics*, *100*(2), 212–222. doi:10.1016/j.ijpe.2004.11.012

Tayur, S., Ganeshan, R., & Magazine, M. (1999). *Quantitative Models for Chain supply Management*. Kluwer's International Series. doi:10.1007/978-1-4615-4949-9

Thapa, G. (2015). *Optimization of just-in-time sequencing problems and supply chain logistics*. Mälardalen University Press Dissertations No. 184. Recovered from https://www.diva-portal.org/smash/get/diva2:851513/FULLTEXT01.pdf

Urek, B., & Kaya, O. (2016). A mixed integer nonlinear programming model and heuristic solutions for location, inventory and pricing decisions in a closed loop supply chain. *Computers & Operations Research*, *65*, 93–103. doi:10.1016/j.cor.2015.07.005

Voss, S., & Woodruff, D. (2006). Connecting MRP, MRP II and ERP- supply chain production planning via optimization models. In Tutorials on Emerging Methodologies and Applications in Operations Research. Springer Science & Business Media.

Wang, R. C. H., & Liang, T. F. (2004). Application of fuzzy multi-objective linear programming to aggregate production planning. *Computers & Industrial Engineering*, *46*(1), 17–41. doi:10.1016/j.cie.2003.09.009

Wazed, M. A., Ahmed, S., & Nukman, Y. (2011). Mathematical models for multistage production under commonality and uncertainty. *International Journal of Physical Sciences*, *6*(22), 5202–5222.

Wei, C. C., & Wang, J. J. (2004). A comprehensive framework for selecting an ERP system. *International Journal of Project Management*, *22*(2), 161–169. doi:10.1016/S0263-7863(02)00064-9

What are automated manufacturing systems? (n.d.). Recovered from https://connell-ind.com/what-are-automated-manufacturing-systems/

What is agile manufacturing? (n.d.). Recovered from https://translate.google.com/?hl=es#en/es/WHAT%20IS%20AGILE%20MANUFACTURING%3F%20Why%20is%20it%20an%20effective%20strategy%3F

What is inventory management? (n.d.). Recovered from https://www.tradegecko.com/learning-center/what-is-inventory-management

What is Supply Chain Management (SCM)? (n.d.). Recovered from https://scm.ncsu.edu/scm-articles/article/what-is-supply-chain-management-scm

Yadav, A., & Jayswal, S. C. (2017). Modelling of flexible manufacturing systems: A review. *International Journal of Production Research*, *56*(7), 2464–2487. doi:10.1080/00207543.2017.1387302

Yazgan, H. R., Boran, S., & Goztepe, K. (2019). An ERP software selection process with using artificial neural network based on analytic network process approach. *Expert Systems with Applications, 36*(5), 9214–9222. doi:10.1016/j.eswa.2008.12.022

Yazıcı, E., Büyüközkan, G., & Baskak, M. (2016). A New Extended MILP MRP Approach to Production Planning and Its Application in the Jewelry Industry. *Mathematical Problems in Engineering, 2016*, 7915673. doi:10.1155/2016/7915673

Yilmaz, O. F., & Suleiman, T. (Eds.). (2018). *Handbook of Research on Applied Optimization Methodologies in Manufacturing*. IGI-Global. doi:10.4018/978-1-5225-2944-6

Yousefi, M., Hosseinioun, S. S., & Taghikhah, F. (2012). Determining optimum number of kanbans for workstations in a kanban-based manufacturing line using discrete-event simulation: a case of study. *Proceedings of the 2012 International Conference on Industrial Engineering and Operations Management.*

Zhang, W., & Chen, M. (2010). A mathematical programming model for production planning using CONWIP. *International Journal of Production Research, 39*(12).

Zhenyuan, J., Xiaohong, L., & Wei, W. (2011). Design and implementation of lean facility layout system of a production line. *International Journal of Industrial Engineering, 18*, 260–269.

Zhou, J., Yang, G., & Tang, G. (2015). The mathematical modeling of the mobile recharging facility vehicles' scheduling and its genetic algorithm solution. *IEEE International Conference on Cyber Technology in Automation, Control, and Intelligent Systems (CYBER)*. 10.1109/CYBER.2015.7288262

Zhou, M., & Dicesare, F. (1993). *Petri Net Synthesis for Discrete Event Control of Manufacturing Systems*. Springer Science. doi:10.1007/978-1-4615-3126-5

Ziukov, S. (2015). A literature review on models of inventory management under uncertainty. *Business Systems & Economics, 1.*

Compilation of References

Ahmad, F. & Alam, M. S. (2018), Economic and ecological aspects for microgrids deployment in India. *Sustainable cities and society*, *37*, 407-419

Remote food and plant growth monitoring with IoT technology. (n.d.). https://www.soracom.io/customers/agriculture/food-plant-growth-monitoring-iot/

Gangwar, K. K., & Singh, R. (2010). *Guidebook on Energy Education for Students, Published by Punjab ENVIS Centre, Punjab State Council For Science & technology*. Chandigarh and Punjab Energy Development Agency.

Schipani, S. (2019). *What Is Micro Farming?* https://hellohomestead.com/what-is-micro-farming/

Climate-Smart Agriculture. (n.d.). https://www.fao.org/climate-smart-agriculture/knowledge/en/

Garv. (2017, April 03). *Rural households electrification*. Garv.gov: http://garv.gov.in/dashboard/ue

HREDA. (2008). *Energy Conservation Hand Book for Buildings and home*. Haryana Renewable Energy Development Agency.

McKinsey & Company. (2010). India's urban awakening: building inclusive cities, sustaining economic growth. McKinsey.

MNRE. (2007). In: Akshay Urja. Ministry of New & Renewable Energy Government of India, 3(1), 48.

MNRE. (2007-08). In *Annual Report, Published by Ministry of New & Renewable Energy* (p. 112). Government of India.

MOEF. (2010). *India: Greenhouse Gas Emissions 2007*. Ministry of Environment and Forests, Government of India.

Phadke A., Sathaye J. and Padmanabhan, S. (2005) Economic benefits of Reducing Maharashtra's electricity shortage through end-use efficiency improvement. *LBNL Report 57053.*

Prayas Energy Group. (2005). Demand-side Management (DSM) in the Electricity Sector: Urgent Need for Regulatory Action and Utility-Driven Programs. Prayas Group.

Bureau of Energy Efficiency - BEE. (2010). *BEELINE, Quarterly Newsletter*, 1. Bureau of Energy Efficiency.

Shalimov, A. (2020). *IoT in agriculture. 8 technology use cases for smart farming (and challenges to consider).* https://easternpeak.com/blog/iot-in-agriculture-technology-use-cases-for-smart-farming-and-challenges-to-consider/

Jerath, N. Saxena, S.K., gangwar, K.K. & Singh, R. (Eds.). (2008). Renewable Energy. Punjab ENVIS Centre Newsletter on Status of Environment and related Issues, 6(2-3), 32. Punjab State Council for Science & Technology.

Robichaud, L., & Anantatmula, V. (2010). —Greening Project Management Practices for Sustainable Construction||. *Journal of Management Engineering*, *27*(1), 48–57. doi:10.1061/(ASCE)ME.1943-5479.0000030

Sathaye, J. & Bouille, D. (2001). Barriers, Opportunities, and Market Potential of Technologies and Practices. In Metz, O. Davidson, and R. Swart Cambridge (eds.) Climate Change Mitigation. BUniversity Press for the Intergovernmental Panel on Climate Change.

Sathaye J. & Gupta, A. (2010). Reducing Electricity Deficit through Energy Efficiency in India: An Evaluation of Economic and Emissions Benefits. *LBNL Report.*

Sivaharan, T, Blair, G., & Coulson, G (2005). GREEN: A Configurable and Re-configurable Publish Subscribe Middleware for Pervasive Computing. *Lecture Notes in Computer Science.* Springer

TERI *(2003-04*). Beacon, Published by The Energy and Resource Institute.

TERI. (2015). *Energy Security Outlook: Defining a secure and sustainable energy future for India.* TERI Press.

UNDP (2011). *Improving energy efficiency in the Indian Railways System.* UNDP India.

US Department of Energy. (2008). Energy Savers. National Renewable Energy Laboratory, a DOE National Laboratory, Pp.32.

VCFEE BEE. (2015). Venture Capital Fund for Energy Efficiency (VCFEE). BEE India. https://beeindia.gov.in/sites/default/files/VCFEE.PDF

Bureau of Energy Efficiency (BEE). (2004). ESCO: Cashless Alternative for Energy Conservation. *IEEMA Journal.*

Internet of Things in Agriculture. What is IoT and how is it implemented in agriculture? (n.d.). https://www.cropin.com/iot-internet-of-things-applications-agriculture/

Bureau of Energy Efficiency (BEE). (2008a). *Awards Booklet: Municipalities.* BEE. www.emtindia.net/eca2008/AwardBooklet2008/f_Municipalities.pdf

Ravindra, S. (2020). *IoT Applications in Agriculture*. https://www.iotforall.com/iot-applications-in-agriculture/

Bureau of Energy Efficiency (BEE). (2008). *Roadmap for Energy Efficiency in Buildings*. Australia- India Technical & Policy Exchange Power.

Canning, P., Rehkamp, S., Waters, A., & Etemadnia, H. (2017), *The Role of Fossil Fuels in the U.S. Food System and the American Diet, Economic Research, Economic Research Report no. 224*. United States Department of Agriculture.

Winn, M. (2020). *How Drones Are Keeping the Lights On*. https://www.iotforall.com/drones-keeping-lights-on/

CEA. (2016, December 30). *Draft National Electricity Plan*. CEA. http://www.cea.nic.in/reports/committee/nep/nep_dec.pdf

Crop Health Management. (n.d.). https://www.iotforall.com/use-case/crop-health-management/

Central Energy Regulatory Commission (CERC). (2010). *The Electricity (Supply) Act, 1948*. Government of India. gov.in/ElectSupplyAct1948.pdf

Crop Irrigation Management. (n.d.). https://www.iotforall.com/use-case/crop-irrigation-management/

Centre for Environment Education – CEE (2010). *Low Carbon Life Styles - A trainers Guide to low carbon life styles*. CEE.

IoT enabled smart farm. (n.d.). https://www.soracom.io/customers/agriculture/iot-enabled-smart-cows/

Agriculture IoT Solutions and Cloud Connectivity. (2019). https://www.soracom.io/blog/agriculture-iot-solutions-and-cloud-connectivity/

Chaturvedi, P. (1997). *Energy Management: Policy, Planning Utilization*. Concept Publishing Company Ltd.

Abad, M., Noguera, P., Puchades, R., Maquieira, A., & Noguera, V. (2002). Physico-chemical and chemical properties of some coconut coir dusts for use as a peat substitute for containerised ornamental plants. *Bioresource Technology*, *82*(3), 241–245. doi:10.1016/S0960-8524(01)00189-4 PMID:11991072

Abualigah, L., Zitar, R. A., Almotairi, K. H., Hussein, A. M., Elaziz, M. A., Nikoo, M. R., & Gandomi, A. H. (2022). Wind, Solar, and Photovoltaic Renewable Energy Systems with and without Energy Storage Optimization: A Survey of Advanced Machine Learning and Deep Learning Techniques. *Energies*, *15*(2), 578. doi:10.3390/en15020578

Acharya, M. (2020). Floating Solar Photovoltaic (FSPV): A Third Pillar to Solar PV Sector? *Lecture Notes in Civil Engineering*, *15*(3), 68. doi:10.1007/978-981-13-8743-2_21%0Ahttps://

AgriSolar. (2022). *Agrisolar SolarPower Europe*. Agrisolar. https://www.solarpowereurope.org/ interests/ agrisolar

Ahamed, R., McKee, K., & Howard, I. (2020). Advancements of wave energy converters based on power take off (PTO) systems: A review. *Ocean Engineering, 204*, 107248. doi:10.1016/j.oceaneng.2020.107248

Ah-kioon, S., Bulgak, A. A., & Bektas, T. (2009). Integrated cellular manufacturing systems design with production planning and dynamic system reconfiguration. *European Journal of Operational Research, 192*(2), 414–428. doi:10.1016/j.ejor.2007.09.023

Akia, M., Yazdani, F., Motaee, E., Han, D., & Arandiyan, H. (2014). A review on conversion of biomass to biofuel by nanocatalysts. *Biofuel Research Journal, 1*(1), 16–25. doi:10.18331/BRJ2015.1.1.5

Akole, H. U., & Jadhav, P. S. (2018). Floating Solar Plant. *3rd Int. Conf. Converg. Technol. I2CT* (pp. 7–10). IEEE. 10.1109/I2CT.2018.8529479

Albadawi, Z., Bashir, H. A., & Chen, M. (2005). A mathematical approach for the formation of manufacturing cells. *Computers & Industrial Engineering, 48*(1), 3–21. doi:10.1016/j.cie.2004.06.008

Algin, V., Ishin, M., Siarhei Paddubka, S., Starzhinsky, V., Shil'ko, S., Rackov, M., & Čavić, M. (2020), Development of information model of power transmissions in the light of Industry 4.0. Mathematical Modelling of Technological Processes and Systems, 4(2), 54-63.

Almeida, R., Schmitt, R., Grodsky, S., Flecker, A., Gomes, C., Zhao, L., Liu, H., Barros, N., Kelman, R., & McIntyre, P. (2022). Floating solar power could help fight climate change— Let's get it right. *Nature, 606*(7913), 246–249. doi:10.1038/d41586-022-01525-1 PMID:35672509

Alonso, D. M., Bond, J. Q., & Dumesic, J. A. (2010). Catalytic conversion of biomass to biofuels. *Green Chemistry, 12*(9), 1493–1513. doi:10.1039/c004654j

Al-Tahat, M. D., & Bataineh, K. M. (2012). *Statistical analyses and modeling of the implementation of agile manufacturing tactics in industrial firms*. Academic Press.

Al-Widyan, M., Khasawneh, M., & Abu-Dalo, M. (2021). Potential of floating photovoltaic technology and their effects on energy output, water quality and supply in Jordan. *Energies, 14*(24), 1–13. doi:10.3390/en14248417

Amores, I., Ballesteros, I., Manzanares, P., Sáez, F., Michelena, G., & Ballesteros, M. (2013). Ethanol Production from Sugarcane Bagasse Pretreated by Steam Explosion. *Electronic Journal of Energy and Environment, 1*(1), 25–36. doi:10.7770/ejee-V1N1-art519

Anagnostis, A., Papageorgiou, E., & Bochtis, D. (2020). Application of artificial neural networks for natural gas consumption forecasting. *Sustain., 12*(16), 6409. doi:10.3390u12166409

Ananthapadmanaban, D. (2018). Summary of Some Selected Characterisation Methods of Geopolymers. In Mazen Alshaaer and Han-Yong Jeon (Ed), *Geopolymers and Other Geosynthetics IntechOpen,* doi:. https://www.intechopen.com/chapters/64656 doi:10.5772/intechopen.82208

Andrade, A., Santoro, M., & De Tomi, G. (2014). Mathematical model and supporting algorithm to aid the sequencing and scheduling of mining with loading equipment allocation. *Rem: Rev. Esc. Minas, 67*(4).

Andreas Hees, A., & Reinhart, G. (2015). Approach for production planning in reconfigurable manufacturing systems. *9th CIRP Conference on Intelligent Computation in Manufacturing Engineering - CIRP ICME '14.*

Ariafara, S., & Ismail, N. (2009). An improved algorithm for layout design in cellular manufacturing systems. *Journal of Manufacturing Systems, 28*(4), 132–139. doi:10.1016/j.jmsy.2010.06.003

Arnett, E. B., Johnson, G. D., Erickson, W. P., & Hein, C. D. (2013). *A Synthesis of Operational Mitigation Studies to Reduce Bat Fatalities at Wind Energy Facilities in North America.* Bat Conservation International.

Arostegui, M. Jr, Kadipasaoglu, S., & Khumawala, B. (2006). An empirical comparison of Tabu Search, Simulated Annealing and Genetic Algorithms for facilities location problems. *Journal of Production Economics, 103*(2), 742–754. doi:10.1016/j.ijpe.2005.08.010

Azam, M., Jahromy, S. S., Raza, W., Raza, N., Lee, S. S., Kim, K. H., & Winter, F. (2020). Status, characterisation, and potential utilisation of municipal solid waste as renewable energy source: Lahore case study in Pakistan. *Environment International, 134*, 105291. doi:10.1016/j.envint.2019.105291 PMID:31730999

Azran, Z., & Majid, A. (2014). Study on Performance of 80 Watt Floating Photovoltaic Panel. *Journal of Mechanical Engineering Science, 7*, 1150–1156. doi:10.15282/jmes.7.2014.14.0112

Barla, S. B. (2003). A case study of supplier selection for lean supply by using a mathematical model. *Logistics Information Management, 16*(6), 451–459. doi:10.1108/09576050310503420

Baruah, S. (2015). A scheduling model inspired by control theory. *Proceedings of the 6th International Real-Time Scheduling Open Problems Seminar (RTSOPS).*

Bekele, G., & Tadesse, G. (2012). Feasibility study of small Hydro/PV/Wind hybrid system for off-grid rural electrification in Ethiopia. *Applied Energy, 97*, 5–15. doi:10.1016/j.apenergy.2011.11.059

Belyakov, N. (2019). Sustainable power generation: Current status, future challenges, and perspectives. *Sustainable Power Generation: Current Status, Future Challenges, and Perspectives*, 1–620. doi:10.1016/C2018-0-01215-3

Bidaoui, H., El Abbassi, I., El Bouardi, A., & Darcherif, A. (2019). Wind speed data analysis using Weibull and Rayleigh distribution functions, case study: Five cities northern Morocco. *Procedia Manufacturing, 32*, 786–793. doi:10.1016/j.promfg.2019.02.286

Blackburn, K., & Stuard, S. B. (2014). A framework to facilitate consistent characterisation of read across uncertainty. *Regulatory Toxicology and Pharmacology, 68*(3), 353–362. doi:10.1016/j.yrtph.2014.01.004 PMID:24457134

Boopathi, S. (2019). Experimental investigation and parameter analysis of LPG refrigeration system using Taguchi method. In SN Applied Sciences (Vol. 1, Issue 8). doi:10.100742452-019-0925-2

Boopathi, S. (2021). Improving of Green Sand-Mould Quality using Taguchi Technique. *Journal of Engineering Research.* doi:10.36909/jer.14079

Boopathi, S., and Myilsamy, S. (2021). Material removal rate and surface roughness study on Near-dry wire electrical discharge Machining process. *Materials Today: Proceedings, 45*, 8149–8156. doi:10.1016/j.matpr.2021.02.267

Boopathi, S., Saranya, A., Raghuraman, S., & Revanth, R. (2022). Design and Fabrication of Low Cost Electric Bicycle. *International Research Journal of Engineering and Technology, 5*(July), 5–7. www.irjet.net

Boopathi, S., Venkatesan, G., & Anton Savio Lewise, K. (2023). Mechanical Properties Analysis of Kenaf–Grewia–Hair Fiber-Reinforced Composite. In *Lecture Notes in Mechanical Engineering* (pp. 101–110). Springer. doi:10.1007/978-981-16-9057-0_11

Boxwell, M. (2017). *Solar electricity handbook.* Greenstream Publishing Limited. [http://solarelectricityhandbook.com/ solar-irradiance.html]

Braglia, M., Grassi, A., & Montanari, R. (2004). Multi-attribute classification method for spare parts inventory management. *Journal of Quality in Maintenance Engineering, 10*(1), 55–65. doi:10.1108/13552510410526875

Brah, S., Hunsucker, J., & Shah, J. (1991). Mathematical Modeling of Scheduling Problems. *Journal of Information and Optimization Sciences, 12.*

Brenner, S. C., & Carstensen, C. (2017). *Finite Element Methods.* Wiley Online Library.

Bukari, D., Tuokuu, F. X. D., Suleman, S., Ackah, I., & Apenu, G. (2020). *Ghana's energy access journey so far: a review of key strategies.* International Journal of Energy Sector Management., doi:10.1108/IJESM-02-2020-0008

Bunt, A. C. (1988). Media and Mixes for Container-Grown Plants. In *Media and Mixes for Container-Grown Plants.* British Library Cataloguing in Publication Data Bunt. doi:10.1007/978-94-011-7904-1

Burnhual, S., & Deb, S. (2013). Scheduling optimization of flexible manufacturing systems using cuckoo search-based approach. *The International Journal of Advance Manufacturing Technology, 64(*5-8).

Buzacott, J. A., & Shanthikumar, J. G. (1993). Stochastic Models of Manufacturing Systems. Prentice Hall.

Calone, R., & Orsini, F. (2022). Aquaponics: A Promising Tool for Environmentally Friendly Farming. *Frontiers for Young Minds, 10*, 707801. doi:10.3389/frym.2022.707801

Camargo, L. A., Pereira, S. C., Correa, A. C., Farinas, C. S., Marconcini, J. M., & Mattoso, L. H. C. (2016). Feasibility of Manufacturing Cellulose Nanocrystals from the Solid Residues of Second-Generation Ethanol Production from Sugarcane Bagasse. *BioEnergy Research, 9*(3), 894–906. doi:10.100712155-016-9744-0

Canilha, L., Carvalho, W., De Almeida Felipe, M. D. G., De Almeida E., Silva, J. B., & Giulietti, M. (2010). Ethanol production from sugarcane bagasse hydrolysate using Pichia stipitis. *Applied Biochemistry and Biotechnology, 161*(1–8), 84–92. doi:10.1007/s12010-009-8792-8

Cao, D., & Chen, M. (2005). A mixed integer programming model for a two-line CONWIP-based production and assembly system. *International Journal of Production Economics, 95*(3), 317–326. doi:10.1016/j.ijpe.2004.01.002

Cárdenas, G. E., Pérez-Lechuga, G., Tuoh-Mora, J. C., & Medina-Marín, J. (2012). Obtaining the optimal short-term hydrothermal coordination scheduling: A stochastic view point. *Applied Mathematical and Computational Sciences, 3*(4), 361-379.

Carzaniga, A., Fuggetta, A., Hall, R. S., Heimbigner, D., Van Der Hoek, A., & Wolf, A. L. (1998). A characterisation framework for software deployment technologies. Colorado State Univ Fort Collins, Dept of Computer Science. doi:10.21236/ADA452086

Cassandras, C. G., & Lafortune, S. (2010). *Introduction to Discrete Event Systems*. Springer.

Cavazos, J., Moss, J. E. B., & O'Boyle, M. F. P. (2006). Hybrid Optimizations: Which Optimization Algorithm to Use. In A. Mycroft & A. Zeller (Eds.), Lecture Notes in Computer Science: Vol. 3923. *Compiler Construction. CC 2006*. Springer. doi:10.1007/11688839_12

Chamadugu, R., & Stecke, K. E. (1994). Classification and review of FMS scheduling procedures. *Production Planning & Control, The Management of Operations, 5*(1).

Chandel, A. K., Kapoor, R. K., Singh, A., & Kuhad, R. C. (2007). Detoxification of sugarcane bagasse hydrolysate improves ethanol production by Candida shehatae NCIM 3501. *Bioresource Technology, 98*(10), 1947–1950. doi:10.1016/j.biortech.2006.07.047 PMID:17011776

Chang, C., Teng, J.-T., & Goyal, S. K. (2008). Inventory lot-size models under trade credits: A review. *Asia-Pacific Journal of Operational Research, 25*(01), 89–112. doi:10.1142/S0217595908001651

Chavan, S. A., & Anwar, P. M. A., & R, P. S. B. (2017). Review of Effect of Process Parameters on Mechanical Properties of 3D Printed Components. *Internation Journal of Advance Research in Science and Engineerng, 6*(5), 461–467.

Cheng, M. H., Wang, Z., Dien, B. S., Slininger, P. J. W., & Singh, V. (2019). Economic analysis of cellulosic ethanol production from sugarcane bagasse using a sequential deacetylation, hot water and disk-refining pretreatment. *Processes (Basel, Switzerland), 7*(10), 642. doi:10.3390/pr7100642

Chico Hermanu, B. A., & Santoso, B., & Wicaksono, F. X. R. (2019). Design of 1 MWp floating solar photovoltaic (FSPV) power plant in Indonesia. *AIP Conference Proceedings*. doi:10.1063/1.5098188

Chin, C, Chen, W., Chien, F., & Wang, J.J. (2005). An AHP-based approach to ERP system selection. *International Journal of Production Economic, 96*(1), 47-62.

Choudhary, S. K. (2016), Study about the Types of Information Technology Service for Supply Chain. *Proceedings of the World Congress on Engineering and Computer Science 2016, 2.*

Coumou, D., & Rahmstorf, S. (2012). A decade of weather extremes. *Nature Climate Change, 2*(7), 491–496. doi:10.1038/nclimate1452

Culot, G., Nassmbeni, G., Orzes, G., & Sartor, M. (2020). Behind the definition of Industry 4.0: Analysis and open questions. *International Journal of Production Economics, 226*, 107617. doi:10.1016/j.ijpe.2020.107617

Da Silveira Dos Santos, D., Camelo, A. C., Rodrigues, K. C. P., Carlos, L. C., & Pereira, N. (2010). Ethanol production from sugarcane bagasse by Zymomonas mobilis using simultaneous saccharification and fermentation (SSF) Process. *Applied Biochemistry and Biotechnology, 161*(1–8), 93–105. doi:10.1007/s12010-009-8810-x

Das, S., & Sahu, G. (2020). Soil-Less Farming- An Innovative Way Towards Sustainability. *Agricultural & Food: E-Newsletter, 2*(10), 82–85.

De Carlo, F., & Arlo, M.A. (2013). Layout Design for a low-capacity manufacturing line: A case study. *International Journal of Engineering Business.*

Ekşioğlu, S. D., Acharya, A., Leightley, L. E., & Arora, S. (2009). Analyzing the design and management of biomass-to-biorefinery supply chain. *Computers & Industrial Engineering, 57*(4), 1342-1352.

Elamouri, M., & Amar, F. B. (2008). Wind energy potential in Tunisia. *Renewable Energy, 33*(4), 758–768. doi:10.1016/j.renene.2007.04.005

Elangovan, R., Santhanakrishnan, D. N., Rozario, R., & Banu, A. (2018). Tomen:A Plant monitoring and smart gardening system using IoT. *International Journal of Pure and Applied Mathematics, 119*(March), 703–710. https://www.researchgate.net/publication/323811801_TomenA_Plant_monitoring_and_smart_gardening_system_using_IoT

EnergyLive (2019). Netherlands' largest floating solar array switched on. *Energy Live News.*

Espinal, C. A., & Matulić, D. (2019). Recirculating Aquaculture Technologies. *Aquaponics Food Production Systems, 01*, 35–76. doi:10.1007/978-3-030-15943-6_3

Etter, P. C. (2017). Underwater Acoustic Modeling and Simulation, (Fourth Edition). In Underwater Acoustic Modeling and Simulation, Fourth Edition. CRC press. doi:10.1201/b13906

EUScienceHub. (2021). Floating solar panels in African hydropower reservoirs: a win-win for electricity generation and water savings. EU Science Hub. https://joint-research-centre.ec.europa.eu/jrc-news/floating-solar-panels-african-hydropower-reservoirs-2021-02-26_en,

Ezzahoui, I., Abdelhouahid, R. A., Taji, K., Marzak, A., & Ghanimi, F. (2022). The Aquaponic Ecosystem Using IoT and IA Solutions. *International Journal of Web-Based Learning and Teaching Technologies*, *17*(5), 1–15. doi:10.4018/IJWLTT.20220901.oa2

Faizrahnemoon, M. (2012). *Mathematical modelling of the scheduling of a production line at SKF* [Master's Thesis]. University of Gothenburg.

Fantahun, M. D., & Chen, M. (2006). A comprehensive mathematical model for the design of cellular manufacturing systems. *International Journal of Production Economics*, *103*(2), 767–783. doi:10.1016/j.ijpe.2005.10.008

Fanti, M. P., & Zhou, M. (2004). Deadlock control methods in automatic manufacturing systems. IEEE Transactions on Systems, 34(1).

Faranda, R. S., Hafezi, H., Leva, S., Mussetta, M., & Ogliari, E. (2015). The optimum PV plant for a given solar DC/AC converter. *Energies*, *8*(6), 4853–4870. doi:10.3390/en8064853

Felix, M., & Gheewala, S. H. (2011). A Review of Biomass Energy Dependency in Tanzania. *Energy Procedia*, *9*(2), 338–343. doi:10.1016/j.egypro.2011.09.036

Feng, S., Kang, S., Huo, Z., Chen, S., & Mao, X. (2008). Neural networks to simulate regional ground water levels affected by human activities. *Ground Water*, *46*(1), 80–90. doi:10.1111/j.1745-6584.2007.00366.x PMID:18181867

Fornes, F., Belda, R. M., Abad, M., Noguera, P., Puchades, R., Maquieira, A., & Noguera, V. (2003). The microstructure of coconut coir dusts for use as alternatives to peat in soilless growing media. *Australian Journal of Experimental Agriculture*, *43*(9), 1171–1179. doi:10.1071/EA02128

Future Technology Trends. (2016). https://read.oecd-ilibrary.org/science-and-technology/oecd-science-technology-and-innovation-outlook-2016/future-technology-trends_sti_in_outlook-2016-5-en#page1

Gagandeep, A. D., & Saini, H. S. (2018). Design and implementation of an automatic irrigation feedback control system based on monitoring of soil moisture. *Proceedings of the International Conference on Inventive Computing and Informatics, ICICI 2017*, (pp. 540–544). IEEE. doi:10.1109/ICICI.2017.8365190

Ganoe, R. D., Stackhouse, P. W. Jr, & DeYoung, R. J. (2014). *RET Screen Plus Software Tutorial. Technical Report.* International Clean Energy Decision Support Centre.

García, R., Pizarro, C., Lavín, A. G., & Bueno, J. L. (2012). Characterisation of Spanish biomass wastes for energy use. *Bioresource Technology*, *103*(1), 249–258. doi:10.1016/j.biortech.2011.10.004 PMID:22050839

Ghapanchi, A., Jafarzadeh, M. H., & Khakbaz, M.-H. (2008). Fuzzy-Data Envelopment Analysis approach to Enterprise Resource Planning system analysis and selection. *International Journal of Information Systems and Change Management*, *3*(2), 157. doi:10.1504/IJISCM.2008.020693

Ghosha, T., Senguptaa, S., Chattopadhyayb, M., & Dana, P. K. (2011). Meta-heuristics in cellular manufacturing: A state-of-the-art review. *International Journal of Industrial Engineering Computations*.

Ghosh, I. (2014). An Intelligent Hybrid Multi Criteria Decision Making Technique to Solve a Plant Layout Problem. *International Journal of Industrial Engineering Research and Development*, *5*(3), 13–23. doi:10.34218/IJIERD.5.3.2014.002

Giri, S., Ahmad, A., & Katiyar, V. (2019). Avoiding Power Clipping Losses by Inverter having High DC-to-AC Loading Ratio in Grid Connected Solar PV Plant Using Battery Energy Storage System, 213-219. PEEIC. doi:. doi:10.1109/PEEIC47157.2019.8976581

Goddek, S., Joyce, A., Wuertz, S., Körner, O., Bläser, I., Reuter, M., & Keesman, K. J. (2019). Decoupled Aquaponics Systems. In Aquaponics Food Production Systems. doi:10.1007/978-3-030-15943-6_8

Govindaraju, R. S. (2000). Artificial Neural Networks in Hydrology. II: Hydrologic Applications. *Journal of Hydrologic Engineering*, *5*(2), 124–137. doi:10.1061/(ASCE)1084-0699(2000)5:2(124)

Gozálvez, J. J. F. (2012). Covering Reservoirs With a System of Floating Solar Panels: Technical and Financial Analysis. *Sel. Proc. from 16th Int. Congr. Proj. Eng.*, (pp. 177–187). IEEE.

Gulati. Jacobs, M., Joostem, A., Naidoob, D., & Fakir, S. (2013). The Water–energy–food Security Nexus: Challenges and Opportunities for Food Security in South Africa. *Aquatic Procedia*, *1*, 150–164. doi:10.1016/j.aqpro.2013.07.013

Gumus, A. T., & Guneri, A. F. (2009). A multi-echelon inventory management framework for stochastic and fuzzy supply chains. *Expert Systems with Applications, 36*(3), 5565-5575.

Ha, G. S., El-Dalatony, M. M., Kurade, M. B., Salama, E. S., Basak, B., Kang, D., Roh, H. S., Lim, H., & Jeon, B. H. (2020). Energy-efficient pretreatments for the enhanced conversion of microalgal biomass to biofuels. *Bioresource Technology*, *309*(April), 123333. doi:10.1016/j.biortech.2020.123333 PMID:32305842

Haghiabi, A. H., Nasrolahi, A. H., & Parsaie, A. (2018). Water quality prediction using machine learning methods. *Water Qual. Res. J.*, *53*(1), 3–13. doi:10.2166/wqrj.2018.025

Haluschak, P. (2006). Laboratory Methods of Soil Analysis. Canada-Manitoba Soil Survey. Manitoba.

Hao, Q., & Shen, W. (2008). Implementing a hybrid simulation model for a Kanban-based material handling system. *Manufacturing*, *24*(5), 635–646. doi:10.1016/j.rcim.2007.09.012

Hassan, Z., Hossain, G. J., & Islam, M. M. (2020). Internet of Things (IoT) based water quality monitoring system. *Educ. Res*, *2*(4), 168–180.

Hauwanga, A. S., & Musti, K. S. S. (2022). Evaluation of Energy Losses and Costs Based on the Location of PCC in Smart Grids. IEEE 11th International Conference on Communication Systems and Network Technologies (CSNT), (pp. 76-82). IEEE. 10.1109/CSNT54456.2022.9787642

Heuristic Optimization. (n.d.). Recovered from file:///Users/gilbertoperezlechuga/Downloads/9780387258522-c2.pdf

Hong, M.M., Payandeh, S., & Gruver, W.A. (1996). Modeling and analysis of flexible fixturing systems for agile manufacturing. *IEEE International Conference on Systems, Man and Cybernetics. Information Intelligence and Systems.*

How does additive manufacturing work? (n.d.). Recovered from https://www.ge.com/additive/additive-manufacturing

Iacono, F., Ferretti, S., Mezzadra, A., Magni, L., & Toffanin, Ch. (2019). Industry 4.0: Mathematical model for monitoring sterilization processes. *2019 IEEE* International Conference on Systems, Man and Cybernetics (SMC), 610-615. 10.1109/SMC.2019.8914206

Immanuel, J., Patil, R., Sudheer, L. S., Parvathi, C. S., & Bhaskar, P. (2018). Web based Weather Monitoring System using Wireless Sensor Node. *International Journal of Computer Applications, 181*(11), 47–52. doi:10.5120/ijca2018917747

Ivanov, D., Dolgui, A., & Sokolov, B. (2016). A dynamic model and an algorithm for short-term supply chain scheduling in the smart factory industry 4.0. *International Journal of Production Research, 54*(2), 386-402. doi:10.1080/00207543.2014.999958

Jackson de Moraes Rocha, G., Martin, C., Soares, I. B., Souto Maior, A. M., Baudel, H. M., & Moraes de Abreu, C. A. (2011). Dilute mixed-acid pretreatment of sugarcane bagasse for ethanol production. *Biomass and Bioenergy*, *35*(1), 663–670. doi:10.1016/j.biombioe.2010.10.018

Jarimo, T., & Pulkkinen, U. (2005). A multi-criteria mathematical programming model for agile virtual organization creation. In *Collaborative Networks and Their Breeding Environments*. Springer. https://link.springer.com/content/pdf/10.1007%2F0-387-29360-4_13.pdf

Jha, S., Nanda, S., Acharya, B., & Dalai, A. K. (2022). A Review of Thermochemical Conversion of Waste Biomass to Biofuels. *Energies*, *15*(17), 1–23. doi:10.3390/en15176352

Jin, J. M. (2014). *The Finite Element Method in Electromagnetic* (3rd ed.). Wiley & Sons.

Johnston, R. Z., & Matlock, M. D. (2011). Geospatial Climate Data. University of Arkansas. The Sustainability Consortium, Food, Beverage and Agriculture.

Junianto, B., Dewi, T., & Sitompul, C. R. (2020). Development and Feasibility Analysis of Floating Solar Panel Application in Palembang, South Sumatra. *Journal of Physics: Conference Series*, *1500*(1), 012016. doi:10.1088/1742-6596/1500/1/012016

Just-in-Time manufacturing. (n.d.). https://translate.google.com/?hl=es#en/es/justtime%20manufacturing%20(JIT%20manufacturing)

Kahimba, F. C., Mbaga, S., Mkoko, B., Swai, E., Kimaro, A. A., Mpanda, M., Liingilie, A., & Germer, J. (2015) Rain-fed crop, livestock, and agroforestry systems in Semi-arid and sub-humid Tanzania: A Baseline report, Trans- SEC Project, ZALF. http://project2.zalf.de/trans-sec/public/media/ upload/ product/pdf/83905803646d86cb3ba323cdc201860b.pdf.

Kampker, A., Kreisköther, K., & Reinders, C. (2017). Material and Parameter Analysis of the PolyJet Process for Mold Making Using Design of Experiments. *International Journal of Materials and Metallurgical Engineering*, *11*(3), 242–249.

Kanyeka, E., Kamala, R., & Kasuga, R. (2007). *Improved Agricultural Research Technologies Recommended in Tanzania*. Ministry of Agriculture Food Security and Cooperatives.

Karsak, E. E., & Özogul, C. O. (2009). An integrated decision-making approach for ERP system selection. *Expert Systems with Applications*, *36*(1), 660–667. doi:10.1016/j.eswa.2007.09.016

Khan, A. A., Minai, A. F., Pachauri, R. K., & Malik, H. (2022). Optimal Sizing, Control, and Management Strategies for Hybrid Renewable Energy Systems: A Comprehensive Review. *Energies*, *15*(17), 6249. doi:10.3390/en15176249

Khanna, N., Mikkilineni, A. K., Martone, A. F., Ali, G. N., Chiu, G. T. C., Allebach, J. P., & Delp, E. J. (2006). A survey of forensic characterisation methods for physical devices. *Digital Investigation*, *3*, 17–28. doi:10.1016/j.diin.2006.06.014

Khoo, K. S., Wen, Y. C., Doris, Y. Y. T., & Pau, L. S. C. (2020). Nanomaterials Utilization in Biomass for Biofuel and Bioenergy Production. *Energies*, *13*(4), 892. doi:10.3390/en13040892

Khusna, D., Zawiah, S., Jamasri, J., & Aoyama, H. (2010). A Proposed Study on Facility Planning and Design in Manufacturing Process. *Proceedings of the International Multiconference of Engineers and Computers Scientists 2010*, 3.

Kia, R., Baboli, A., Javadian, N., Tavakkoli-Moghaddam, R., Kazemi, M., & Khorrami, J. (2012). Solving a group layout design model of a dynamic cellular manufacturing system with alternative process routings, lot splitting and flexible reconfiguration by simulated annealing. *Computers & Operations Research*, *39*(11), 2642–2658. doi:10.1016/j.cor.2012.01.012

Kilic, H. S., Zaim, S. M., & Delen, D. (2014). Development of a hybrid methodology for ERP system selection: The case of Turkish Airlines. *Decision Support Systems*, *66*, 82–92. doi:10.1016/j.dss.2014.06.011

Kiobia, D. O. (2015). *Design and development of a low-cost acoustic device to detect pest infestation in stored maise*. Virginia Tech. [http://hdl.handle.net/10919/56588]

Kohnert, D. (2007). On the Articulation of Witchcraft and Modes of Production among the Nupe, Northern Nigeria. *Hexenglauben Im Modernen Afrika: Hexen, Hexenverfolgung Und Magische Vorstellungswelten (Witchcraft in Modern Africa: Witches, Witch-Hunts and Magical Imaginaries), 48*(2), 39–61. doi:10.1227/01.NEU.0000349921.14519.2A

Kumar, P., Sampath, B., Kumar, S., Babu, B. H., & Ahalya, N. (2023). Hydroponics, Aeroponics, and Aquaponics Technologies in Modern Agricultural Cultivation. In Trends, Paradigms, and Advances in Mechatronics Engineering (pp. 223–241). IGI Global.

Lakhiar, I. A., Gao, J., Syed, T. N., Ali Chandio, F., Tunio, M. H., Ahmad, F., & Ali Solangi, K. (2020). Overview of the aeroponic agriculture – An emerging technology for global food security. *International Journal of Agricultural and Biological Engineering, 13*(1), 1–10. doi:10.25165/j.ijabe.20201301.5156

Lall, V., & Teyarachakul, S. (2016). Enterprise Resource Planning (ERP) System Selection: A Data Envelopment Anaysis (DEA) Approach. *Journal of Computer Information Systems, 47*(1).

Lean Manufacturing. (n.d.). http://www.leansolutions.co/wp-content/uploads/2011/07/Indicadores-Lean-manufacturing.png

Lee, Y. H., Golinska, P., & Wu, J. Z. (2016). Mathematical Models for Supply Chain Management. *Mathematical Problems in Engineering.*

Lenny Koh, S. C., & Mike Simpson, M. (2007). Could enterprise resource planning create a competitive advantage for small businesses? *Benchmarking, 14*(1), 59–76. doi:10.1108/14635770710730937

Li, P., Shi, K., & Zhang, J. (2000). *Modeling Agile Manufacturing Cell using Object-Oriented Timed Petri Net.* Recovered from: http://keshi.ubiwna.org/paper/Modeling%20Agile%20Manufacturing%20Cell%20using%20Object-Oriented%20Timed%20Petri%20net.pdf

Li, Z., & Wu, N. (2012). Deadlock control of automatic manufacturing systems based on Petri Nets- A literature review. IEEE Transactions on Systems, 42(4).

Liao, Y., Deschamps, F., Loures, E.F., & Ramos, L.F.P., (2017). Past, present and future of Industry 4.0- a systematic literature review and research agenda proposal. *International Journal of Production Research, 55*(12).

Liao, X., Li, Y., & Lu, B. (2007). A model for selecting an ERP system based on linguistic information processing. *Information Systems, 32*(7), 1005–1017. doi:10.1016/j.is.2006.10.005

Li, X., Guo, S., Liu, Y., Du, B., & Wang, L. (2017). A production planning model for make-to-order foundry flow shop with capacity constraint. *Mathematical Problems in Engineering, 2017*, 6315613. doi:10.1155/2017/6315613

López-Ortega, O., & Pérez-Lechuga, G. (2001). An express meta model for sharing and exchanging flexible manufacturing resources. In *World Congress Manufacturing*. Rochester Institute of Technology.

Luo, J., Fang, Z., & Smith, R. L. Jr. (2014). Ultrasound-enhanced conversion of biomass to biofuels. *Progress in Energy and Combustion Science*, *41*(1), 56–93. doi:10.1016/j.pecs.2013.11.001

Macrelli, S., Zacchi, G., & Mogensen, J. (2012). Techno-economic evaluation of 2nd generation bioethanol production from sugar cane bagasse and leaves integrated with the sugar-based ethanol process. *Biotechnology for Biofuels*, *5*(1), 152. . doi:10.1186/1754-6834-5-22 PMID:22502801

Madavhi, I., Aalaei, A., Paydar, M. M., & Solymanpur, M. (2010). *Designing a mathematical model for dynamic cellular manufacturing systems considering production planning and worker assignment*. Academic Press.

Madavhi, I. (2009). Genetic algorithm approach for solving a cell formation problem in cellular manufacturing. *Expert Systems with Applications*, *36*(3), 6598–6604. doi:10.1016/j.eswa.2008.07.054

Madhav, P. A. (1986). *Mathematical Modelling of Flexible Manufacturing Systems*. University of Windsor.

Malu, P. R., Sharma, U. S., & Pearce, J. M. (2017). Agrivoltaic potential on grape farms in India. *Sustainable Energy Technologies and Assessments*, *23*, 104–110. doi:10.1016/j.seta.2017.08.004

Mandal, N. K., Roy, T. K., & Maitic, M. (2005). Multi-objective fuzzy inventory model with three constraints: A geometric programming approach. *Fuzzy Sets and Systems*, *150*(1), 87–106. doi:10.1016/j.fss.2004.07.020

Maniya, K., & Bhatt, M. G. (2011). An alternative multiple attribute decision making methodology for solving optimal facility layout design selection problems. *Computers & Industrial Engineering*, *61*(3), 542–549. doi:10.1016/j.cie.2011.04.009

Manufacturing resource planning (MRP-II). (n.d.). http://www.businessdictionary.com/definition/manufacturing-resource-planning-MRP-II.html

Manzhirov, A. V. (2015). Mechanical design of viscoelastic parts fabricated using additive manufacturing technologies. *Proceedings of the World Congress on Engineering 2015, 2.*

Manzhirov, A. V., & Lychev, S. A. (2014). Mathematical Modeling of Additive Manufacturing Technologies, *Proceedings of the World Congress on Engineering 2014, 2.*

Material requirements planning (MRP). (n.d.). https://searcherp.techtarget.com/definition/material-requirements-planning-MRP

Maucieri, C., Nicoletto, C., van Os, E., Anseeuw, D., Van Havermaet, R., & Junge, R. (2019). Hydroponic Technologies. In Aquaponics Food Production Systems. doi:10.1007/978-3-030-15943-6_4

McKendry, P. (2002). Energy production from biomass (part 1): Overview of biomass. *Bioresource Technology*, *83*(1), 37–46. doi:10.1016/S0960-8524(01)00118-3 PMID:12058829

Meneghel, D., Bortoni, E., & Karimi, A. (2018). *Boosting DC/AC Ratio of PV Plant for BESS Integration on DC side,* 1-4. Semantic Scholar. doi:. doi:10.1109/SusTech.2018.8671354

Mesa, L., Morales, M., González, E., Cara, C., Romero, I., Castro, E., & Mussatto, S. I. (2014). Restructuring the processes for furfural and xylose production from sugarcane bagasse in a biorefinery concept for ethanol production. *Chemical Engineering and Processing: Process Intensification, 85*, 196–202. doi:10.1016/j.cep.2014.07.012

Mishra, D., Khan, A., Tiwari, R., & Upadhay, S. (2018). Automated Irrigation System-IoT Based Approach. *Proceedings - 2018 3rd International Conference On Internet of Things: Smart Innovation and Usages*. IEEE. 10.1109/IoT-SIU.2018.8519886

Mks, S. (2005). Simplified algorithm to determine break point realys & relay coordination based on network topology [for realys read relays. 2005 IEEE International Symposium on Circuits and Systems (ISCAS), (*vol. 1,* pp. 772-775). IEEE. doi:10.1109/ISCAS.2005.1464702

MNRC, Ministry for Natural Resources Canada. (2015). *Clean Energy Project Analysis: RETScreen Engineering and Cases* (3rd ed.). Natural Resources.

Moberg, J. P. (2001). Soil and Plant Analysis Manual. Revised Edition, the Royal Veterinary and Agricultural University. Chemistry Department, Copenhagen University.

Mohammed, Y. S., Mustafa, M. W., & Bashir, N. (2013a). Status of renewable energy consumption and developmental challenges in Sub-Sahara Africa. *Renewable & Sustainable Energy Reviews, 27*, 453–463. doi:10.1016/j.rser.2013.06.044

Mohammed, Y. S., Mustafa, M., Bashir, N., & Mokhtar, A. S. (2013b). Renewable energy resources for distributed power generation in Nigeria: A review of the potential. *Renewable & Sustainable Energy Reviews, 22*, 257–268. doi:10.1016/j.rser.2013.01.020

Mula, J., Peidro, D., Diaz-Madroñero, M., & Vicens, E. (2010). Mathematical programming models for supply chain production and transport planning. *European Journal of Operational Research, 204*(3), 377–390. doi:10.1016/j.ejor.2009.09.008

Mula, J., Poler, R., & García, J. P. (2006a). MRP with flexible constraints: A fuzzy mathematical programming approach. *Fuzzy Sets and Systems, 157*(1), 74–97. doi:10.1016/j.fss.2005.05.045

Mula, J., Poler, R., & García, J. P. (2006b). Models for production planning under uncertainty: A review. *International Journal of Production Economics, 103*(1), 271–285. doi:10.1016/j.ijpe.2005.09.001

Muñoz, E., & Capón-García, E. (2019). Intelligent Mathematical Modelling Agent for Supporting Decision-Making at Industry 4.0. In J. Mejia, M. Muñoz, Á. Rocha, A. Peña, & M. Pérez-Cisneros (Eds.), *Trends and Applications in Software Engineering. CIMPS 2018. Advances in Intelligent Systems and Computing* (Vol. 865). Springer.

Musti, K. S. (2020). Circular economy in energizing smart cities. In Handbook of Research on Entrepreneurship Development and Opportunities in Circular Economy (pp. 251-269). IGI Global. doi:10.4018/978-1-7998-5116-5.ch013

Musti, K. S., Iileka, H., & Shidhika, F. (2020). Industry 4.0-based enterprise information system for demand-side management and energy efficiency. In Novel Approaches to Information Systems Design (pp. 137-163). IGI Global. doi:10.4018/978-1-7998-2975-1.ch007

Musti, K. S. (2021). *Industry 4.0-Based Large-Scale Symbiotic Systems for Sustainable Food Security in Namibia.: Impacts of Climate Change on Agriculture and Aquaculture*. IGI Global. doi:10.4018/978-1-7998-3343-7.ch009

Musti, K. S. S. (2020). Quantification of Demand Response in Smart Grids. *IEEE Int. Conf. (INDISCON)*, (pp. 278-282). IEEE. 10.1109/INDISCON50162.2020.00063

Mwinuka, L., Schneider, I., Maeda, C., Mutabazi, K. D., Makindara, J., Graef, F., Sieber, S., Swai, E., Mbwana, H., & Swamila, M. (2015). Comparing stakeholder views for mutual acceptable food value chain upgrading strategies in Tanzania. *African Journal of Agricultural Research*, *10*(12), 1376–1385. doi:10.5897/AJAR2014.8901

Naveed, N. (2021). Investigate the effects of process parameters on material properties and microstructural changes of 3D-printed specimens using fused deposition modelling (FDM). *Materials Technology*, *36*(5), 317–330. doi:10.1080/10667857.2020.1758475

Norani, M. N. M., & Abdollah, M. F., Abdullah, M. I. H. C., Amiruddin, H., Ramli, F. R., & Tamaldin, N. (2021). 3D printing parameters of acrylonitrile butadiene styrene polymer for friction and wear analysis using response surface methodology. *Proceedings of the Institution of Mechanical Engineers, Part J: Journal of Engineering Tribology, 235*(2), 468–477. doi:10.1177/1350650120925601

Norouzi, G., Heydari, M., Noory, S., & Bagherpour, M. (2015). Developing a mathematical model for scheduling and determining success probability of research projects considering complex-fuzzy networks. *Journal of Applied Mathematics*, *2015*, 809216. doi:10.1155/2015/809216

Odetoye, O. A., Ibikunle, F. A., Olulope, P. K., Onyemenam, J. O., & Okeke, U. N. (2022). Large-Scale Solar Power in Nigeria: The Case for Floating Photovoltaics. *IEEE Int. Conf. on Disruptive Technologies for Sustainable Development (NIGERCON)*, (pp. 1-5). IEEE. 10.1109/NIGERCON54645.2022.9803156

Omar, M., Sarker, R., & Othman, W. A. M. (2013). A just-in-time three-level integrated manufacturing system for linearly time-varying demand process. *Applied Mathematical Modelling*, *37*(3), 1275–1281. doi:10.1016/j.apm.2012.03.041

Ong, H. C., Chen, W. H., Singh, Y., Gan, Y. Y., Chen, C. Y., & Show, P. L. (2020). A state-of-the-art review on thermochemical conversion of biomass for biofuel production: A TG-FTIR approach. *Energy Conversion and Management*, *209*. Advance online publication. doi:10.1016/j.enconman.2020.112634

Oracle, E. R. P. (n.d.). What Is ERP? https://translate.google.com/?hl=es#en/es/Oracle%20ERP%0AWhat%20Is%20ERP%3F

Osman, A. I., Mehta, N., Elgarahy, A. M., Al-Hinai, A., Al-Muhtaseb, A. H., & Rooney, D. W. (2021). Conversion of biomass to biofuels and life cycle assessment: a review. In Environmental Chemistry Letters (Vol. 19, Issue 6). Springer International Publishing. doi:10.100710311-021-01273-0

Ouimet, Charu, & Chaubal. (2001). *No. 6308163 B1*. United States Patent. https://translate.google.com/?hl=es#es/en/Method%20for%20controlled%20optimization%20of%20enterprise%20planning%20models

Oyedepo, S. O. (2014). Towards achieving energy for sustainable development in Nigeria. *Renewable & Sustainable Energy Reviews*, *34*, 255–272. doi:10.1016/j.rser.2014.03.019

Papadimitriou, G., Kaliorakis, M., Chatzidimitriou, A., Gizopoulos, D., Favor, G., Sankaran, K., & Das, S. (2021). A system-level voltage/frequency scaling characterisation framework for multicore CPUs.

Parmar, B., Chokhalia, J., & Desarda, S. (2019). Terrace Garden Monitoring System Using Wireless Sensor Networks. *51 | Research. Journal of Engineering and Technology Management*, *02*(01), 2582–2610. www.rjetm.in/

Pearce, J. M., Harshavardhan Dinesh. (2016). The potential of agrivoltaic systems. *Renewable & Sustainable Energy Reviews*, *308*, 299–308. doi:10.1016/j.rser.2015.10.024

Pérez, L. G., Bujari, A. A., & Venegas, M. F. (2017). Economic aspects of the production planning in lines of manufacturing under conditions of uncertainty. In Modeling of Economic and Financial Phenomena: A Contemporary Vision. Editorial Castdel.

Pérez, L. G., Gress, E. S. H., Karelyn, A., & Orán, M. G. M. (2010). Some efficiency measures in the operation of flexible manufacturing systems: A Stochastic Approach. *Proceedings of the 2009 SIAM Conference on Mathematics for Industry*, 31-36.

Pérez, L. G., Pérez, R. E., & Venegas, M. F. (2015). Stochastic Optimization of Manufacture Systems by Using Markov Decision Processes. In Handbook of Research on Modern Optimization Algorithms and Applications in Engineering and Economics. IGI Global.

Pérez-Lechuga, G., Flores-Rivera, C., Karelyn, A., & Tarasenko, A. (2011). A multi-period stochastic model to link the supply chain production planning in a manufacturing organization. *Advances in Science and Statistics Applications, 6*(4), 255-283.

Pérez-Lechuga, G., Madrid F. K. N., & Fiore, H. (2021). Hybrid Optimization and Artificial Intelligence Applied to Energy Systems: A review. In *Advances of Artificial Intelligence in a Green Energy Environment*. IGI Global.

Pérez-Lechuga, G. (2018). Optimal logistics strategy to distribute medicines in clinics and hospitals. *Journal of Mathematics in Industry, Springer*, *8*(1), 2. doi:10.118613362-018-0044-5

Peters, J., Combs, S. M., Hoskins, B., Jarman, J., Kovar, J. L., Watson, M. E., Wolf, A. M., & Wolf, N. (2003). *Recommended Methods of Manure Analysis*. Publication No. 3769. Coop. Ext. Publishing.

Pickering, H. W., Menzies, N. W., & Hunter, M. N. (2002). Zeolite/rock phosphate - A novel slow release phosphorus fertiliser for potted plant production. *Scientia Horticulturae*, *94*(3–4), 333–343. doi:10.1016/S0304-4238(02)00006-7

Pi, R., Paul, S., Antony, A., & Aswathy, B. (2017). Android Based Home Automation Using Raspberry Pi. *International Journal of Advance Engineering and Research Development*, *4*(06), 143–147. doi:10.21090/ijaerd.58329

Pitra, G. M., & Musti, K. S. S. (2021). Duck Curve with Renewable Energies and Storage Technologies. 13th International Conference on Computational Intelligence and Communication Networks (CICN), (pp. 66-71). IEEE. 10.1109/CICN51697.2021.9574671

Pizarro, F., Salazar, R., Rajo-Iglesias, E., Rodriguez, M., Fingerhuth, S., & Hermosilla, G. (2019). Parametric Study of 3D Additive Printing Parameters Using Conductive Filaments on Microwave Topologies. *IEEE Access: Practical Innovations, Open Solutions*, *7*, 106814–106823. doi:10.1109/ACCESS.2019.2932912

Pochet, Y., & Wolsey, L. A. (2006). *Production Planning by Mixed Integer Programming*. Springer.

PRNewswire. (2021) Where the Sun Meets the Sea: Offshore Floating-PV Powers Singapore's Journey Toward Carbon Neutrality. PR Newswire.

Production planning. (n.d.). Recovered from http://www.businessdictionary.com/definition/production-planning.html

Radhiansyah, T. D., & Hamdani, D. (2021). Performance Analysis of Offshore Floating PV Systems in Isolated Area. *2021 3rd Int. Conf. High Volt. Eng. Power System,* (pp. 651–655). IEEE. doi:10.1109/ICHVEPS53178.2021.9600926

Rahmad, I. F., Nababan, E. B., Tanti, L., Triandi, B., Ekadiansyah, E., & Fragastia, V. A. (2019). Application of the Alcohol Sensor MQ-303A to Detect Alcohol Levels on Car Driver. *2019 7th International Conference on Cyber and IT Service Management, CITSM 2019*, (pp. 247–255). IEEE. doi:10.1109/CITSM47753.2019.8965395

Rai, J. K., & Xirouchakis, P. (2008). Finite element method-based machining simulation environment for analyzing part errors induced during milling of thin-walled components. *International Journal of Machine Tools & Manufacture*, *48*(6), 629–643. doi:10.1016/j.ijmachtools.2007.11.004

Rajkumar, M. N., Abinaya, S., & Kumar, V. V. (2017). Intelligent irrigation system - An IOT based approach. *IEEE International Conference on Innovations in Green Energy and Healthcare Technologies,* (pp. 1–5). IEEE. 10.1109/IGEHT.2017.8094057

Raj, S. A., Sudheer, A., Vinodh, S., & Anand, G. (2013). A mathematical model to evaluate the role of agility enablers and criteria in a manufacturing environment. *International Journal of Production Research*, *51*(19).

Ramos, S., Soares, J., Cembranel, S. S., Tavares, I., Foroozandeh, Z., Vale, Z., & Fernandes, R. (2021). Data mining techniques for electricity customer characterisation. *Procedia Computer Science, 186*, 475–488. doi:10.1016/j.procs.2021.04.168

Ranjbaran, P., Yousefi, H., Gharehpetian, B., & Astaraei, F. R. (2019). A review on floating photovoltaic (FPV) power generation units. *Renewable & Sustainable Energy Reviews, 110*, 332–347. doi:10.1016/j.rser.2019.05.015

Raposo, F., De la Rubia, M. A., Fernández-Cegrí, V., & Borja, R. (2012). Anaerobic digestion of solid organic substrates in batch mode: An overview relating to methane yields and experimental procedures. *Renewable & Sustainable Energy Reviews, 16*(1), 861–877. doi:10.1016/j.rser.2011.09.008

REN21. (2017). Renewable Energy Policy Network for the 21st Century. Renewables Global Status Report. REN21 Secretariat.

Roberts, I. A., Wang, C. J., Esterlein, R., Stanford, M., & Mynors, D. J. (2009). A three-dimensional finite element analysis of the temperature field during laser melting of metal powders in additive layer manufacturing. *International Journal of Machine Tools & Manufacture, 49*(12-13), 916–923. doi:10.1016/j.ijmachtools.2009.07.004

Rosen, D. W. (2007). *Design for additive manufacturing: a method to explore unexplored regions of the design.* Recovered from: https://pdfs.semanticscholar.org/101a/8f03ed20725ae71d30537e90f783a398e86f.pdf

Rouphael, Y., & Colla, G. (2005). Growth, yield, fruit quality and nutrient uptake of hydroponically cultivated zucchini squash as affected by irrigation systems and growing seasons. *Scientia Horticulturae, 105*(2), 177–195. doi:10.1016/j.scienta.2005.01.025

Rui, L., Xiaofei, Z., & Guiwu, W. (2014). Models for selecting an ERP system with hesitant fuzzy linguistic information. *Journal of Intelligent & Fuzzy Systems, 26*(5), 2155–2165. doi:10.3233/IFS-130890

Sadrazadeh, A. (2012). A genetic algorithm with the heuristic procedure to solve the multi-line layout problem. *Computers & Industrial Engineering, 62*(4), 1055–1064. doi:10.1016/j.cie.2011.12.033

Safari, B., & Gasore, J. (2010). A statistical investigation of wind characteristics and wind energy potential based on the Weibull and Rayleigh models in Rwanda. *Renewable Energy, 35*(12), 2874–2880. doi:10.1016/j.renene.2010.04.032

Sahu, A., Yadav, N., & Sudhakar, K. (2016). Floating photovoltaic power plant: A review. *Renewable & Sustainable Energy Reviews, 66*, 815–824. doi:10.1016/j.rser.2016.08.051

Sampath, B., Sasikumar, C., & Myilsamy, S. (2023). Application of TOPSIS Optimization Technique in the Micro-Machining Process. In Trends, Paradigms, and Advances in Mechatronics Engineering (pp. 162–187). IGI Global.

Sampath, B., & Myilsamy, S. (2021). Experimental investigation of a cryogenically cooled oxygen-mist near-dry wire-cut electrical discharge machining process. *Strojniski Vestnik. Jixie Gongcheng Xuebao*, *67*(6), 322–330. doi:10.5545v-jme.2021.7161

Sampath, B., Pandian, M., Deepa, D., & Subbiah, R. (2022). Operating parameters prediction of liquefied petroleum gas refrigerator using simulated annealing algorithm. *AIP Conference Proceedings*, *2460*(1), 070003. doi:10.1063/5.0095601

Samuel, A. L. (2000). Some studies in machine learning using the game of checkers. *IBM Journal of Research and Development*, *44*(1–2), 207–219. doi:10.1147/rd.441.0206

Sanchez, R. G., Kougias, L., Fahl, F., & Jäger-Waldau, A. (2021). Assessment of floating solar photovoltaics potential in existing hydropower reservoirs in Africa. *Renewable Energy*, *169*, 687–699. doi:10.1016/j.renene.2021.01.041

Sanchez, R.G., Seliger, R, Fahl, F, De Felice, L., & Taha, B.M.J., Fabio. (2020). Freshwater use of the energy sector in Africa. *Applied Energy*, *270*. doi:10.1016/j.apenergy.2020.115171

Sandeep, K., & Rakesh, K. (2011). CLIPS based decision support system for water distribution networks. *Drinking Water Engineering and Science*, *4*(1), 37–50. doi:10.5194/dwes-4-37-2011

Sathyanarayana, S. R., Gangadhar, W. V., Badrinath, M. G., Ravindra, R. M., & Shriramrao, A. U. (2022). Hydroponics: An Intensified Agriculture Practice to Improve Food Production. In Reviews in Agricultural Science (Vol. 10, pp. 101–114). doi:10.7831/ras.10.0_101

Savsar, M., & Jawini, A. N. (1995). Simulation analysis of just-in-time production systems. *International Journal of Production Economics*, *42*(1), 67–78. doi:10.1016/0925-5273(95)00166-2

Sawik, T. (2011). *Scheduling in Supply Chains Using Mixed Integer Programming*. Wiley.

Scheid, A., Hafner, J., Hoffmann, H., Kächele, H., Sieber, S., & Rybak, C. (2018). Fuelwood scarcity and its adaptation measures: An assessment of coping strategies applied by small-scale farmers in Dodoma region, Tanzania. *Environmental Research Letters*, *13*(9), 095004–095015. doi:10.1088/1748-9326/aadb27

Schnurer, A., & Jarvis, A. (2010). Microbiological Handbook for Biogas Plants. Swedish Waste Management U2009:03.

Shahabudeen, P., Krishnaiah, K., & Narayanan, M. T. (2003). Design of a Two-Card Dynamic Kanban System Using a Simulated Annealing Algorithm. *International Journal of Advanced Manufacturing Technology*, *21*(10–11), 754–759. doi:10.100700170-002-1387-y

Shah, K., Ripon, N., & Glette, K. (2012). Adaptive variable neighborhood search for solving multi- objective facility layout problem with unequal area facilities. *Swarm and Evolutionary Computation.*

Shapiro, J. F. (1993). Chapter 8 Mathematical programming models and methods for production planning and scheduling. *Management Science*, *4*, 371–443.

Sharma, N., Acharya, S., Kumar, K., Singh, N., & Chaurasia, O. P. (2018). Hydroponics as an advanced technique for vegetable production: An overview. *Journal of Soil and Water Conservation*, *17*(4), 364. doi:10.5958/2455-7145.2018.00056.5

Shen, G., Yuan, X., Chen, S., Liu, S., & Jin, M. (2022). High titer cellulosic ethanol production from sugarcane bagasse via DLCA pretreatment and process development without washing/detoxifying pretreated biomass. *Renewable Energy*, *186*(January), 904–913. doi:10.1016/j.renene.2022.01.062

Sherif, M. (2013). *Agile manufacturing system scheduling using genetic algorithms and simulated* [Master Degree Thesis]. American University in Cairo. http://dar.aucegypt.edu/handle/10526/3397

Singh, J., Kuchroo, P., Bhatia, H., & Sidhu, E. (2016). Floating TEG based solar energy harvesting system. *Doi*, *763-766*, 763–766. doi:10.1109/ICACDOT.2016.7877689

Six Reasons Manufacturing is Central to the Economy. (2011). Recovered from http://rooseveltinstitute.org/six-reasons-manufacturing-central-economy/

Solomin, E., Sirotkin, E., Cuce, E., Selvanathan, S. P., & Kumarasamy, S. (2021). Hybrid Floating Solar Plant Designs: A Review. *Energies*, *14*(10), 2751. doi:10.3390/en14102751

Solymanpur, M., Vrat, P., & Shankar, R. (2004). Ant colony optimization algorithm to the inter-cell layout problem in cellular manufacturing. *European Journal of Operational Research*, *157*(3), 592–606. doi:10.1016/S0377-2217(03)00248-0

Somerville, C., Cohen, M., Pantanella, E., Stankus, A., & Lovatelli, A. (2014). Small-scale aquaponic food production. In *FAO Fisheries and Aquaculture*. https://www.fao.org/3/contents/1dea3c92-1faa-47bb-a374-0cf4d9874544/i4021e00.htm

Sreenath, S., Sudhakar, K., & AF, Y. (2021). 7E analysis of a conceptual utility-scale land-based solar photovoltaic power plant. *Energy*, *219*, 119610. doi:10.1016/j.energy.2020.119610

Suberu, M. Y., Mustafa, M. W., Bashir, N., Muhamad, N. A., & Mokhtar, A. S. (2013). Power sector renewable energy integration for expanding access to electricity in sub-Saharan Africa. *Renewable & Sustainable Energy Reviews*, *25*, 630–642. doi:10.1016/j.rser.2013.04.033

Sullivan, K. M., Dean, A., & Soe, M. M. (2009). On Academics: OpenEpi: A web-based epidemiologic and statistical calculator for public health. *Public Health*, *124*(3), 471–474. PMID:19445426

Sun, L.L., Heragu, S.S., Chen, L., Spearman, M.L. (2009). Simulation analysis of a multi-item MRP system based on factorial design. *Winter Simulation Conference (WSC), Proceedings of the 2009*, 2107–2114.

Suri, M., Remund, J., Cebecauer, T., Dumortier, D., Wald, L., Huld, T., & Blanc, P. (2008). *First steps in the cross-comparison of solar resource spatial products in Europe.* In EUROSUN 1st International Conference on Solar Heating, Cooling and Buildings, Lisbon, Portugal.

Taha, M. F., ElMasry, G., Gouda, M., Zhou, L., Liang, N., Abdalla, A., Rousseau, D., & Qiu, Z. (2022). Recent Advances of Smart Systems and Internet of Things (IoT) for Aquaponics Automation: A Comprehensive Overview. *Chemosensors (Basel, Switzerland), 10*(8), 303. doi:10.3390/chemosensors10080303

Talluri, S., Narasimhan, R., & Nair, A. (2006). A., Vendor performance with supply risk: A chance-constrained DEA approach. *International Journal of Production Economics, 100*(2), 212–222. doi:10.1016/j.ijpe.2004.11.012

Tarigan, E. (2021). Simulation of the Floating PV System to Supply Electricity Demand for the City of Surabaya, Indonesia. *Int. conf. on on Renewable Energy: Generation and Applications (ICREGA),* (pp. 48–53). IEEE. doi: . doi:10.1109/ICREGA50506.2021.9388219

Tayur, S., Ganeshan, R., & Magazine, M. (1999). *Quantitative Models for Chain supply Management.* Kluwer's International Series. doi:10.1007/978-1-4615-4949-9

Thapa, G. (2015). *Optimization of just-in-time sequencing problems and supply chain logistics.* Mälardalen University Press Dissertations No. 184. Recovered from https://www.diva-portal.org/smash/get/diva2:851513/FULLTEXT01.pdf

Urek, B., & Kaya, O. (2016). A mixed integer nonlinear programming model and heuristic solutions for location, inventory and pricing decisions in a closed loop supply chain. *Computers & Operations Research, 65*, 93–103. doi:10.1016/j.cor.2015.07.005

Vandaele, N., & Porter, W. (2015). Renewable energy in developing and developed nations: Outlooks to 2040. *Journal of Undergraduate Research, 15*(3), 1–7.

Verdonck, O., Penninck, R., & De Boodt, M. (1984). The Physical Properties of Different Horticultural Substrates. In Acta Horticulturae (Issue 150, pp. 155–160). doi:10.17660/actahortic.1984.150.16

Voss, S., & Woodruff, D. (2006). Connecting MRP, MRP II and ERP- supply chain production planning via optimization models. In Tutorials on Emerging Methodologies and Applications in Operations Research. Springer Science & Business Media.

Wald, L. (2007). Solar radiation energy (fundamentals and theory). Solar Energy Conversion and Photoenergy Systems. In J. Blanco and S. Malato (eds.) Encyclopedia of Life Support Systems (EOLSS). Eolss Publishers.

Wang, R. C. H., & Liang, T. F. (2004). Application of fuzzy multi-objective linear programming to aggregate production planning. *Computers & Industrial Engineering, 46*(1), 17–41. doi:10.1016/j.cie.2003.09.009

Wawa, A. I. (2012). *The Challenges of Promoting and Adopting Biogas Technology in Semi-arid areas of Tanzania: The Case of Kongwa and Bahi Districts in Dodoma Region.* [Doctoral Thesis, The Open University of Tanzania].

Wazed, M. A., Ahmed, S., & Nukman, Y. (2011). Mathematical models for multistage production under commonality and uncertainty. *International Journal of Physical Sciences, 6*(22), 5202–5222.

Wei, C. C., & Wang, J. J. (2004). A comprehensive framework for selecting an ERP system. *International Journal of Project Management*, *22*(2), 161–169. doi:10.1016/S0263-7863(02)00064-9

Weiland, P. (2010). Biogas production: Current state and perspectives. *Applied Microbiology and Biotechnology*, *85*(4), 849–860. doi:10.100700253-009-2246-7 PMID:19777226

What are automated manufacturing systems? (n.d.). Recovered from https://connell-ind.com/what-are-automated-manufacturing-systems/

What is agile manufacturing? (n.d.). Recovered from https://translate.google.com/?hl=es#en/es/WHAT%20IS%20AGILE%20MANUFACTURING%3F%20Why%20is%20it%20an%20effective%20strategy%3F

What is inventory management? (n.d.). Recovered from https://www.tradegecko.com/learning-center/what-is-inventory-management

What is Supply Chain Management (SCM)? (n.d.). Recovered from https://scm.ncsu.edu/scm-articles/article/what-is-supply-chain-management-scm

Wong, Y. C., & Sanggari, V. (2014). Bioethanol production from sugarcane bagasse using fermentation process. *Oriental Journal of Chemistry*, *30*(2), 507–513. doi:10.13005/ojc/300214

Yadav, A., Gupta, R., & Garg, V. K. (2013). Organic manure production from cow dung and biogas plant slurry by vermicomposting under field conditions. *Recycling of Organic Waste in Agriculture*, *2*(1), 1–21. doi:10.1186/2251-7715-2-21

Yadav, A., & Jayswal, S. C. (2017). Modelling of flexible manufacturing systems: A review. *International Journal of Production Research*, *56*(7), 2464–2487. doi:10.1080/00207543.2017.1387302

Yanes, A. R. (2020). Towards Aquaponics 4.0 : A framework for automated, IoT and smart implementations in indoor farming. *Information Processing in Agriculture, 9*(4).

Yanes, A. R. (2020). *Towards Aquaponics 4.0 : A framework for automated, IoT and smart implementations in indoor farming*. SORACOM.

Yazgan, H. R., Boran, S., & Goztepe, K. (2019). An ERP software selection process with using artificial neural network based on analytic network process approach. *Expert Systems with Applications*, *36*(5), 9214–9222. doi:10.1016/j.eswa.2008.12.022

Yazıcı, E., Büyüközkan, G., & Baskak, M. (2016). A New Extended MILP MRP Approach to Production Planning and Its Application in the Jewelry Industry. *Mathematical Problems in Engineering*, *2016*, 7915673. doi:10.1155/2016/7915673

Yilmaz, O. F., & Suleiman, T. (Eds.). (2018). *Handbook of Research on Applied Optimization Methodologies in Manufacturing*. IGI-Global. doi:10.4018/978-1-5225-2944-6

Yousefi, M., Hosseinioun, S. S., & Taghikhah, F. (2012). Determining optimum number of kanbans for workstations in a kanban-based manufacturing line using discrete-event simulation: a case of study. *Proceedings of the 2012 International Conference on Industrial Engineering and Operations Management*.

Zareef, L., Jalali, M., & Sadat, S. M. (2021). Analysis of the potential for use of floating solar panels on Naghlo hydropower dam. *Int. Jnl of Innovative Research and Scientific Studies*, *4*(2), 126–132. doi:10.53894/ijirss.v4i2.66

Zhang, W., & Chen, M. (2010). A mathematical programming model for production planning using CONWIP. *International Journal of Production Research, 39*(12).

Zheng, A., & Casari, A. (2018) *Feature Engineering: Principle and Techniques for Data Scienctists*. O'Reilly Media, Inc. https://www.oreilly.com/library/view/feature-engineering-for/9781491953235/

Zhenyuan, J., Xiaohong, L., & Wei, W. (2011). Design and implementation of lean facility layout system of a production line. *International Journal of Industrial Engineering*, *18*, 260–269.

Zhou, J., Yang, G., & Tang, G. (2015). The mathematical modeling of the mobile recharging facility vehicles' scheduling and its genetic algorithm solution. *IEEE International Conference on Cyber Technology in Automation, Control, and Intelligent Systems (CYBER)*. 10.1109/CYBER.2015.7288262

Zhou, K., & Yang, S. (2016). Understanding household energy consumption behaviour: The contribution of energy big data analytics. *Renewable & Sustainable Energy Reviews*, *56*, 810–819. doi:10.1016/j.rser.2015.12.001

Zhou, M., & Dicesare, F. (1993). *Petri Net Synthesis for Discrete Event Control of Manufacturing Systems*. Springer Science. doi:10.1007/978-1-4615-3126-5

Zhu, Y., Merbold, L., Leitner, S., Pelster, D. E., Okoma, S. A., Ngetich, F., Onyango, A. A., & Butterbach-Bahl, K. (2020). The effects of climate on decomposition of cattle, sheep and goat manure in Kenyan tropical pastures. *Plant and Soil*, *451*(1), 325–343. doi:10.100711104-020-04528-x

Ziukov, S. (2015). A literature review on models of inventory management under uncertainty. *Business Systems & Economics, 1*.

About the Contributors

Pandian Vasant is a Research Associate at Modeling Evolutionary Algorithms Simulation and Artificial Intelligence, Faculty of Electrical and Electronics Engineering, Ton Duc Thang University, Ho Chi Minh City 700000, Vietnam, and an Editor in Chief of International Journal of Energy Optimization and Engineering (IJEOE). He holds PhD in Computational Intelligence (UNEM, Costa Rica), MSc (University Malaysia Sabah, Malaysia, Engineering Mathematics) and BSc (Hons, Second Class Upper) in Mathematics (University of Malaya, Malaysia). His research interests include soft computing, hybrid optimization, innovative computing and applications. He has co-authored research articles in journals, conference proceedings, presentations, special issues Guest Editor, chapters (310 publications indexed in Research-Gate) and General Chair of the EAI International Conference on Computer Science and Engineering in Penang, Malaysia (2016) and Bangkok, Thailand (2018). In the years 2009 and 2015, he was awarded top reviewer and outstanding reviewer for the journal Applied Soft Computing (Elsevier). He has 35 years of working experience at the universities. Currently, he is General Chair of the International Conference on Intelligent Computing and Optimization () and Research Associate at Modeling Evolutionary Algorithms Simulation and Artificial Intelligence, Faculty of Electrical & Electronics Engineering, Ton Duc Thang University, HCMC, Vietnam.

* * *

Sampath Boopathi () completed his undergraduate in Mechanical engineering and postgraduate in the field of Computer-Aided Design. He completed his Ph.D. from Anna University, and his field of research includes Manufacturing and optimization. He published 60 more research articles in Internationally Peer-reviewed journals, one Patent grant, and three published patents. He has 16 more years of Academic and research experiences in the various Engineering Colleges.

Sasikymar C. completed his undergraduate in Mechanical Engineering and postgraduate in the field of Engineering Design. He completed his Ph.D. from Anna University, Chennai, Tamil Nādu, India.

Radhika K. is the Professor in the Department of Electronics and Communication Engineering at Muthayammal Engineering College. She has around 15 years of Teaching experience. She has published papers in reputed International Journals ,Books/Chapters and published Patents. She is the life member of ISTE, IETE. Her area of interest includes VLSI, Embedded Systems, and Instrumentation.

Matheus is presently with Department of Electrical and Computer Engineering, Namibia University of Science and Technology. His research interests include renewable energy resource, micro and smart grid planning, symbiotic systems and agrivoltaics.

K. S. Sastry Musti is presently with Namibia University of Science and Technology, Windhoek, Namibia. He has than 30 years of professional experience. He obtained his BTech in Electrical Engineering from JNTU, Kakinada (A.P.) in 1990, MTech and PhD from NIT, Warangal in the years 1996 and 2002 respectively. Some of his earlier employments include University of the West Indies, St Augustine, Trinidad, ESIGELEC, Rouen, France and Sophia University, Tokyo, Japan. His research interests include Power and Energy Systems, Information Systems and Engineering Education.

Puviarasi R. received her PhD in Saveetha Institute of Medical and Technical Sciences in 2018 and completed M. E (Electrical Drives and Embedded Control) from Anna University Coimbatore in 2011. Her research area includes Power Electronics, Renewable Energy systems, Image Processing and 3D Printing. She has published 65 papers in various indexed journals (IEEE transaction, WoS, PubMed, Scopus Google Scholar etc) and 25 papers are presented in National/International conferences. She has received Best Young Faculty Award from Tamil Nadu Association of Intellectuals and Faculty (TAIF) (2022), Women Researcher Award from 1st International Conference and Research Awards on Science, Health & Engineering (2020), Inspiring Young Teacher Award 2017-18-IET CLN International Awards - IET on 29th Sep 2018 and Excellent Researcher award of RULA International Awards on 15th Aug 2018. currently she is working as Associate Professor in Saveetha School of Engineering, Saveetha Institute of Medical and Technical Sciences, Chennai, Tamil Nadu, India.

Valerian C. K. Silayo received his PhD (Sun Drying of Grains) from Newcastle upon Tyne, UK in December 1995. He has been working at the Sokoine University of Agriculture from 1985 to date. Currently, he is a professor in the School of Engineering and Technology at Sokoine University of Agriculture, Tanzania. Silayo has written articles covering issues of process engineering, renewable energy, upscaling strategies for ensuring food security, and post-harvest processing.

I**Pawan Kumar Singh** is working as a software architect with Honeywell Technology Solutions Lab. He holds Master's in Technology in Mobile Computing from Jadavpur University. He has more than a decade of extensive and diverse working experience in mobile application development and solution design. His area of interest are IOT, Data Science, Cyber Security, Mobile Application and Industrial Automation.

Vartika Singh, Assistant Professor, AIGWES, Amity University Noida, holds Ph.D. degree in Hydrogeology and Remote Sensing, M.Phil. degree in Hydrogeology from Vikram University and Master degree in App. Geology from Kurukshtera University. She has GATE Qualified all India rank 47. She has 15 years of research experience with Defence Terrain Research Laboratory (DTRL), DRDO, Delhi and Amity University Noida, Uttar Pradesh. Her specialization of research is in Hydrogeology, Remote Sensing, Watershed; Spring shed Management, Environmental Geology & Artificial Intelligence. She has more than 60 publications, 3 patents, 15 book chapters in various national, international journals. She is one of the founder member of Computer Intelligence Research Group and Magazine. She is in Editorial Board of several national and international journals.

Lakshmana Kishore T. is working as an Assistant Professor in the Department of Mechanical Engineering at University College of Engineering, JNTUK Kakinada, India. His areas of research include Heat Transfer, Computational Fluid Dynamics, Bio-fuels, Renewable energy sources and Nano-Fluidics.

N. Suthanthira Vanitha is presently working as a Professor in the Department of Electrical and Electronics Engineering and Dean-Quality Assurance at Muthayammal Engineering College, Rasipuram. She has 21 years of experience in Academic, Research and Administration. She has published 170 papers in National/International Journals and Conferences,12 Books/Chapters and 13 Patents. She is the Life Member in IEEE, ISTE and Fellow Member in IE(I). Her area of specialization includes, Embedded Systems, Biomedical Instrumentation and Renewable Energy Systems.

Yusto Mugisha Yustas received his PhD in Renewable Energy from the Sokoine University of Agriculture (SUA) in November 2021. He worked on developing a predictive model for integrated renewable energy system performance. In 2010, as the District Agriculture Engineer, he was employed by the Rufiji District Council-Coast region, Tanzania and is currently the academic member of staff in the School of Engineering and Technology at SUA. Yusto writes and presents widely on issues of renewable energy, upscaling strategies for ensuring food security, agricultural mechanisation and post-harvest processing.

Index

F

G

H

I

M

N

O

P

R

S

V

W

www.ingramcontent.com/pod-product-compliance
Lightning Source LLC
LaVergne TN
LVHW080901300926
849162LV00003B/513